MW01518630

Web Offset
Press Operating

Web Offset
Press Operating
Fifth Edition

by Daniel G. Wilson
and PIA/GATF Staff

PIA/GATF*Press*
PITTSBURGH

PIA/GATF*Press* books are widely used by companies, associations, and schools for training, market-ing, and resale. Quantity discounts are available by contacting 800/910-4283.

 Weyerhaeuser
Printed on Weyerhaeuser 60-lb. Cougar® Opaque Smooth Offset

PIA/GATF*Press*
Printing Industries of America/
    Graphic Arts Technical Foundation
200 Deer Run Road
Sewickley, PA 15143-2600
Phone: 412/741-6860
Fax: 412/741-2311
Email: AWoodall@piagatf.org
Internet: www.gain.net

Orders to:
Online: www.gain.net
PIA/GATF Orders
200 Deer Run Road
Sewickley, PA 15143
Phone (U.S. and Canada): 866/855-4283
Phone (all other countries): 301/393-8624
Fax: 301/393-2555

Contents

Foreword

Designed to supplement press operating manuals and formal apprenticeship programs, the fifth edition of PIA/GATF's *Web Offset Press Operating* provides both novice and experienced press operators with valuable information to improve productivity and print quality. Theoretical and practical how-to information have been combined in a single volume. The concepts presented in this book are applicable to most web offset presses. Actual press adjustments, however, must be made following the press manufacturer's recommendations.

In the preface to the first edition of this book published in 1974, Charles Shapiro, then GATF publications editor, said:

> Web offset lithography, as a major printing method has "come of age." For some time now, a definitive reference text has been sorely needed. But, until the web offset field passed through its "growing-pains" stage—1955–1970—any text would have been obsolete before the ink dried on the author's manuscript.

Here we are almost 30 years later, and web offset lithography is undergoing a new round of growing pains: the rapid proliferation of web press automation has made the need for skilled press operators ever more critical. To address this need for training, GATF unveiled its comprehensive Web Offset Training Curriculum in the late 1980s. Well over 1,000 press operators have gone through this program, which is revised and updated periodically.

Along with *Solving Web Offset Press Problems*, this book is one of the two keystone publications supporting the training curriculum. Based on feedback from the users of the both this book and the training program, the fifth edition of *Web Offset Press Operating* has been divided into eleven main sections that correspond to the eleven tasks in the training program. Further, the chapters within each section now more closely parallel the knowledge sections of the corresponding training program task. The safety precautions and operations from the training program are now incorporated into the book where appropriate.

Although all chapters have been updated, expanded, and heavily illustrated, one chapter deserves special mention. Many users of the training curriculum requested enhanced treatment of the maintenance function. To achieve this, the new maintenance chapter now tips the scales at 53 pages, up from 8 in the fourth edition. Much of the new material in the chapter has been adapted from *Maintaining Printing Equipment*, written by Herschel L. Apfelberg and published by GATF in 1984.

Thomas M. Destree
Editor in Chief

Acknowledgments

A special thanks goes to MAN Roland Druckmaschinen AG and MAN Roland, Inc. for the donation of their state-of-the-art, shaftless ROTOMAN heatset web offset press to the Foundation. The press was dedicated on September 19, 2002, at an open house that also recognized the contributions of over twenty technology partners that contributed auxiliary equipment, services, and supplies.

PIA/GATF and the author would also like to thank the following companies, especially MAN Roland and the technology partners, for contributing information, photographs, and/or line illustrations for the fifth edition of *Web Offset Press Operating*:

- AEC, Inc.
- American Ed-Co, Inc.
- Applied Web Systems, Inc.
- Bacou-Dalloz
- Baldwin Technology Co., Inc.
- Bel-Art Products
- Brantjen & Kluge, Inc.
- Buhler Inc.
- Cascade Corporation
- Catalytic Products International
- Cole-Parmer Instrument Co.
- Converter Accesory Corp.
- Dahlgren USA, Inc.
- DAY International, Inc.
- R.R. Donnelley & Sons Co., Inc.
- Double E Co., Inc.
- Epic Products International Corp.
- Fisher Scientific
- Flint Ink Corp.
- Glatfelter
- Goss Graphic Systems, Inc.
- Graphics Microsystems, Inc.
- Heidelberg Prepress
- Heidelberg USA, Inc.
- Heidelberger Druckmaschinen AG
- Hurletron Incorporated
- Imation

- INNOTECH Graphic Equipment Corp.
- Institute of Paper Science and Technology
- Jardis Industries, Inc.
- KBA North America, Web Press Division
- Lincoln Industrial
- Lithco, Inc.
- MAN Roland Druckmaschinen AG
- MAN Roland, Inc.
- Mannesmann Rexroth Corporation
- Martin Automatic, Inc.
- MEGTEC Systems
- Myron L Company
- Network Industrial Services
- Plexus Pacific Industries
- Printcafe Software, Inc.
- Procam Controls, Inc.
- PTC Instruments
- QTI
- Raytek Corporation
- RTE
- SIMCO Industrial Static Control, an Illinois Tool Works Co.
- Solna Web USA, Inc.
- SPEC
- Sun Chemical/GPI Division
- Systems Technology, Inc.
- Tobias Associates, Inc.
- Toray Marketing & Sales (America), Inc.
- Valco Cincinnati, Inc.
- Valmet Paper Machinery, Inc.
- Vits-America, Inc.
- Web Printing Controls Co., Inc.
- Web Systems, Inc.
- Western Printing Machinery Co.

The author would also like to thank the following PIA/GATF staff members for their assistance:

- Tom Destree
- Christy Holstead Semple
- William Farmer, Jr.
- Peter Oresick
- Amy Woodall

Section I

Orientation to the Web Offset Press

1 History of Lithography

The modern offset lithographic web press is the product of a wide range of historic technological innovations and to present them all is beyond the scope of this book. However, a few of the most critical developments are presented to provide a historical perspective on the modern web press and also to pay credence to the many important inventors who contributed directly and indirectly to this extraordinary machine's development.

The Invention of Lithography

Figure 1-1. *Alois Senefelder.*

Alois Senefelder, the inventor of lithography, came into his role as a technological innovator indirectly. He saw himself as a dramatist despite his preparation for a law career. Lacking sufficient personal funds, a sponsor, or a publisher, he set out to discover an inexpensive method for printing his own plays. Copperplate engraving was too costly; the copper could be used only once and the process was too slow.

In 1796, he began experimenting with Bavarian limestone on which he produced a relief image by chemical etching. In 1798, as a result of discoveries made during the course of these experiments, he invented the lithographic process of printing, a flat-surface principle rather than an intaglio principle.

According to Senefelder (*The Invention of Lithography*), he had "just ground a stone plate smooth in order to treat it with etching fluid and to pursue on it my practice in reverse writing, when my mother asked me to write a laundry list for her. Without bothering to look for writing materials, I wrote the list hastily on the clean stone,

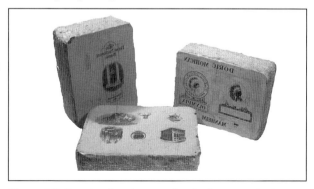

Figure 1-2. *Bavarian limestone used as the image carrier for early stone lithography.*

with my prepared stone ink of wax, soap, and lampblack, intending to copy it as soon as paper was supplied."

Later, Senefelder became curious to see what would happen if the stone were etched with nitric acid. As he suspected, the acid ate into that part of the plate that was not protected by his specially developed ink. Thus, the writing stood about "the thickness of a playing card in relief." This became known as shallow-relief printing.

Senefelder's lack of skill in writing on stone, which is much more difficult than writing on paper due to the requisite reverse writing and the material itself, led him to experiment with transferring. He developed an ink of linseed oil, soap, and lampblack with which he wrote on paper; then, he placed the paper face down on the stone plate and passed it through a press, producing an accurate, reversed image on the stone. "But," he wondered, "why could I not invent an ink that would serve on the stone without making it necessary to trace over it with the stone ink? Why not make an ink that would leave the paper under pressure and transfer itself to the stone entirely? Could one give the paper itself some property so that it would let go of the ink under given conditions?"

This led to more experiments, in the course of which he observed that "if there happened to be a few drops of oil in the water into which I dipped paper inscribed with my greasy stone-ink, the oil would distribute itself evenly over all parts of the writing, whereas the rest of the paper would take no oil, and especially so if it had been treated with gum solution or very thin starch paste."

Next, Senefelder drew a sheet of an old book through a thin gum solution and then sponged it with a thin oil color, which adhered only to the printed letters. He laid a clean sheet on top of the page and put both through a press, obtaining a "very good transfer, in reverse, of course." This transfer, when treated in the same manner,

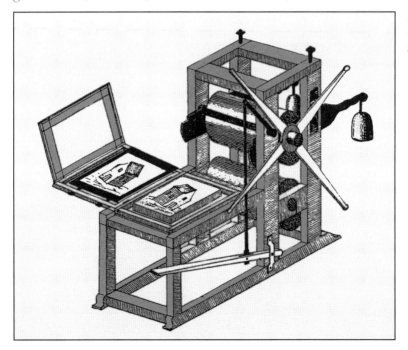

Figure 1-3.
Senefelder's press.

could produce right-reading copies. What was significant, though, was that "this process, depending solely on chemical action, was totally, fundamentally different from all other processes of printing." Senefelder also found that this process worked well with the limestone plate, which readily absorbed the fatty ink and was much sturdier than paper. It is from this stone that lithography gets its name (meaning, literally, stone writing).

The life of Senefelder was devoted to lithography. In 1817, he submitted a model of a lithographic press, with automatic inking and dampening, to the Royal (Bavarian) Academy, which gave him a gold medal in recognition of this invention. At about the same time, he invented the first paper plates with which he wanted to replace the heavy lithographic stones.

"Chemical printing." The principle on which lithography is based is the mutual repellence of water and greasy substances. The limestone used by Senefelder was porous and particularly well suited to lithography. Stone, however, was not the essence of the process. Senefelder preferred the term "chemical printing" for his invention and used this term as a generic one; stone printing (in German Steindruck), or lithography, was for him only one branch of many possible in chemical printing.

Terminology has been neither generally accepted nor consistent throughout history. At certain times and in certain countries, a distinction was made, for example, between lithography and zincography (lithography from zinc plates). The term chemical printing was less used in the Anglo-American orbit than planography, and planography, too, has been used by different people with differing connotations. In the course of the twentieth century, the term lithography has lost its meaning as stone printing and is in our day more or less generally used in the same sense in which Senefelder used chemical printing.

Lithography and copperplate engraving. The original method of stone printing was slow and cumbersome compared with contemporary lithographic techniques, but it was very fast and simple compared with copperplate engraving, the method with which lithography competed most vigorously.

Copperplate engraving was a widely used printing technique at the time, but limestone was a much cheaper material and was also reusable. The superiority of lithography—at least costwise—is set forth in a cost comparison of lithography and copperplate engraving in the English translation of Antoine Raucourt's treatise on lithography, published in 1821. There, it is shown that "lithography is about seven times cheaper than copperplate engraving, if plate costs of both processes are compared. The printing costs show much less of a difference, but lithographic printing costs only a little more than half of the printing of copperplate engravings."

Stone lithography in America. The first lithograph to be published in the United States was by Bass Otis and appeared in the *Analectic Magazine* of August 1819. This magazine also published a comparatively detailed description of the process and summarized the uses and advantages of lithography in the following eight points:

Figure 1-4. *The first American lithograph.*

1. It is a perfect facsimile: there can be no mistake or miscopy.

2. It supersedes all kinds of engraving: when the drawing is finished, it is now sent to the engravers, and no impression can be taken until the engraving is finished; in lithography, impressions can be taken the instant the drawing is dry, more perfect than any engraving can possibly produce.

3. It can imitate not only drawings in crayon and Indian ink, but etching, mezzo-tinto, and aqua tinta.

4. The plate is never worn out as in copperplate engraving. In France, 70,000 impressions of a circular letter were taken, before the engraving was finished of a similar letter written on paper.

5. Maps, large prints, calico printing, etc., can be executed in this way on rollers of stone, turned and the design drawn, etched or aqua tinted, on the stone roller itself. For roller work in calico-printing, it would be inestimable.

6. All works of science may now be freed from the prodigious expense attending numerous engravings.

7. Any man who can draw can take off any number of impressions of his own designs, without trusting to any other artist.

8. The advantage of expedition in the process now recommended is beyond all calculation.

The publication of Bass Otis's lithograph did not lead to an immediate growth of lithography in the United States. Only a handful of people were active in this field during the 1820s and '30s. Barnett and Doolittle opened shop in New York in 1821 and therewith became the first lithographers in the United States.

The real spread and growth of lithography in the United States did not set in until the 1840s. In the following decade, chromolithography, the making of lithographic prints in many colors, became established. In these chromos, stone lithography reached its highest and, in a sense, final form. The next great changes, whereby modern lithography—lithography without stone—was created, did not take place before the beginning of the twentieth century.

The Development of the Rotary Press

The first rotary web presses were in use long before the invention of lithography. The engineering of these presses, which included developments in such mechanisms as ink systems, infeed systems, and in-line finishing systems paved the way for the development of the web offset press. Historians are unsure of the exact origin of the rotary press, but they do agree that the first ones were developed around the mid-17th century. There are patents on record from this time period for presses that employed engraved cylinders used to print on sheets of vellum, parchment, and cloth. Various developments over the ensuing 100 years led to the first multicolor rotary press, invented by Thomas Bell in 1783, who is considered to be the principal founder of the rotogravure press. These presses were originally designed to print patterns and designs on cloth and paper.

Though the rotogravure press was in use in the late 1700s, the major printing method at this time was letterpress, which involved printing from flat forms composed of movable type. Invented by Johannes Gutenberg in 1442, movable type involved placing individual, interchangeable type pieces in the right order to compose words, sentences, paragraphs, and finally pages. This resulting flat form (called a galley), which was made up of thousands of type pieces, was placed in the flat bed of a press, the surface was inked, and the paper was placed against the inked form and put under pressure to make the impression. By today's standards, this was a very slow and tedious method of printing indeed, and while engraved-cylinder rotary presses were incredibly fast by comparison, letterpress printers had not yet invented a means to wrap their flat type forms around a cylinder.

Stereotype cylinders. The limitation of flat letterpress forms was finally overcome with the invention of stereotyping, credited to two Frenchmen, M. Worms and M. Philippe, who patented the process in 1845. Their invention essentially involved the process of making a papier-maché mold (called a "flong") of the flat type form. This flexible mold was then wrapped in a cylindrical carrier and cast into a relief metal cylinder by pouring molten metal into the mold. Worms and Philippe also demonstrated a steam-powered rotary letterpress that printed a web of paper on both sides at 10,000 impressions per hour, an absolutely fantastic speed at that time. The press design never held up in production environments, and was therefore never successfully marketed. However, the demonstration sparked the development of high-speed rotary printing.

Manufacture of paper rolls. The development of a webfed rotary printing press would not have been possible without the invention of a machine that could manufacture rolls of paper. Paper was originally made into sheets by hand, dipping a wire mold into a vat of a cellulose fiber and water mixture. The wet sheets would be removed from the mold, pressed, and dried. Paper production was extremely labor-intensive and slow, resulting in very high costs. In 1798 the Frenchman Nicholas-Louis Robert invented the "endless paper machine," a device that could manufacture rolls of paper by automatically pouring the fiber/water mixture onto a continuously rotating wire mat. The further development of this machine was funded in England by Henry and Sealy Fourdrinier and engineered by Bryan Donkin. By 1820, a working machine, powered by water, was being sold to papermakers who would forevermore efficiently mass-manufacture paper in rolls, rather than by tedious sheet-by-sheet methods.

William Bullock's press. The manufacture of rolls of paper led to the development of printing presses that could print rolls. The first of these were called "reel-fed" presses, rather than web presses. In 1863, William Bullock, an American, patented the first automatic, reel-fed rotary press working

Figure 1-5. Bullock's press of 1865.

from stereotype plates. This early press cut the web into sheets before impression. (Note: Bullock died four years after his invention in an accident involving the gears of the press.) This press was first installed to print the *Philadelphia Enquirer.*

As key enhancements to this press, Walter Scott invented the first in-line folder for Bullock's press in 1869. In 1870, John William Kellberg patented a gripperless system to feed the web directly into the cylinders, cutting the paper into sheets after impression, an enhancement that is said to have doubled production of the press from 10,000 to 20,000 impressions per hour.

Direct rotary lithographic presses. At the beginning of the twentieth century, lithography within the graphic arts was considerably weakened. The advance in photo-engraving made good-quality picture reproduction by the letterpress method possible. The advances in printing press technology made it possible to print much more efficiently by letterpress than by lithography.

The flatbed stone press was well adapted to the requirements of stone lithography, but it was absolutely unsuitable for the use of photomechanically made printing image carriers. It seemed only logical to replace the stone press with a press that would operate on the rotary principle and use a thin metal plate in place of the lithographic stone. The rotary principle had proved its soundness in newspaper press design. Zinc plates had been suggested by Senefelder and had been successfully used for quite some time, and aluminum had also been introduced into lithography in the early 1890s.

An additional, very important reason for the replacement of limestone plates was the scarcity, enormous cost, and unwieldiness of large stones. Such sizes were a necessity for the printing of large posters. Replacing the stones with sheet metal, which was available in the desired dimensions, was an economic necessity as well as a convenience.

In the 1890s, rotary lithographic presses began to be used. They were valuable in the poster field and in other large-sheet applications. For many purposes, the deficiencies of the early metallic lithographic plate (e.g., its relatively delicate image and the inability of the level-surfaced metal to push the image into the contours of rougher papers) didn't permit direct rotary lithography to compete strongly with other printing processes. Two main factors helped overcome lithography's weaknesses and bring the process to predominance—(1) the growing effect of photography on printing and (2) the unexpected development of the offset press.

Role of Photography

Modern lithography is unthinkable without photography. Photography supplies a considerable part of the subject matter to be reproduced. It is also essential to lithographic reproduction. The scientific basis of photography lies in physics and chemistry. The same sciences are of great importance to lithography; they are essential for all phases of the lithographic process, especially in the areas in which photography plays a dominant role.

Here, we are not concerned with photography as a picture-making process but only with such aspects of photography as are used in the lithographic reproduction of pictures. However, these two functions are closely related.

Historically, photography was first used in lithography for the making of the print image, the stone or the plate. The next important contribution of photography to lithography had to do with the reproduction of tonal images. At this point, photography became a decisive step in the conversion into a press plate of the subject matter to be lithographed. Process color printing depends completely on photography.

The development of the halftone process. Two problems dominate the history of picture reproduction. One is the rendering of tonal values; the other is the combination of reading matter and pictures for printing in one pressrun. Invention of the halftone process was the first effective solution. It was developed for letterpress printing and made possible the reproduction of photographs and other tonal pictures together with reading matter.

In the 1880s, when the revolution in photography occurred, lithography was utterly unsuitable for mass production printing. Nothing illustrates this backwardness better than the story of the New York *Daily Graphic*. On March 4, 1873, the *Graphic* published the first issue of the first illustrated daily newspaper in the world. It was an eight-page paper. The pictorial four pages were printed first, by a lithographic stone transfer process, and the inside four pages were run off by letterpress. Production on the lithographic presses was between 700 and 800 sheets per hour, a rate much too slow to satisfy the demand. Lithographic stone presses were completely outdistanced by letterpress machinery. At that time, letterpress had no satisfactory means for reproducing continuous-tone pictures. However, the development of the photo-engraved halftone process overcame this deficiency of letterpress reproduction and became an excellent process of reproducing pictures.

The development of photoengraving may not appear to belong in this history, but since the use of halftone dots was later adapted to lithography, photoengraving was a prelude to modern means of reproducing continuous-tone pictures lithographically. Many people worked on the problem of making tonal pictures reproducible by breaking them up into very small units. Frederick Eugene Ives (1856–1937) of Philadelphia solved the problem of tonal reproduction in 1886 by introducing the glass cross-line screen. Max Levy of Philadelphia succeeded in 1890 in developing a precision manufacturing process for these screens. These contributions were by no means the only ones that made photoengraving into the leading reproduction process. Lithography stagnated technically until it received its next great impulse through the offset press, as described in the next section, "The Offset Press."

The offset press revived the interest in photolithography; however, conditions had vastly changed compared to the 1870s, when photolithography had been a contact printing transfer process for line work. During the time of the broad new interest in photolithography, which may be placed arbitrarily at the period shortly after the first World War, graphic arts photography had immensely advanced. Photolithography acquired a different meaning. It stood for offset lithographic printing in which the camera and, where necessary, the halftone screen were used for converting line and tone originals by means of photography into printable form.

Photography: going from letterpress to lithography. The need to proof the flat type forms (galleys) before going through the time and expense of making the relief cylinders with the stereotype method led to the development of the proof press. The operation of this simple press involved rolling ink over the form, placing a piece of paper over the form, and cranking an impression roller by hand over the paper to make an impression. In the early 1900s, Robert Vandercook of Chicago became the best known builder and innovator of proof presses. The developers of the proof presses could not have realized it at the time, but these machines became an important building block in the development of lithography as an industrial process. With the development of photographic techniques in the 1920s and 1930s, a proof made with a proofing press began to be used as a reproduction proof, placed on a camera to create photographic film images of the page. These films would then be used to expose the flexible metal lithographic printing plates, which could then be wrapped around a press cylinder.

The Offset Press

The offset press and photomechanics are the foundation of contemporary lithography. Of these two, the offset press is the more important contribution because without it, photomechanics would not have been nearly as successful. Direct rotary presses established themselves for general use but died out once the offset press had found acceptance. Direct printing put a much too heavy burden on the photomechanical plates. Only indirect printing made modern long-run lithography from photomechanical plates possible.

The origin of the offset press. The basic offset principle was known for a long time before the paper offset press was introduced. In fact, offset presses were used for metal lithography long before they were adapted to paper use.

The first metal decorating presses were lithographic stone presses equipped with an intermediate cylinder. In 1875, an English patent was granted to R. Barclay, of Barclay & Fry, for such a press. The intermediate cylinder had a surface of specially treated cardboard for transferring the inked design to the sheet metal. Shortly thereafter a rubber blanket was substituted for the cardboard. Such presses were made in England around 1880.

There were also offset presses that could be used for either paper or metal, but strangely enough the offset method was never used for paper lithographing by the owners of these machines.

Ira Rubel develops the first paper offset press. It is generally agreed that Ira W. Rubel was the first to develop an offset lithographic press specifically for the printing of paper. At the time Rubel developed his offset press, about 1904, lithographic stone presses had a rubber blanket on the surface of their impression cylinder. Whenever the feeder operator missed feeding a sheet when the press was operating, the ink image was transferred to the rubber blanket from the stone. The following sheet would then be printed on both sides because the rubber blanket transferred the ink image to the back side of the sheet. This unintentionally made transfer produced a print superior to that made directly from stone, because the soft blanket conformed to the contours of the paper surface while the hard stone did not.

Figure 1-6. *An early rotary offset press.*

Rubel noticed this and decided to use it as the basis of a printing press that employed an intermediate blanket cylinder to transfer the image. Thus, "offset" lithography was born.

This new press, and the introduction of the photocomposing machine in 1950, made lithography a direct competitor with letterpress. As a result, from the middle of the 20th century, offset lithography became the fastest growing printing process.

Commercial and color lithography. The offset press was originally intended for commercial lithography but not for color lithography. According to the then-accepted definitions, commercial lithographic originals were "engraved on stone or sheet," whereas color lithography was the appropriate designation when the original was "drawn on stone, zinc, or aluminum."

Commercial lithography produced business and bank stationery, mainly in black and white; color lithography produced greeting cards, labels, posters, etc. It wasn't very long until the offset press was very successfully used for color work. Its advantage over the stone and direct rotary presses was thereby established in both branches of lithography. As time passed, the offset press became the dominant lithographic printing machine.

The lithographic plate and the founding of LTF. At first, it may seem odd that the lithographic plate and the founding of the Lithographic Technical Foundation (LTF), predecessor of the Graphic Arts Technical Foundation, are discussed under the same heading, but these subjects are very closely related.

The plate was the most important source of problems in lithography during the first half of the twentieth century. The plate was also the problem source closest to the printer. All other elements in lithography were either not at all or only to a very small degree subject to change by the average printer. Presses, paper, ink, photo graphic materials, rollers, and blankets are all manufactured by specialized industries, but the plate was made in the lithographic shop. It was possible to purchase plates from firms specializing in platemaking, but these firms were not much better equipped to cope with the problems of the lithographic plate than many printing shops.

The quality of the plate is important in many respects. For one, it is decisive for the quality of the final lithographic print. Another point is plate life, which has a great influence on run length and costs. Photomechanical platemaking is a complex process. The effort necessary to control this process could never be made by an individual lithographic shop of average size. The solution to the plate problem had to wait until the industry began to do cooperative research—until LTF was in existence.

Two kinds of photomechanically made plates were used in lithography up to the end of the Second World War: albumin plates and deep-etch plates. Bichromated albumin was first used by Poitevin in 1855, and deep-etch plates have a long history in which many inventors participated. J. S. Mertle, in *Photolithography and Offset Printing*, credits the Reverend Hannibal Goodwin with having made "one of the first attempts toward the reversal of photomechanical images on metal plates," which he patented in 1881. Many other inventors did further work on deep-etch plates.

The real change in platemaking techniques, however, had to wait for LTF, which was founded in the early 1920s. Not all plate problems were eliminated from lithography, but the plate ceased to be the problem child of lithography through the efforts of LTF. In the middle of the twentieth century, lithographic platemaking became a field of activity for the supply industry.

Chronologically, deep-etch and bimetallic plates preceded presensitized plates, which appeared in the 1950s. In the 1960s, "wipe-on" plates came into commercial use, and it was in the 1960s that lithography overtook letterpress to become the predominant printing process. The invention of the offset press in the early 1900s, coupled with the great strides in photography and photomechanical platemaking, made lithography much more competitive than letterpress.

Web Offset Development

As for web offset, the mid-1930s saw an increasing demand for heavily illustrated mass-market magazines, which inspired the development of a heatset ink drying system for high-speed rotary letterpress printing. At this time web offset lithography was beset by technical problems and was not yet ready for high-speed, large-volume production. But by the late 1940s, lithographic platemaking had been perfected and web offset manufacturers adapted the inks and drying systems developed for high-speed magazine letterpress printing—this about thirty-five years after the first web offset press had been built.

By 1950, web offset was ready for serious competition with letterpress, as well as with sheetfed lithography. The ability of the web offset method to perfect (print on two sides of the sheet simultaneously) was unique, and high productivity was the reward for any printer who made the switch to web offset. But because the industry's workforce was primarily geared for letterpress printing, initial growth in web offset was slow, so it was not until the 1960s that web offset truly caught on. In the 1950s and early '60s, printing companies buying their very first web offset presses accounted for most of the growth, and as these firms continued purchasing additional presses, growth continued and web offset began to cut heavily into letterpress sales.

Newspapers began converting to web offset during the 1960s, when over half the weeklies and 20% of the dailies switched from letterpress to web offset. The newspaper's switch to web offset was influenced mainly by the growth of each plant's operations. As demand grew, most large daily newspaper operations quickly replaced slower letterpress equipment with high-speed web offset lithographic presses. National and international newspaper production and distribution have been made possible by the use of satellite transmission and digital telephony to connect publishers to printing plants in varied geographic locations around the world. Prepress production for lithography has been well suited to the application of digital data transmission required in such an arrangement.

Currently, about 70–80% of all books are printed on web offset presses. It was during the 1960s that periodical publishers turned more and more to web offset, increasing the value of web-offset-produced periodical shipments from $50 million to

$210 million. Today, many large-edition weekly magazines are printed in regional editions similar to newspapers produced with digital content moving by way of satellite transmission and telecommunications. The bulk of web offset printing markets today include direct mail, magazines and inserts, newspapers, catalogs, books, and other forms of general commercial printing. With all of these markets turning to web offset during the '60s, this printing method experienced a dramatic increase in the value of printed products, starting out the decade at $60 million in shipments and ending it at $830 million.

The continued growth of web offset has depended upon improvements in equipment and supplies. The quality of the paper, inks, plates, blankets, and other press supplies has constantly improved to meet the ever-increasing requirements of great speed and quality demanded by the web offset printing market. Along with this, improvements in press and related equipment including infeed, dryers, and folders have kept pace with the growing demand for increased speed and quality.

In the past, much of the growth of web offset printing was due to the economics involved. Savings due to fast makereadies and high production speeds alone motivated many printers to switch to web offset, regardless of other factors. Today, there are many buyers of printed products who depend upon print quality as a foremost consideration, seeking the ability of heatset web offset to produce printing that is exceedingly sharp, with high gloss and heavy ink coverage.

Current Trends in Web Offset

Color plays an important role in web offset printing today. In the late 1980s and throughout the 1990s, the average number of units per press purchased was five, allowing for full-color printing on both sides of the web with one pass through the press. This reflects the pervasive amount of process color printing currently produced on web offset presses.

Problems with air pollution control have increased due to the effluents produced by heatset printing coupled with stricter U.S. Environmental Protection Agency (EPA) regulations. Because of this, an increasing amount of work is being printed today by nonheatset web offset presses using ultraviolet (UV) inks. This will likely continue in the future, with much growth sure to continue in nonheatset UV printing.

Since the inception of web offset printing, the sixteen-page blanket-to-blanket press has been the standard of the industry. However, printers are employing an increasing number of smaller eight-page and larger thirty-two-page presses. The eight-page press competes in sheetfed lithographic markets, and the thirty-two-page format press is moving into markets that have traditionally been served by rotogravure presses. Both of these press configurations are expected to play a role in the continued expansion of the web offset segment of the printing industry.

Waterless printing is catching on quickly with web offset printers today. Printing without dampening solution has environmental benefits and the added benefit of reproducing more brilliant color images. While there are some drawbacks, many printers have successfully made the switch from traditional lithography to waterless printing, a trend that is likely to increase in the future.

There is a trend toward the manufacture of web presses without drive shafts, but rather with independent motors to power each printing unit. These presses, called shaftless presses, have the potential to be more productive and reduce some of the problems associated with tension control and color-to-color registration.

More computer controls and automation are being integrated into web offset presses. Digital prepress functions are being tied to speedy press makeready via higher levels of automation. New presses are being manufactured today that make ready faster, run faster, and produce the highest quality printing possible. As many experimental systems are currently being developed by manufacturers, web offset will surely continue to grow and improve in the 21st century.

2 Terminology and Principles

Lithographic printing may be done on two types of printing press: the ***sheetfed offset press*** and the ***web offset press.*** As the names suggest, sheetfed presses print on precut sheets of paper while web presses print on rolls of paper. This chapter provides a foundation for understanding web offset presses by presenting basic principles and terminology associated with the printing method. In-depth information on sheetfed presses may be found in a variety of GATF*Press* publications including *Sheetfed Offset Press Operating* and the *Lithography Primer.*

Lithography is a printing process that works essentially on the principle that water will not readily mix with oil. The ***printing plate*** that receives the ink and water is usually a thin rectangular sheet of pliable aluminum, which can be wrapped around and fastened to a press cylinder. In simple terms, image areas on the printing plate are ***oleophilic*** (meaning "oil attracting") and readily accept oil-based inks. The nonimage areas are ***hydrophilic,*** and so attract water. The chemical repulsion principle keeps the oily ink film on the plate from moving into the water-coated nonimage areas of the plate. Lithography is a planographic printing process, meaning that the image and nonimage areas are essentially on the same level (or plane).

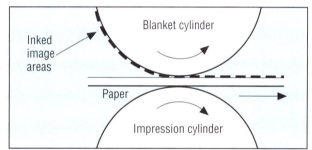

Figure 2-1. *Transfer of the inked image from the blanket to the printing substrate.*

Offset lithography refers to the action of transferring the inked image from the plate cylinder to an intermediate synthetic-rubber covered blanket cylinder, thus "offsetting" the image (figure 2-2). The inked image on this blanket cylinder then comes into contact with the paper to transfer the image. As described in the previous chapter, Ira Rubel discovered the offset principle around 1905. He found that offsetting the inked image before transferring it to the paper resulted in a cleaner, sharper image.

Following are the basic steps involved in the most common form of offset lithography:

1. Plate with photochemically produced image and nonimage areas is mounted on a cylinder.

2. Plate is dampened with a mixture of chemical concentrates in a water-based solution, which adheres to the nonimage areas of the plate.

3. Plate surface is contacted by inked rollers, which apply ink only to the image areas of a properly dampened printing plate.

4. Right-reading inked image on the printing plate is transferred under pressure to a rubber-like blanket, on which it becomes reversed (wrong-reading).

5. Inked image on the blanket is transferred under pressure to a sheet of paper or other printing substrate, producing an impression of the inked image on the paper.

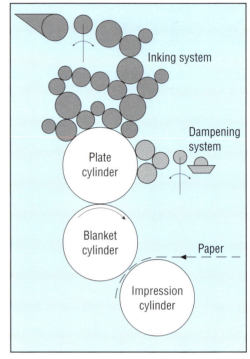

Figure 2-2. *The offset principle. The ink on the printing plate is transferred first to the intermediate blanket and then to the paper. Shown above is a typical sheetfed printing unit.*

An *offset press,* then, is a mechanical device that dampens and inks the printing plate and transfers the inked image to the blanket and then to the printing substrate. A *web offset press* (or webfed press) prints on a continuous web, or ribbon, of paper fed from a roll and threaded through the press at speeds four or five times greater than that of a sheetfed press. A *sheetfed offset press* feeds and prints on individual sheets of paper (or other substrate) at speeds of approximately 10,000–15,000 impressions per hour (iph). Compared to sheetfed presses, web presses have much smaller gaps on the plate and blanket cylinders, which means that ink and water flow much more continuously. Blanket-to-blanket web offset presses lack a hard impression cylinder, which is inherent in sheetfed press designs. This manual deals with (to use the most accurate expression) *lithographic web offset presses.*

Waterless offset lithography is a variation of offset lithography that does not require the use of a water-based dampening solution. The process uses an offset press equipped with temperature-controlled inking systems. The process requires special inks and special waterless plates.

Direct lithography, or *direct printing,* is the process of printing directly from a printing plate to a substrate. It is accomplished in web offset by running paper through a blanket-to-blanket unit so that the paper contacts one plate and takes ink from it. In this case, the blanket cylinder of the couple acts as an impression cylinder.

Basic Systems of Web Offset Presses

One can gain a better grasp on the web offset press by analyzing the press in terms of its systems. We'll use a typical four-color blanket-to-blanket heatset press as an example. For the purposes of this book, think of the press as consisting of several sections. The *infeed* of the press is where the unprinted rolls of paper are mounted. The *delivery* is where the final printed material comes out. Going from the infeed to the delivery, the elements of a heatset web offset press are, in order, *infeed, printing units* (press), *dryer, chill rolls,* and *delivery* (either a folder, sheeter, or rewinder). A nonheatset web press does not have a dryer or chill rolls.

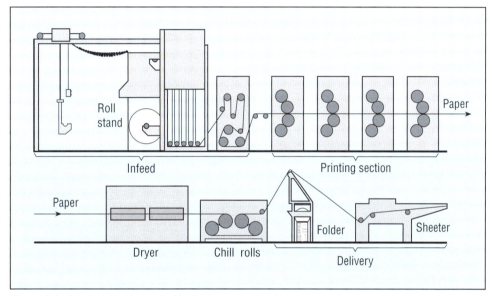

Figure 2-3. *A four-color blanket-to-blanket heatset web offset press with optional delivery to a folder or sheeter. The press configuration shown here is the most common, but variations are possible. For example, with some presses having more than six units, the folder(s) is placed in the middle of the total press configuration and the infeeds are placed at each end.*

The *infeed system* of the press consists of the mechanisms required to guide the web of paper into the first unit, where printing will take place. It includes all equipment from roll stand to the first printing unit and controls the speed, tension, and lateral positioning of the web. It keeps the web taut and flat as it enters the first printing unit.

The *printing section* of the press consists of several special cylinders and inking and dampening systems. Each printing couple has the following components:

- *Plate cylinder.* A cylinder that carries the printing plate, which is a flexible image carrier with ink-receptive image areas and, when moistened with a water-based solution, ink-repellent nonimage areas.

- *Blanket cylinder.* A cylinder that carries the offset blanket, which is a fabric coated with synthetic rubber that transfers the image from the printing plate to the substrate.

- *Impression cylinder.* A cylinder, found on in-line web presses, that forces the paper against the inked blanket and the paper.

- *Dampening system.* A series of rollers that carry a metered film of water mixed with other important chemicals (together called ***damp-ening solution),*** "dampening" the printing plate. The water-based dampening solution contains additives such as acid, gum arabic, and isopropyl alcohol or other wetting agents.

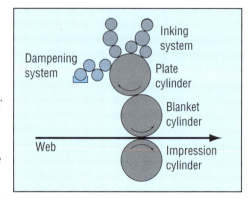

Figure 2-4. *An in-line offset lithographic press unit that prints a single color on one side of a web.*

- *Inking system.* An ink fountain that functions as an ink reservoir and a series (set) of rollers that transport and mill the ink, finally depositing a metered film of ink onto the printing plate.

The ***dryer and chill rolls*** evaporate volatile solvents and cool the heat-softened binding resins. Final solidification or setting of the ink occurs with the cooling of the binding resins. Web offset presses may be either heatset or nonheatset. ***Heatset*** presses

Figure 2-5. *Goss C700E heatset web offset press. (Courtesy Goss Graphic Systems, Inc.)*

Figure 2-6. *Sunday 2000 web offset press. (Courtesy Heidelberger Druckmaschinen AG)*

employ a hot-air dryer to evaporate solvents from inks, leaving a semidry ink film. Chill rolls are needed in this type of system to cool the hot web after drying. **Nonheatset** web presses may have no dryer at all or may employ other types of dryers, such as ultraviolet lamps for curing UV inks.

The **delivery system** of the press, where the final printed material comes out, can perform a variety of potential **in-line finishing** functions. Presses vary tremendously in capabilities and sophistication of in-line finishing from one press to another. However, a few in-line finishing capabilities are commonplace on web presses. A **folder** delivers folded signatures ready for mailing or for binding with other signatures to form a magazine or book. On many presses, folding is the final press operation, whether it is a "former fold," "jaw fold," "chopper fold," or "quarter fold." A **sheeter** cuts the web and delivers flat, printed sheets. A **rewinder,** as the name implies, rewinds the printed web back into roll form; this is done when the web is to be finished off line. A folder produces signatures; a rewinder produces rolls. The bulk of web offset work involves folding and producing signatures. Presses may also be equipped with remoistenable glue applicators, inkjet labeling devices, plow folders, perforators, variable-cutoff units, as well as many other finishing devices.

Orientation to Press Direction

The two sides of the press have specific designations. One side of the press houses the drive shaft and gears that power the press. This side of the press is called the **gear side.** The crew always works on the other side, because this is where all of the press controls are located. This is called the **operator side** of the press. The ends of the press are referred to as the **infeed** (or **front)** and the **delivery.** The terms **sheeter** and **folder** are also commonly used to refer to the delivery end of the press.

Web Offset Press Configurations

Web offset presses are available in a variety of special configurations or arrangements of printing couples within the printing units. A ***printing couple*** is an assembly that includes an inking system, a dampening system, a plate cylinder, a blanket cylinder, and an impression cylinder, all of which are required to apply one color of ink to one side of a substrate. (See figure 2-4.) If another color is to be printed, the paper must go through another printing couple. A ***printing unit,*** on the other hand, is the housing in which a number of printing couples may be mounted. The exact number of couples within the unit determines the number of colors of ink that may be printed as the web moves through the unit.

Another important difference among press configurations is the ability or inability to print on both sides of the web as it moves through the press unit. ***Perfecting*** is the term that describes printing on two sides of the web with one pass of the paper through the press. The increasing need for high-speed printing created by the booming newspaper industry had inspired the invention of the web perfecting press in 1856. Some press configurations are ***nonperfecting;*** these designs permit printing on only one side of the paper as it moves through the printing units.

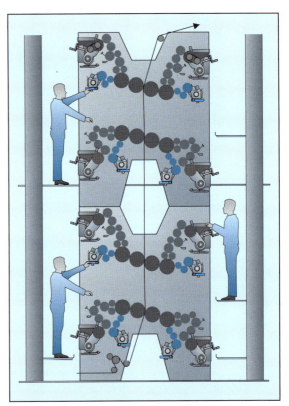

Figure 2-7. *The vertically arranged printing unit of a CROMOMAN web press for four-page newspaper printing. (Courtesy MAN Roland Druckmaschinen AG)*

There are three classifications of web offset presses in use today: (1) in-line, (2) common-impression-cylinder (CIC), and (3) blanket-to-blanket (BTB). Each of these types of presses varies in the impression system, more specifically in how the printing couples are arranged within the units of the press.

In-line presses. Most products made with uncoated paper and printed on one or perhaps both sides are printed on ***in-line presses.*** Business forms are a common example, so much so that these presses are sometimes called ***business forms presses.*** As the paper moves through the press, ink is printed on one side of the web only, thus in-line presses are nonperfecting. Perfecting may be accomplished on these presses by using a turn bar to flip the web over between units, resulting in two-sided printing. Each unit contains a single printing couple consisting of an inking system, a dampening system, a plate cylinder, a blanket cylinder, and an impression cylinder (see figures 2-4, 2-8, and 2-9). Such presses are generally smaller than CIC and BTB presses and are also typically equipped with auxiliary devices matched to business forms products, like imprinters, numbering devices, perforators, and punches.

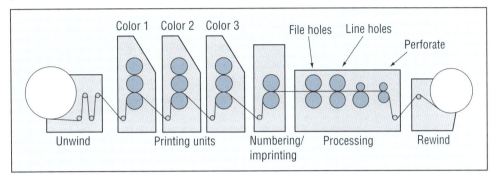

Figure 2-8. *An in-line press.*

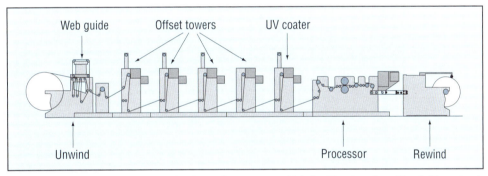

Figure 2-9. *A web press system for business documents and labels. (Courtesy Brandtjen & Kluge, Inc.)*

Common-impression-cylinder presses. Each printing unit of a common-impression-cylinder (CIC) press has one very large impression cylinder with the blanket cylinders of four or five printing couples arranged around it. The web of paper wraps around the surface of the impression cylinder. Because of the arrangement of the couples and the size of the impression cylinder, these presses are also termed *satellite presses.*

A CIC press can be made to perfect in two ways. The most common means is to turn the web over between units, in much the same manner as

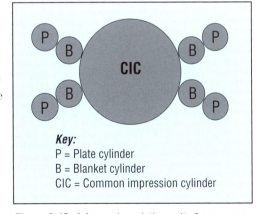

Figure 2-10. *A four-color printing unit of a common-impression-cylinder press (CIC) with four printing couples arranged around the large common impression cylinder. Note: For clarity, the inking and dampening systems are not shown.*

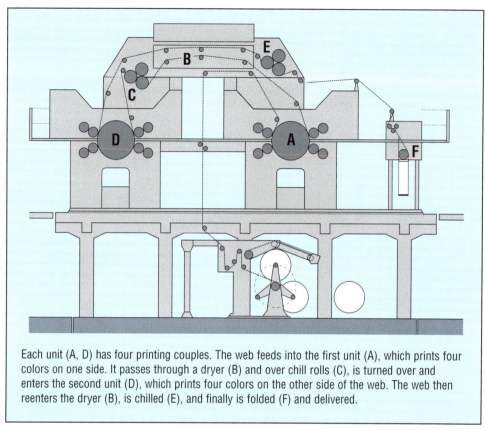

Each unit (A, D) has four printing couples. The web feeds into the first unit (A), which prints four colors on one side. It passes through a dryer (B) and over chill rolls (C), is turned over and enters the second unit (D), which prints four colors on the other side of the web. The web then reenters the dryer (B), is chilled (E), and finally is folded (F) and delivered.

Figure 2-11. *A two-unit common-impression-cylinder (CIC) press.*

in-line presses. The second method involves printing half-webs of paper, where the width of the web is only half the width of the plate cylinder. The web is threaded so that it runs along one side of the printing unit, resulting in the top of the half-web being printed. The web then moves through a turning bay where it is flipped over, and returned to the printing unit, where it runs on the other side of the cylinder. On its second pass through the printing unit, the other side of the web is printed. During operation, paper runs continuously on both sides of the press. This method of perfecting on CIC web presses is called ***double-ending.***

The large size and longer makereadies are disadvantages inherent in the design of CIC presses, but their speed is higher than that of blanket-to-blanket presses.

Blanket-to-blanket presses. The most common commercial web offset press in the United States is the blanket-to-blanket (BTB) press, which is a perfecting press. Each printing unit on a blanket-to-blanket press has two blanket cylinders simultaneously printing both sides of the web, with each blanket cylinder serving as the other's impression cylinder. Each blanket cylinder is part of a printing couple that comprises a dampening system, inking system, and plate cylinder. Each unit of the web press contains two printing couples, typically with one on the top and one on the bottom. These presses have no impression cylinders; the blanket cylinder of the top couple acts as the impression cylinder for the bottom couple, and vice versa. Perfecting capability is inherent in the BTB design

Blanket-to-blanket press units come in two basic configurations. The most common arrangement is horizontal, where the web runs through the printing units in a horizontal plane; the printing couples are therefore stacked vertically, one on top of the other. Vertical blanket-to-blanket presses, common in newspaper printing, have the web running vertically between blankets, with the couples organized horizontally and with the folder located in the center of the press. One advantage of this arrangement is easier access to the printing couples. With a vertical web press, the rolls are usually located on a level below the pressroom.

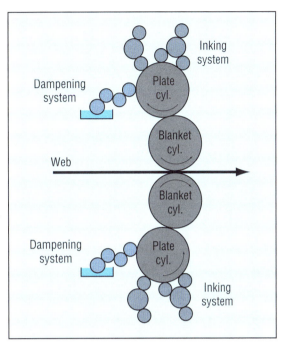

Figure 2-12. *A typical blanket-to-blanket printing unit (horizontal). Each blanket cylinder acts as an impression cylinder for the other couple in the printing unit. Blanket-to-blanket units have two couples, arranged to simultaneously print both sides of the web.*

Figure 2-13. *CROMOMAN web press for four-page newspaper printing. (Courtesy MAN Roland Druckmaschinen AG)*

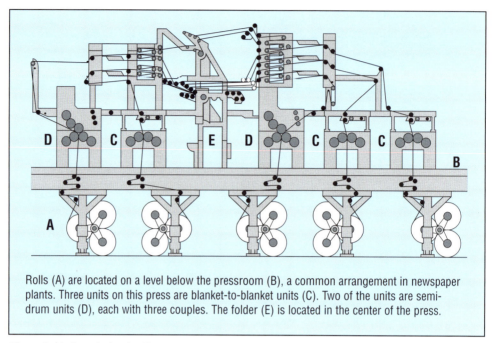

Rolls (A) are located on a level below the pressroom (B), a common arrangement in newspaper plants. Three units on this press are blanket-to-blanket units (C). Two of the units are semi-drum units (D), each with three couples. The folder (E) is located in the center of the press.

Figure 2-14. A vertical web offset press.

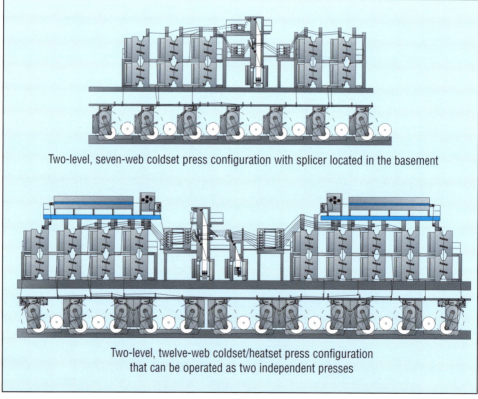

Two-level, seven-web coldset press configuration with splicer located in the basement

Two-level, twelve-web coldset/heatset press configuration
that can be operated as two independent presses

Figure 2-15. Two possible configurations of the CROMOMAN web press. (Courtesy MAN Roland Druck-maschinen AG)

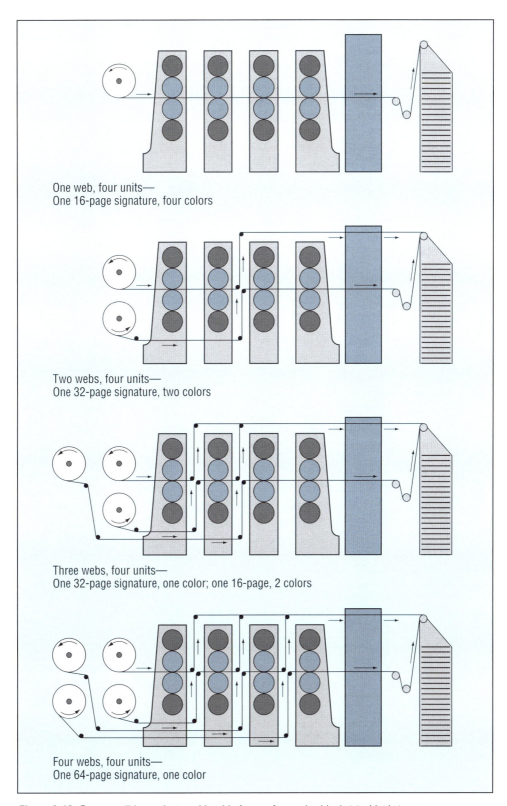

One web, four units—
One 16-page signature, four colors

Two webs, four units—
One 32-page signature, two colors

Three webs, four units—
One 32-page signature, one color; one 16-page, 2 colors

Four webs, four units—
One 64-page signature, one color

Figure 2-16. *Four possible products achievable from a four-color blanket-to-blanket press.*

The popularity of the blanket-to-blanket press is due to several factors:

- It is capable of producing a high-quality product.
- Makereadies are efficient.
- Paper passes through a dryer and cooling system once.

The BTB press is also flexible in terms of web configurations and the number of colors run in one pass. Many of these presses have the capability of feeding one or multiple webs simultaneously. In this way, many different variations of product may be achieved. For example, with a four-unit press, one web can be run straight through, with four colors printed on each side. Alternately, the press might be set up so that four webs are fed into the press, each through one unit resulting in only one color printed per side, but four times the output as a single web. Many variations in between are possible. This flexibility combined with high speed accounts for web offset's popularity. Figure 12-16 shows four possible products produced from a four-unit blanket-to-blanket press.

In rare cases, BTB web printers may employ an "s-wrap" to print by **direct lithography,** rather than offset. In this way, the inked image is transferred directly from the plate to the paper. This enables the press operator to lay an additional color of ink on one side of the web.

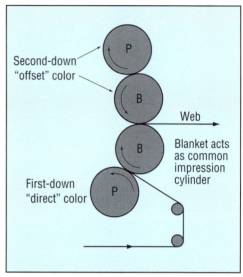

Figure 2-17. *A blanket-to-blanket offset press unit that has been so threaded that the web comes in direct contact with one plate and takes ink from it. Two colors are printed on the same side of the web, one by offset lithography and one by direct lithography.*

Shaftless Presses

Web offset presses have traditionally been constructed with a mechanical line shaft running the length of the press. Each unit is driven off this single shaft, which is powered by a main drive motor. Gears and timing belts make up the gear train, which provides synchronous motion of all driven rollers and cylinders, from the reel stand, through the printing units and folder.

Shaftless presses are becoming more popular with press manufacturers in recent times, and some experts predict that all new presses will be shaftless in years to come. With shaftless presses, manufacturers are attaching individual, independent AC servo motors to each driven element in the press, including printing couples, reel stand, chill rolls, and folder. Each electric motor is controlled via a fiber optic link to a central controller. As such, all units are electronically synchronized and independently controlled for speed, position, and torque. Following are several advantages to shaftless presses:

Figure 2-18. ROTOMAN shaftless web offset press. (Courtesy MAN Roland Druckmaschinen AG)

Reduction of mechanical errors. As already stated, all components of a mechanical line shaft press are synchronized via a line shaft, gears, chains, clutches, belts, and pulleys. Because all of these elements are connected end-to-end, any slight gear backlash (caused by any slight play in the gears) in one unit is passed down the line to the units that follow. The mechanical error introduced, resulting in a slight acceleration or deceleration, is cumulative and can worsen as it moves through the press if more error is introduced. These accumulated errors can cause problems with color register, web tension control, and print faults such as slur.

By contrast, each independent motor on a shaftless press is synchronized electronically, rather than mechanically, capable of maintaining synchronization with other motors to within an arc minute (or 0.008°). Errors are not passed down the line, and a tighter machine stiffness is achieved.

Dynamic synchronization. Each press unit can also be adjusted independently of its companions. Because of this, plates can be changed and blankets washed on one unit without driving or affecting any other unit. This is also possible on mechanical line presses, but it requires clutches and other mechanical parts that make the manufacture of the units much more complicated.

Figure 2-19. LITHOMAN shaftless web offset press. (Courtesy MAN Roland Druckmaschinen AG)

Figure 2-20. *The KBA COMPACTA 215, a shaftless 16-page commercial web offset press. (Courtesy KBA North America, Web Press Division)*

With a five-unit press, the press can be running at full production speed through four of the units while plate changes are made on the fifth unit. When the plate change is complete, the fifth unit can be brought up to press speed, synchronized with the other units, and the impression can be turned on. This capability is known as dynamic synchronization, which is only possible on shaftless presses.

Velocity variations. The speed of each unit is controlled independently, allowing for precise web tension between units. Variations in velocity allow the motor driving a unit to run at a fixed speed percentage above or below press speed. This allows a positive or negative draw of almost any amount to be established between any two units in the press. The draw can be dynamically controlled based on feedback from a tension load cell. This allows for much higher level of tension control on the running web. With mechanical line shafts, any draw must be built into the gear system, with little capacity for variation.

Simplified cylinder timing. Each print couple's circumferential register adjustments can be made directly at the servo motor. Unit-to-unit timing (synchronization) is simple because each unit can advance or retard with respect to the main press synchronization.

Ease of press modification. Because electronically timed presses are designed and built in sections, presses can be modified more easily to meet new production requirements. When adding a new component, there is no drive train to tie into and no gears to align. A new section can be tested and installed and then electronically synchronized to all other press sections.

Factors Affecting the Productivity of Web Offset Presses

Web offset presses are highly productive machines, though they vary widely in cost and capability. There are several key measurements that can be analyzed to determine a press's productivity, including press size, press rate, makeready requirements, and in-line finishing capabilities.

Press size. To analyze press size, both the width of the press and cutoff must be specified. As the paper is printed, the image on the plate cylinder is repeated over and over again along the length of the web. The length of the repeated image is called the *repeat length.* The gaps between the repeated image will be cut into sheets at the delivery end of the press, thus the term *cutoff.* The circumference of the plate cylinder determines cutoff. A larger-diameter plate cylinder will produce a longer repeat length.

Besides web cutoff length, web presses sizes are also determined by the across-the-cylinder direction, which is usually the maximum roll width that can be handled. A wider press is capable of printing a wider web of paper, resulting in more image area per cutoff. It is important to note that a wider press could also print narrower webs, providing the printer with more possible product variations.

The size classifications of web offset presses are typically measured by the size of the signature that can be printed, which are traditionally stratified into 32-page, 24-page, 16-page, 8-page, and narrow web. Signatures are large sheets of paper with multiple pages printed on them. A 32-page signature would contain 32 sequential pages when folded down to the size of a single page. (See figure 2-21.) These terms are based around the number of 8.5×11-in. (216×279-mm) pages or, in metric, A4 pages that can be arranged on a single cutoff.

The larger web offset presses can print a 32-page signature, with sixteen 8.5×11-in. pages on one side and sixteen identical-sized pages on the back. These presses might be about 60 in. (1,524 mm) in width with cutoffs of about 35 in. (889 mm). Presses categorized as *full-sized presses* can produce either 24-page signatures (12 equal sized pages on each side) or in some cases 16-page signatures, with eight equal sized pages printed on each side of the signature. These presses tend to be about 40 in. (1016 mm) to 60 in. in width with cutoffs of about 23 in. (584 mm). *Half-sized presses* print 8-page signatures, with four equal-sized pages on each side. These presses are commonly about 25 in. (635 mm) in width with cutoffs of about 17 in. (432 mm). *Quarter-sized presses* and *narrow webs* vary in size considerably, but typically will print webs from about 20 in. (508 mm) wide and narrower, with cutoffs of about 18 in. (457 mm) and less. This size allows for the printing of a four-page 8.5×11-in. signature or smaller 8.5×11-in. forms.

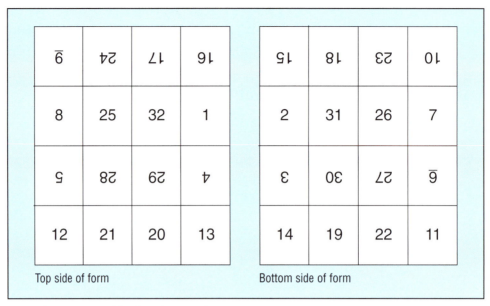

Figure 2-21. An imposition for a 32-page signature.

Many web presses being manufactured today are ***variable-cutoff presses.*** This means that the press design allows for a plate and blanket cylinder assembly of one cutoff to be removed from the press and replaced with a plate and blanket cylinder assembly of a longer or shorter cutoff. This capability allows the printer to match the product with the most efficient cutoff, providing greater flexibility and minimizing paper waste.

Press speed. The rate or speed of a web offset press is typically measured in feet per minute (fpm) or, in metric, meters per second (mps). This measurement refers to the length of the web that is removed from the roll in one minute or one second. Typical web offset press speeds range from 8 mps to about 15 mps in very fast presses.

Sheetfed press rates are always measured in impressions per hour (iph), which is the number of sheets that can be delivered in one hour. Web press rates can be measured in much the same manner. However, on web presses the rate is defined as cutoffs per hour (cph). This rate measurement refers to the number of cutoff signatures that are produced each hour. It is important to note that a variable-cutoff press moving at 10 mps will have various cph ratings, depending upon the cutoff length. Conversions between mps to cph can be calculated as follows:

- Number of millimeters per cutoff ÷ 1000 = meters per cutoff
- Speed in mps ÷ meters per cutoff = cutoffs per second
- Cutoffs per second × 3600 = cph

Makeready requirements. *Makeready* is defined as all of the preparations performed on the press in order to produce the first salable sheet or signature coming off the delivery. This may include changing blankets, inking the press, mounting a roll and webbing the press, hanging plates, cleaning impression cylinders, running pre-

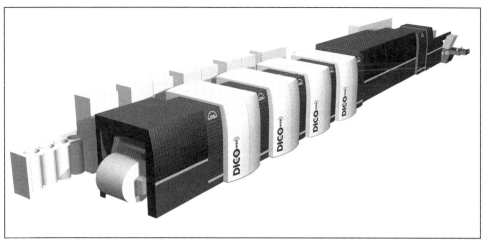

Figure 2-22. The DICOweb press, a computer-to-press offset printing system. This system is based on DCO (digital changeover) technology that allows printers to image, erase, and reimage printing forms directly on the press. (Courtesy MAN Roland Druckmaschinen AG)

liminary impressions, checking ink film thickness and color, checking and correcting registration problems, and setting up in-line finishing devices. These functions can require a considerable amount of time—time that will increase the cost of the job. Web presses vary greatly in the amount of automation equipment incorporated into the design. Newer presses tend to incorporate a great deal of efficient automation features, including automatic plate hanging devices, automatic registration, and automatic blanket and impression cylinder washers. These devices cut down considerably on the time required for makeready. However, presses with these features also tend to cost more.

Auxiliary Equipment

Auxiliary equipment for a web press is composed of those devices that perform specialized operational functions. These devices can significantly improve print quality and increase productivity. The following auxiliary equipment may be found on a web offset press:

- **Remote control console,** a free-standing computerized device that enables the press operator to control numerous press functions without leaving the inspection table. A computerized (remote) console allows the operator to adjust inking, dampening, and circumferential register; control ink density and ink trapping; and monitor dot gain.
- **Plate scanner,** a device that measures the image area percentages at selected increments across the printing plate prior to mounting the plate on press.
- **Scanning densitometer,** a computerized quality control table that measures and analyzes printed color bars using a densitometer.
- **Web preconditioners,** located in the infeed before the infeed metering roller. They moisture-condition the web and burn off paper lint and slitter dust. They also help to reduce paper stretch and blistering problems.

- *Sheet cleaners,* mounted on the press in front of or in place of preconditioners. They remove loose paper, lint, and dust from the surfaces and edges of the web before it reaches the first printing unit.

- *Ink agitators,* which keep ink in the fountain in constant motion, lowering its viscosity and allowing it to flow to the forward part of the fountain and transfer to the fountain roller.

- *Water-cooled ink oscillators,* which maintain the temperature of the ink oscillator at a level that will not set the ink.

- *Fountain solution recirculation systems,* which pump fountain solution into each fountain pan from one or more central tanks.

- *Fountain solution refrigeration unit systems,* which keep the solution at constant, low temperature to avoid problems due to fluctuating dampening solution temperature; e.g., changes in viscosity. Chemical changes occur more slowly at lower temperatures.

- *Fountain solution mixers,* usually incorporated into recirculation systems. These units mix in small quantities, constantly adding fresh fountain solution and maintaining greater uniformity.

- *Automatic blanket washers,* which consist of a row of nozzles, mounted over the blanket, from which solvent is sprayed onto the rubber surface. Solvent and loosened material are carried off by the web. Some automatic blanket washers also utilize a cloth and brush system.

- *Automatic plate changers,* devices that automatically mount a new printing plate on the plate cylinder.

- *Side-lay sensors,* which detect minute lateral web movement and signal steering devices to adjust the paper's position.

- *Web break detectors,* devices that mechanically, electronically, or pneumatically detect a break in a web of paper, subsequently stopping the press.

- *Remoisturizers,* devices that apply moisture to the surface of the web after it is chilled in order to replace lost moisture during the drying process and to minimize dimensional change after binding and trimming.

- *Antistatic devices,* which turn the air immediately adjacent to the web into a conductor by ionizing the air. Antistatic devices employ electricity, ultraviolet light, and even a weak radioactive field to neutralize static charges.

- *Imprinters,* usually a flexographic unit with a rubber letterpress plate, an impression cylinder, and a simple inking system. Used to print a small amount of copy such as names and addresses for varying pressruns.

These auxiliary devices are discussed in detail later in this book.

Section II
Safety

3 Pressroom Safety

General Safety

Technological advancements in printing presses have led to increased press speeds. Faster presses pose additional safety problems; however, improved training methods and guarding devices have reduced potential hazards. A systematic safety approach can further reduce the number of accidents. Press operators should work under the basic principle that all accidents in the pressroom can be prevented. Nearly 85% of all accidents are caused by unsafe acts, while the other 15% are caused by unsafe conditions. Chemicals, for example, that are not handled properly can present many hazards. To protect workers and the environment from chemical hazards, federal,

---

Press Safety

- Exercise extreme care and precaution when operating any printing press.
- Observe and practice all safety rules, regulations, and advice given in the press manual and by the facility's hazard communication program and lockout/tagout program.
- Never make major repairs or perform maintenance when the machine is running. Any work on the press, unless considered minor service and maintenance, should be performed only if the power/energy is locked out.
- Obey all verbal and written instructions before operating the press.
- Always wear PPE where necessary or instructed to protect against personal injury.
- Never wear loose clothing that will become entangled in any part of the press equipment.
- Always stand clear of the equipment when the "run" warning signal is sounded.
- Always make sure the press is completely stopped and the safe button is set before touching any of its operating parts.
- Check all safety devices on the press each shift to ensure that they are reliable and working.
- Never switch off or bypass safety devices.
- In stopping a press, make sure that it has been put on "safe" using the "stop (security)" push button.
- Check that all guards, covers, and swiveling footrests are securely fastened or completely locked in place before operating the press.
- Check that stairs, footrests, running boards, gangways, platforms, and other equipment surfaces are clean and free of grease, oil, or debris. Never place tools and supplies on these surfaces.
- Grasp handrails securely when ascending the platforms, standing on the platforms, and before leaving the platforms.
- Only clean the ink fountains while the press is stopped and the safe button depressed to avoid personal injury and press damage.
- Never work on moving parts with shop towels, tools, etc. because of the high risk of accident and personal injury.
- Replace guards immediately after removing any press washup devices.
- Use the reverse button on the press only for plate removal and web splicing, not for cleaning or gumming cylinders, etc.

state, and local agencies have established rules and regulations that help to assure safety, health, and a clean environment. This chapter highlights some of the major safety concerns in the pressroom.

Press Operator Training

The press operator should be properly trained before attempting to operate any equipment. Follow the press manufacturer's recommended operating procedures. Develop supplemental safety procedures to ensure additional personal safety.

Avoid or cover long hair, avoid loose fitting clothing, and do not wear jewelry or ties near moving machinery. Protect yourself from all inherent dangers. Wear a hard hat and ear protection if necessary. Always wear steel-tipped safety shoes to avoid injury by heavy paper rolls or tools; the soles of the safety shoes should provide adequate traction without generating friction sparks. Wear the recommended protective gear and read the Material Safety Data Sheets before handling chemicals.

Caution: Make sure that guards, audible warning devices, flashing warning lights, and personal protection equipment are in use at all times. Alert all crew members before the press is started. Obey all warning devices and labels.

Figure 3-1. Bilsom® Viking heavy-duty ear-muffs attached to safety helmet. (Courtesy Bacou-Dalloz)

Press Location

Installing a press requires careful planning. The efficient use of floor space is essential. The designated location of the press should meet the following requirements:

- The floor should be able to support the weight of the press.
- Aisle space around the press should be large enough to permit the safe transportation of materials and supplies.
- Press operators should have adequate space to safely remove the printed signatures from the delivery.
- The press should be operable and serviceable without interfering with adjacent equipment.
- The location should have sound-absorbing devices to prevent the amplification and transmission of noise generated by the press.
- Space for auxiliary equipment and containers should be provided.
- Lighting conditions and ventilation must conform to local agency requirements.

A good press location must be accompanied, however, by good housekeeping to establish a safe working environment. All spoilage must be placed in proper waste containers that are conveniently located. Solvent-soaked rags should be placed in a closed container.

Production Safety

- Never operate equipment unless properly trained and authorized.
- Ensure that all guards, shields, and panels are in place before operating the press.
- Never release a safe button set by someone else.
- Never restart a machine that has stopped without an apparent reason.
- Check for persons, tools, or pieces of equipment between and around the press before operating it.
- Remove all used plates, tools, and equipment from the press area and alert your co-workers before starting the machinery.
- Wear hearing protection devices when working in areas with high noise levels.
- Never permit workers with jewelry, loose clothing, or exposed long hair near the press.
- Never lean or rest hands on running equipment.
- Avoid carrying tools in pockets to prevent the possibility of dropping them into the press or other hazardous locations.
- When making press adjustments, use only recommended tools that are in good working condition.
- Keep hand and body parts clear of nips, slitters, and all moving parts when operating the press.
- Never reach into the press to make adjustments while it is running.
- Never attempt to remove hickeys from plate or blanket cylinders, or lint and dirt from rollers while the press is running unless a specifically designed safe tool is used.
- Never wipe down cylinders, plates, rollers, or blankets while the press is moving.
- To perform any cleaning or minor service procedure, use the inch-stop-safe method.
- Check plates periodically for any loose edges or cracks.
- Follow instructions carefully when mixing or handling chemicals in the pressroom.
- Maintain a complete and accessible file of all service manuals, instruction manuals, parts lists, and lubrication manuals or charts for each piece of equipment in the pressroom.

Mechanical Safety

Machine Guarding

Machine guards are used as barriers to protect workers from the dangers of pinch points and hazard points on industrial equipment. (A pinch point is where two rollers or cylinders first contact each other.) Manufacturers now enclose all moving parts so that workers cannot harm their fingers or hands. In addition to eliminating direct contact between workers and moving machine parts, guards (1) prevent accidents caused by human error, (2) prohibit worker contact with exposed electrical components, flying metal objects, and splashing machine oils, and (3) permit safe maintenance practices. OSHA says that guards must be machine- and task-specific, prevent access to the safety hazard while the equipment is running, be impossible to remove during equipment operation, not constitute a hazard to the operator, and require a minimum of maintenance.

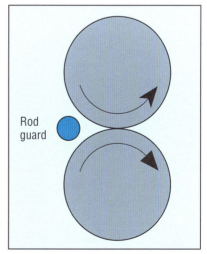

Figure 3-2. The use of guards to protect workers from pinch points. Such guards must be close enough to the rolling surface to prevent passage of a finger.

Figure 3-3. Nip guards between plate and blanket cylinders to prevent the operator's fingers from entering the in-running nip.

In the ideal situation then, a machine should not be able to run without its guard in place. Problems arise, however, when old equipment lacking guarding or equipment with damaged guarding is still in use. In either case, OSHA requires that working guards be installed. OSHA also states that employers should put in place an employee training program covering the proper placement of guards, the correct use of the "stop/safe/ready" button system, and safe cleaning procedures.

Lockout/Tagout Regulations

OSHA's Lockout/Tagout Standard went into effect on January 2, 1990. The standard requires that employees/operators place lockout/tagout devices on their machines whenever they are servicing them and there is a potential for a hazardous release of energy or accidental start-up. In other words, if the equipment is engaged in any activity other than its intended function, with the exception of minor servicing, maintenance, and setup activities that keep the employee's body out of the areas of potential contact with machine components, and/or do not expose the employee to unexpected energization, activation of the equipment, or release of stored energy, lockout/tagout requirements apply.

To comply with the Lockout/Tagout Standard, employers must complete the following:

1. Perform a job hazard analysis to identify activities and potential hazards associated with machinery or equipment.
2. Develop a detailed list of the specific lockout/tagout procedures and when they will be used for each appropriate piece of machinery with a potential energy hazard.
3. Develop a written program.
4. Develop and implement an employee training program.
5. Develop an outside contractor program.

Minor servicing, maintenance, and setup exceptions. Minor servicing and maintenance is defined by OSHA as "those tasks involving operations which can be safely accomplished by employees and where extensive disassembly of equipment is not required."* Minor servicing and maintenance activities include, but are not limited to the following:

OSHA's Lockout/Tagout Regulation and the Printing Industry. GATF et al., pp. 23–29.

- Clearing certain types of paper jams
- Minor cleaning (blanket washing, roller washing)
- Lubricating and adjusting operations
- Paper webbing and paper roll changing

Setup is defined as "any work performed to prepare a machine or equipment to perform its normal operation." Setup activities include the following:

- Mounting a plate
- Setting bearer pressures
- Setting folder adjustments
- Setting rollers

Printers have the option of using "alternative effective protection" for minor servicing and maintenance and setup procedures. OSHA defines an activity as meeting alternative effective protection requirements as they apply to the printing industry if the following conditions are met:

1. Servicing is conducted when the machine or equipment is stopped.
2. Each servicing employee has continuous, exclusive control of the means to start the machine or equipment.
3. Safeguarding is provided to each servicing employee to prevent exposure from the release of harmful, stored, or residual energy.

The inch-safe-service method used in conjunction with a safety system, described in the ANSI standard for printing presses (B65.1-1985), has been recognized by OSHA as one means of providing adequate alternative means of protection.

Servicing and/or maintenance activities that are not performed during normal production and require lockout/tagout procedures include, but are not limited to the following:

1. "Operations where auxiliary motors and pile motors are not disabled by the SAFE button and where the operator cannot maintain exclusive control of the machine or machine elements such as when:
   - Cleaning frames and braces
   - Cleaning the feeder and delivery on sheetfed presses
   - Cleaning the reel stand and other parts of the infeed on web presses
   - Cleaning or replacing air filters used to supply ventilation for toxic or flammable materials or heat-generating electrical equipment."*
2. "Operations that require the machine operator to remove major parts of the equipment such as:
   - Panels or other barriers that restrict access to moving mechanical parts or energized electrical equipment
   - To perform extensive work without removal of such components
   - To perform work requiring the operator to leave the immediate area containing the operation controls where exclusive control by the operator is required."

---

*OSHA's Lockout/Tagout Regulation and the Printing Industry. GATF et al., pp. 23–29.

3. "Roller removal would require Lockout/Tagout when two people are required and/or there are no quick-release sockets which would permit safe roller removal by one person."[*]

4. Gripper bar repair/removal, gear replacement, and electrical work.

Washup. The following procedure is recommended for cleaning blankets and applying gum or finisher to plates on multicolor presses:

- The press operator in charge should lock out all start buttons other than the one being used; all crew members should be clear of the press.

Figure 3-4. Circuit breaker lockout bracket with padlocks attached. Notice that the locks have identification tags attached. (Courtesy American Ed-Co, Inc.)

- The press operator inches the press to expose a section of the blanket and plate and then engages the stop lock button on the first unit.

- The assistant engages the stop lock button on the second unit.

- Both press operators clean the exposed segments of their respective blankets. Gum or finisher is applied to the plates.

- When both operators are clear of the cylinders, the stop lock buttons are disengaged and a signal bell or buzzer is sounded.

- The operator on the first unit inches the press until the next segments of the blanket and plate are in position.

- This procedure is repeated until all blankets have been completely cleaned and all plates have been gummed.

The Roll Stand

Technological changes have automated many of the rolltender's functions; however, spindle-mounted rolls that require some manual handling are widely used. The spindle is a steel shaft that runs through the core of the roll.

Running and braking increases the temperature of the spindle. The rolltender must wear gloves and handle the hot metal with caution. Furthermore, the weight of the spindle requires the rolltender to use the proper lifting technique (i.e., knees bent, back straight).

---

[*]*OSHA's Lockout/Tagout Regulation and the Printing Industry.* GATF et al., pp. 23–29.

Infeed and Delivery Safety

- Read all sections of the press manual discussing specific safety rules, safety guard devices, and safety button operations for the infeed and press delivery areas.
- Make sure the control lights on the operator panel are working correctly.
- Make sure the dryer controls (pyrometer, indicators) are operating properly.
- Identify the following safety devices in the infeed section of the press: (1) main stop (safe) push button. (2) bell/buzzer (warning and signaling device), (3) guards for drive shafts, pulleys, and rollers, (4) emergency stop push buttons, and (5) guards in front of and around the splicer. Consult the sections of the press manual listing specific safety devices and discussing their proper identification and operation.
- Identify the following safety devices located in the delivery section of the press: (1) guards above and/or in front of the folder/sheeter, (2) web travel safety, stopping the press if web breaks or moves beyond the web guides, (3) safety plate above web transport areas, (4) main stop (safe) push button, (5) bell/buzzer (warning and signaling device). Consult the sections of the press manual listing specific safety devices and discussing their proper identification and operation.

Hooks suspended by cables or chains are used to hoist the roll into sockets for running. A roll of stock usually weighs 2,000 lb. (900 kg) or more; therefore, the hooks must securely grasp the spindle before the roll is hoisted into position. The spindle must be long enough to be adequately supported at each end by the sockets. All crew members should remain clear of the area beneath a suspended roll. Roll-tenders should wear leather-palmed gloves to protect their hands from injuries that may be caused by the hoisting cables or chains.

Reel-mounted rolls present fewer potential hazards; however, the rolltender must exercise care when handling and transporting a roll of paper to the infeed.

Avoid contact with the knives that cut the expired roll; make sure that they are in the "rest" position when not in use.

Rolls are usually wrapped in a heavy kraft paper to protect them prior to their use. A sharp knife or razor and a stripper are typically used to remove the wrapper. Push sharp instruments when cutting; never pull them forward.

Inking System

Many rollers comprise the ink train of an offset inking system, creating numerous pinch points wherever two rollers run in contact. These pinch points should be shielded by guards to prevent accidental contact with moving rollers. Metal gratelike guards or angle irons solidly mounted to the press frame should extend across the width of exposed roller nips.

Cylinders

The nip between the plate and blanket cylinders should be guarded to prevent personal injury and mechanical damage. New system standards require the press operator to simultaneously depress the inch and start buttons to rotate these cylinders. This eliminates the possibility of unintentionally starting the press at high speed. These systems should be operable at all times; do not electrically bypass them.

Printing Unit Safety

- Check the guard(s) in front of the inking unit at the infeed side of the press. If this guard is open, the press should not be able to be inched or run in a continuous motion.
- Check the guard(s) in front of the plate and blanket cylinders (delivery side of the press). This guard must be brought back to its original position in order to move the press.
- Check the finger safety nip guard(s) between the last inking form roller and the plate cylinder. If interlocked and the nip guard is moved, the press will immediately stop. The nip guard must be brought back to its original position in order to move the press.
- Check the finger safety nip guard(s) between the plate and blanket cylinder. If this nip guard is moved, the press will immediately stop. The nip guard must be brought back to its original position in order to move the press.
- Check the guard between the dampening and inking rollers. When this guard is open, the press cannot be inched or run. The guard is designed so that hickeys can be removed while the press is running, but only if a specially designed tool is used.

Press operators should stop the press before removing dirt or debris from the plate or blanket. A remotely operated hickey-picking device may be used to perform this job while the press is running.

When the press is initially webbed, or when it is rewebbed, the lead edge of the paper should be tapered to improve visibility of the printing nip during manual paper feeding. The press operator should grasp the paper several inches behind the lead edge.

Folders

Folders perform a variety of folding and cutting operations. The slitters should have guards that cover them as completely as possible. The exposed side that cuts the paper into ribbons is potentially dangerous during rewebbing of the press. Rollers and cutting cylinders in the folder easily grasp fingers, loose clothing, and long hair. Pinch wheels or rolls exert tremendous pressure on the ribbons to feed them into the cutting cylinders; press crew members should avoid contact with all moving machinery.

The tapered lead edge of the paper also improves visibility of roller nips in the folder and allows the press operator to safely feed the paper. Grasping the paper several inches back from the lead edge reduces the risk of injury.

Material Handling

Newly delivered materials should be placed in a specific receiving area and kept clear of aisles and access ways. Always remember that heavy items should not be lifted unless the proper lifting procedure is followed. Following are some of the practices that should be followed when loading or moving a roll of stock:

- Use proper crane/cable and hook lifting equipment.
- Keep clear of the area beneath the roll of stock.
- Roll tenders should wear proper gloves to prevent injury from heated spindles, cables, and chains.
- Lift with leg muscles when lifting spindles.
- Use proper care when handling cutting tools for removing protective kraft paper.

Following are some of the practices that should be observed when handling skids:

- Never lean skids against the press.
- Never leave skids standing on edge as they can fall over and cause personal injury or physical damage to equipment.
- Skids should have a specific storage place until properly discarded.
- Empty skids should always be transported on dollies, not by hand.
- Empty skids should be stacked one up and one down so that the pile is stable. (The maximum pile height is 5 ft. or 1.52 m)
- Always obtain help when handling large skids.
- Never drop skids; always place them properly on the stack.
- In loading skids, make sure the load will not topple or slide if bumped accidentally.
- Remove steel strapping on skids of material in such a way that the strapping will not spring when cut. (Steel strapping cutters with extension handles should be used and safety glasses should be worn. The cutter should stand to the side and place a gloved hand over the cut.)
- Properly and safely dispose of the scrap metal as it is cut away from the skids.

Improper usage of dollies and lift trucks can result in injury. Observing the following precautions will lessen the likelihood of an injury:

- Always wear steel-toe shoes when working around heavy objects and equipment.
- Return lift trucks to designated areas when not in use.
- Never ride or stand on lift trucks at any time.
- Use lift trucks only after proper operating instructions and authorizations are given.
- Report any difficulties noticed while operating the lift truck.
- Check all components (wheels, lights, horns, batteries) of the lift trucks for good operating condition at the beginning of your shift.

Press Decks

Large presses may include three or four levels of decks, which may be located more than 20 ft. (6 m) above the main floor. Catwalks are common on many large presses. Decks, guard rails, stairs, and ladders should conform to state and local agency standards as well as any applicable Occupational Safety and Health Administration (OSHA) standards.

Repairs or maintenance procedures may require the removal of deck components, which should be reassembled before subsequently running the press. All hinged deck sections should be closed during running.

Deck materials should be sturdy and slip-resistant. Metal or wooden deck surfaces require regular cleaning to prevent dirt, oil, or grease from accumulating on them.

Figure 3-5. Platforms with railing on a multilevel press.

Decks should not be used for storage, as containers or tools could easily fall from them. Elevated decks should be furnished with toeboards to prevent objects from rolling over their edges onto people or equipment located below them.

Tools

Tools required for the operation and maintenance of a press should be stored on a tool rack or in a tool box. A stray wrench or screwdriver can easily fall into and damage a printing unit. Furthermore, specific tools should only be used for their intended purpose. Misuse may result in personal injury or mechanical damage. Pliers, for example, should not be used on nuts and bolts, which can easily become stripped and no longer adjustable by the proper tool. Use spark-proof tools in areas where solvents are employed.

Fire Safety

Such familiar materials as isopropanol, ink solvents, and washup solvents like naphtha ignite easily, and they explode under certain conditions. Smoking and open flames must be prohibited when using them. Fire extinguishers are required by law in factories and offices, and they are especially important in the printing plant. They must be clearly labeled and readily accessible. They must be checked at regular intervals to be sure that they are fully charged and in operating condition.

Areas in the plant where flammable or corrosive chemicals are handled should be equipped with safety showers (figure 3-9C).

Different types of fires require different methods of extinguishing:

- **Type A**—paper, wood, solids. Water is a satisfactory extinguishing agent. (Carbon dioxide will also work.)
- **Type B**—liquids, solvents, gasoline. Carbon dioxide is satisfactory. Water may cause the flaming solvent to spread.
- **Type C**—electrical fires. Special extinguishers are required. Water will aggravate the fire.

- **Type D**—igniting metals (magnesium, aluminum, sodium). Although these are not usually found in printing plants, carbon dioxide and water do not extinguish the flame.

Spontaneous ignition or combustion. The oxidative-polymerization reaction by which sheetfed inks dry is exothermic (i.e., it gives off heat). Waste, wiping cloths, and rags that contain drying oils generate heat as the oil slowly reacts with oxygen in the air. As the temperature increases, the reaction speeds up, and more heat is generated faster. The problem is aggravated by the presence of an easily ignited ink solvent. If a pile of wiping cloths or rags soaked in ink or varnish is allowed to stand, it can generate enough heat to ignite the solvent and the cloths. Spontaneous ignition or spontaneous combustion has caused many serious fires.

Figure 3-6. Can for holding used wiping cloths. (Courtesy Fisher Scientific)

Cans (figure 3-6) containing wiping cloths used to clean the press or to wipe up ink must be emptied regularly, preferably once a day.

Electrical ignition: sparks, grounding, and bonding. The printing plant usually has solvent vapors present in the air. Near solvent cans or the press, the concentration of vapors can be high. A spark can ignite the solvent, creating a fire or explosion.

All bulk containers of flammables should be kept in suitable facilities outside the pressroom. They should be both grounded and bonded to prevent sparks from static electricity (figure 3-7). The terms "bonding" and "grounding" are often used

Figure 3-7. Grounding and bonding of solvent containers.

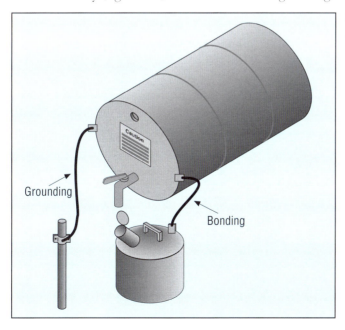

interchangeably, but bonding is the elimination of a difference in electrical potential between objects, such as a solvent drum and a can; grounding eliminates a difference in potential between an object and the ground. Effective grounding and bonding requires that the connections be conductive.

Metal safety cans with a spring-closed lid and metal screen (flame arrester) (figure 3-8) must be used for carrying small amounts of flammable liquids in the pressroom. The screen inside the mouth of the can helps to prevent sparks from entering the can and causing an explosion. The screen acts by diffusing the heat of the spark, in the same way that the screen inside a miner's lamp prevents explosions.

Figure 3-8. Metal safety can for solvents. (Courtesy Fisher Scientific)

Electrical Safety

Because electricity is so common, employees may not be as conscious about its dangers as they should be. Electricity can destroy equipment, cause fires, and seriously burn or fatally shock employees. Care and respect are extremely important when working with or near electrical connections. All employees should be aware of how to handle emergencies associated with electricity.

There should be a master switch in the pressroom that can disconnect power to all machinery, appliances, and outlets at once. This switch ensures that all power is "off" in case of an emergency. In addition, "panic buttons" (emergency disconnect switches) are installed at strategic locations on a press so that an employee may shut off power during an emergency. Additional electrical safety features of printing presses include the warning and signaling devices (flashing lights, bells, whistles) on equipment that indicate when power to the equipment is engaged.

Damaged or frayed electrical cords, plugs, switches, etc., are electrical hazards and should be promptly replaced. All electrical wiring must be inspected by a qualified electrician so that it conforms to all regulations, and only qualified personnel should replace fuses and reset circuit breakers. Additionally, all equipment must be adequately grounded. For example, all convenience outlet wiring and plugs should be three-pronged because the extra prong and wire ensures that a "ground" connection is automatically provided, eliminating accidental electrical shock from defective equipment. Equipment should never be located near sinks, water pipes, or other sources of liquids that may become an electrocution hazard.

All employees who work on electrical systems must be trained and authorized before doing such work. Appropriate lockout/tagout procedures should always be followed; however, if work on a live electrical system is required, the person doing the work needs to be trained in the proper safe procedures for working around live electricity.

Chemical Safety

All chemical substances can be dangerous if mishandled. Even common chemicals and formulations used in printing must be handled in a safe manner.

Every employer must comply with OSHA's Hazard Communication Standard. Employee training, container labeling, and Material Safety Data Sheets (MSDSs) are cornerstones of this standard. The MSDS provides basic information to help employers and employees assess chemical hazards posed by materials they use.

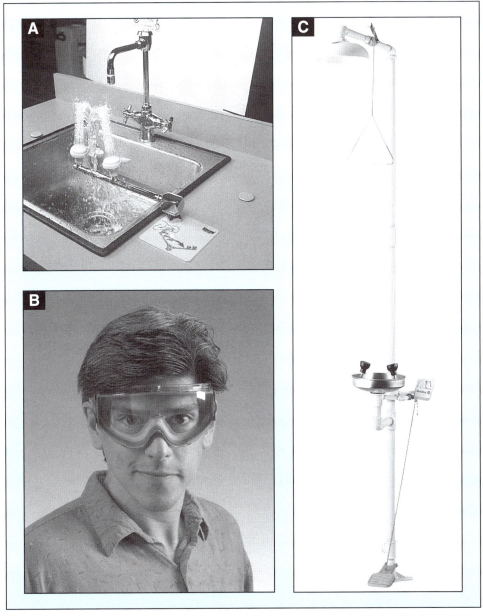

Figure 3-9. (A) Eyewash fountain. (B) Safety goggles for use when working with materials hazardous to the eyes. (C) Safety shower for use when handling flammable or corrosive material in the plant or laboratory. (Courtesy Cole-Parmer Instrument Company)

To comply with the standard, printers must meet the following five requirements:

- Compile a list of all chemicals used in the plant.
- Obtain an MSDS from the manufacturer or distributor for each chemical that appears on the list.
- Properly label each container that contains a hazardous chemical.
- Provide ongoing employee training, including the identification and proper use of all chemicals used in the plant.
- Develop a written hazard communication program.

Under certain conditions, inks, solvents, dampening solutions, and other chemicals used in printing can cause safety and health problems for the workers exposed to them. If any material contains chemical(s) that are defined as health hazards by the Occupational Safety and Health Administration (OSHA), its manufacturer or importer is required by law to provide a Material Safety Data Sheet (MSDS) for that material. An MSDS contains information on the chemical, physical characteristics, safe handling practices, and proper disposal procedures. Employers are required by OSHA regulations to make the MSDSs available to employees. MSDSs for materials no longer in use must be maintained for thirty years.

To help educate employees about the potential hazards associated with using chemicals, the printing industry has adopted the HMIS system. As part of fulfilling the requirements associated with the Hazard Communication Standard, HMIS labels should be affixed to all chemical containers. The label identifies the hazards associated with the chemical and lists the personal protection required when handling that chemical. Employees must be trained in how to read and interpret the HMIS labels.

The Hazardous Materials Identification System (HMIS) uses color coding, numbers, and letters to identify health, flammability, and reactivity hazards. The color blue and the letter "H" indicate a health hazard; 0–4 progressively indicate the severity of the hazard. Red and yellow respectively identify flammability and reactivity hazards; the letter "F" indicates flammability, "R" designates reactivity. The severity of these hazards is also expressed numerically.

The HMIS system also identifies the required personal protection equipment (PPE). Gloves, goggles, an apron, and a respirator are required in various combinations as indicated on the label by alphabetic characters (SA, SB, SC, and SF).

Before a new printing ink or any other new press chemical is used, the technical manager should review the MSDS and product label to evaluate any potential hazards. Some MSDSs contain warnings not present on a product label. The printer must have trained personnel and written emergency handling procedures to deal with accidental exposure and injury caused by hazardous chemicals. For personal injuries, the printer needs to file proper documentation as required by the OSHA.

Solvents

Proper solvent handling is important. Ink solvents present fire, health, and environmental hazards. Training programs together with proper equipment are required to reduce accidents and injuries.

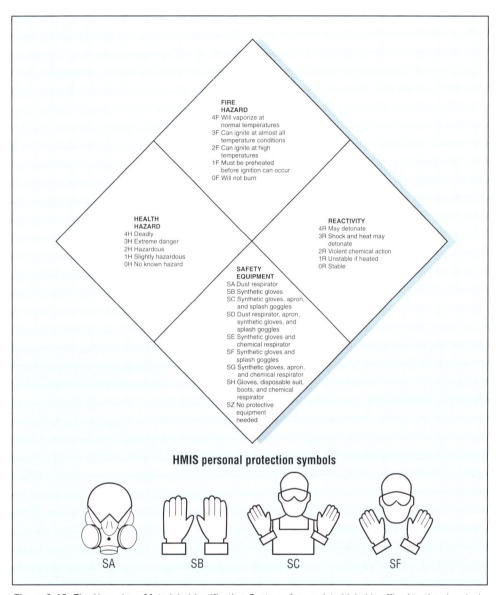

FIRE HAZARD
4F Will vaporize at normal temperatures
3F Can ignite at almost all temperature conditions
2F Can ignite at high temperatures
1F Must be preheated before ignition can occur
0F Will not burn

HEALTH HAZARD
4H Deadly
3H Extreme danger
2H Hazardous
1H Slightly hazardous
0H No known hazard

REACTIVITY
4R May detonate
3R Shock and heat may detonate
2R Violent chemical action
1R Unstable if heated
0R Stable

SAFETY EQUIPMENT
SA Dust respirator
SB Synthetic gloves
SC Synthetic gloves, apron, and splash goggles
SD Dust respirator, apron, synthetic gloves, and splash goggles
SE Synthetic gloves and chemical respirator
SF Synthetic gloves and splash goggles
SG Synthetic gloves, apron, and chemical respirator
SH Gloves, disposable suit, boots, and chemical respirator
SZ No protective equipment needed

HMIS personal protection symbols

SA   SB   SC   SF

Figure 3-10. *The Hazardous Materials Identification System. A completed label is affixed to the chemical container so that employees can be aware of the hazards associated with the chemical and wear proper personal protection equipment.*

Because of the danger of static charges resulting in a flash fire, containers must be grounded and bonded before transferring flammable liquids. Grounding equalizes any charge differential between the container and the ground; bonding equalizes any charge differential between two containers through an electrical conductor.

Containers for press solvents should be approved safety containers, constructed of heavy-gauge metal with lids spring-loaded to the closed position.

Small quantities of solvents are kept at most presses for routine cleaning (figure 3-9). Most solvents are either combustible or flammable; combustible liquids have a flash point at or above 100°F (37.8°C). The flash point of flammable liquids is at or below 100°F.

Potential ignition sources (faulty electrical wiring, static electricity, and friction sparks) should be eliminated. Furthermore, smoking should never be permitted in and around the pressroom.

Liquid	Flash Point	
Hexane	20°F	−6.7°C
Textile spirits	20	−6.7
Mineral spirits	105	40.5
Toluene	45	7.2
Xylene	81	27.2
Kerosene	143	61.6

Table 3-I. Flash point of selected solvents.

Cleaning Rags

Presses often require hand-cleaning, especially after a web break. Cleaning rags should be kept in a designated container when not in use, so as to prevent them from accidentally being pulled into a printing unit. Dirty, solvent-soaked rags present the danger of combustion; therefore, they should be stored in a container that cannot be propped open. The container must remain closed to prevent an accumulation of chemical vapors that could result in a fire. Furthermore, exposure to chemical vapors may be hazardous to pressroom personnel.

Printing Inks

Different printing processes use different kinds of printing inks. Liquid inks used in gravure and flexography usually contain flammable solvents. (Even the water-based inks popular in flexography contain some volatile organic materials to assist in dissolving the binder and to improve performance.) Web offset inks contain large amounts of VOCs (volatile organic compounds). Paste inks, used for offset, screen, and letterpress printing, may or may not be classified as hazardous, depending on the solvent or ink oil, pigments, and additives used in them.

Printing inks rarely cause health or environmental problems when they are used as recommended by the manufacturer. However, MSDSs should be consulted to be sure that inks are properly handled. Traditionally, printing ink manufacturers and chemical suppliers have been responsive to the printing industry's desire to reduce or eliminate the use of highly hazardous materials. In addition, irritating materials have been largely eliminated from inks.

The ink oils or solvents in printing inks are VOCs that can be emitted during the printing process. Other VOCs in the printing plant may include cleaning solvents and solvents used in lithographic dampening solutions. To prevent emissions from exceeding the amount specified by law, VOCs must be controlled or reduced.

Toluene, used as a solvent in many gravure inks, is a toxic, flammable material. It must be handled with great care to prevent hazards to personnel and plant.

Energy-curing inks. Energy-curing inks (UV- and EB-curing inks) deserve special attention because they are chemically different from the more familiar kinds of inks. Energy-curing inks and coatings are promoted as "environmentally friendly." They dry without emission, and they require only about 20–25% as much power to dry as do conventional heatset, web offset inks.

Acrylic materials are intrinsically more toxic and irritating than the vegetable oil and rosin products or the cellulosic, hydrocarbon, and vinyl resins commonly used in inks. To keep eye and skin irritation to a minimum, everyone must take care to avoid getting these inks onto the skin or into the eyes. People working with these inks should wear gloves—nitrile or neoprene gloves, not latex gloves. They must learn not to rub their eyes or their face with gloves that have contacted energy-curing inks.

Barrier creams protect the skin for the short term against inks and many chemicals, but barrier creams do not protect against solvents for energy-curing inks. Clothing soiled with UV/EB-curable materials should be removed and laundered, and the skin should be washed with soap and water.

UV/EB inks and coatings are not highly toxic, but they should not be taken into the mouth or stomach. Eating, drinking, and smoking should be prohibited in the area where these materials are being used.

Storage of printing inks. Small quantities of printing inks should be stored in tightly closed cans in a cool, clean room. Large quantities, stored in tote tanks, drums, or pails, require a secondary containment such as dikes, drip pans, or absorbent material in order to limit any spill to a manageable size, reduce employee exposure to a minimum, and prevent the spill from becoming uncontrollable during an emergency such as fire or explosion.

Dampening System

Dampening solution composition varies from one manufacturer to another; most offer premixed solutions. Printers, however, may find it more economical to mix their own solutions.

Each chemical should be identified on a Material Safety Data Sheet, and the person working with the chemical must be familiar with the proper handling and mixing procedures. Furthermore, suitable protective equipment (rubber apron, rubber gloves, a face shield, and goggles) should be worn when mixing hazardous materials. Eyewash stations should be located in the vicinity.

Figure 3-11. Emergency eyewash station. (Courtesy Bel-Art Products)

Noise Exposure

OSHA requires hearing conservation programs, including annual training on the hazards of noise as well as baseline hearing tests within 30 days of employment and annual audiograms, for all employees exposed to noise levels at or above 85 decibels over an eight-hour time-weighted average. A decibel (dB) is a unit used to express relative intensities of sounds on a scale from zero for the average least-perceptible sound to about 130 for the average level at which severe hearing damage can occur.

Sound Level (dB)	Maximum length of exposure (per day)	Indicators of Sound Level
115	15 min.	Nearly impossible to communicate by voice
110	30 min.	Very difficult to communicate by voice
105	1 hr.	Shout with hands cupped between mouth and other person's ear
102	1.5 hr.	
100	2 hr.	Shout at 0.5 ft. (0.15 m)
97	3 hr.	
95	4 hr.	Shout at 1 ft. (0.3 m)
92	6 hr.	
90	8 hr.	Normal voice at 0.5 ft. (0.15 m), raised voice at 1 ft. (0.3 m), and shout at 2 ft. (0.6 m)
85		Normal voice at 1 ft. (0.3 m), raised voice at 2 ft. (0.6 m), and shout at 4 ft. (1.2 m)
80		Normal voice at 1.5 ft. (0.45 m), raised voice at 3 ft. (0.9 m), and shout at 6 ft. (1.8 m)
75		Normal voice at 2 ft. (0.6 m), raised voice at 4 ft. (1.2 m), and shout at 8 ft. (2.4 m)

Figure 3-12. Maximum exposure lengths per day at various sound levels.

Employers should monitor workplace noise exposure levels with calibrated meters that rate sound in decibels from 0 to 140 and post warnings in areas where the noise level exceeds 90 dB.

If noise levels exceed 90 dB over an eight-hour time-weighted average, workers must wear hearing protection. Personal hearing protection devices include earmuffs, rigid earplugs, and moldable inserts than can be shaped to fit the ear. Properly fitted and used commercial earplugs can reduce noise reaching the ear by 25–30 dB. Earmuffs or cups that cover the external ear provide the best acoustical barrier, reducing noise an additional 10–15 dB. Combining earplugs and earmuffs provides another 3–5 dB of protection.

There are several ways of reducing equipment noise levels. Acoustical ceiling tiles and sound-absorbing baffles and wall coatings, for example, assist in reducing pressroom noise levels. In addition, employees should be encouraged to wear shoes with rubber soles. Floor matting and thick-soled shoes absorb vibrations and reduce noise.

Repetitive Strain Injuries

Even though the pressroom has become increasingly automated, workers are still susceptible to injuries caused by repetitive motion. Each time a worker bends over, picks up a stack of press sheets, fans and jogs it, and places it on the pile table or a pallet, the chance of incurring a repetitive strain injury, also known as a cumulative trauma disorder, increases. Symptoms of repetitive strain injuries include swollen tendons, muscle spasms, numbness, and tingling.

Wrist strain. Carpal tunnel syndrome, the most well-known and debilitating of the repetitive strain injuries, occurs as a result of damage to, or pressure on, the medial nerve, which is located in the arm and travels through the wrist and into the palm of the hand. As the use of computers in the workplace has increased, so too have the reported incidents of wrist strain, carpal tunnel syndrome in particular.

To date, most prevention and treatment efforts have concentrated on developing workstations that lessen the wrist and neck strain on office staff members. The specific concerns of pressroom or other printing industry personnel remain largely unaddressed. For the time being, it is helpful to instruct workers in minimizing wrist strain by observing some basic positioning and lifting requirements. Whenever possible, workers should lift with their elbows down. Elevating unsupported elbows or positioning them too far from the torso makes lifting much more difficult because it requires a physically demanding shoulder swing. Over several hours, the resulting fatigue will reduce efficiency.

Back strain. Back injuries are another common problem caused by repetitive strain. At some firms, employees are required to wear back support belts that are advertised as offering enhanced lumbar and abdominal support while eliminating the need to overstretch muscles. Other companies require employees to wear audible-warning devices on the backs of their shirts. These devices emit 90-dB tones whenever workers bend or lift improperly.

Less controversial and more common are the exercise and training guidelines for avoiding back strain that, unlike the preventive measures for carpal tunnel syndrome, have been in place for sometime.

Employees should size up the weight of a load before lifting or lowering anything. Use two people to lift awkward loads or break the load into smaller groupings if help is not available. During a lift, bend the knees and place the feet close to the load to achieve better balance. Minimize movements of the spine, as even the smallest variations in posture can increase back strain. Keeping the load closer to the body's center of gravity also permits safer movement.

Do not overreach or lift with fast jerking motions, and make certain there are enough places to easily grip the load. Loads with hand grips are ideal. Be especially cautious when lowering a load because the spine is compressed at this time.

Section III
The Infeed System

4 Infeed System Principles

The web offset infeed system encompasses all of the equipment necessary to support the roll and control the speed, tension, and lateral position of the web as it enters the press. A properly setup roll and accurate infeed settings contribute to efficient operation along the entire length of the press. There are many design variations to the web infeed system. However, most systems have the following basic elements:

- **Roll stand.** A roll stand includes a shaft on which to mount the heavy roll of paper. A brake is integrated into the stand to maintain tension on the unwinding web.

- **Splicers.** Almost all large presses have splicers incorporated into the roll stand that function to exchange an expired roll with a new roll without stopping the press. The two variations of splicer are zero-speed and flying paster.

- **Metering rollers.** In addition to the roll stand, most infeeds are equipped with metering rollers that control the rate and tension of paper as it flows into the printing units. The metering rollers are driven from an independent variable-speed drive.

- **Dancer roll.** The speed of the unwinding roll is controlled by a dancer roller (sometimes called a floating roller) that is free to move up and down or forward

---

Safety Precautions

- Wear gloves when handling spindle-mounted rolls to prevent burns from the heat generated while running and braking.
- Observe proper lifting techniques when handling spindles to prevent back injury.
- Make sure hooks are secure and in position in roll spindles before the roll is hoisted into position.
- Stay clear of rolls being hoisted into position.
- Avoid contact with the knives that cut the expired roll, making sure they are in the "rest" position when not in use.
- Push—do not pull—sharp instruments when cutting roll wrappers from fresh roll stock.
- Make sure guards covering folder slitters are in place. The exposed side that cuts paper into ribbons is potentially dangerous during rewebbing.
- Stay clear of rollers, cutting cylinders, and pinch wheels in the folder that can grasp fingers, loose clothing, and long hair.
- Wear leather-palmed gloves when handling wire cables.
- Do not get any part of the body beneath a suspended roll.
- When hoisting a roll, proceed slowly, surely, and safely.

and backward as the press runs. This dancer roller connects to the brake on the roll stand and provides feedback to the brake so that the web does not lose tension. A second dancer roller may be located between the infeed metering rollers and the first printing unit for additional tension control.

- **Lateral positioning guides.** Most infeeds incorporate controls that laterally position the web before it enters the first printing unit. This maintains side-to-side register on the web.

Critical Tension Control

Of primary importance is the ability of the infeed system to maintain consistent tension on the web. A long ribbon of paper responds much like a rubber band to a pulling force. When the ribbon is pulled, the width of the paper decreases and the length increases. This relationship between force and the stretch of paper is called the *modulus of elasticity.* This principle will be presented in detail in chapter 7. Consistent tension will assure that the paper stays at a fixed width so that successive colors can be printed on the web in register. Consistent tension may also help to minimize the potential for web breaks during the pressrun.

Roll Stand and Dancer Roll Principles

There are single-roll stands and multiple-roll stands. A double stand, for example, can feed two webs at once. To further increase the flexibility of the press to produce printed products, an auxiliary stand can be retrofitted. This stand would hold additional rolls (usually two), increasing the number of webs that can be run simultaneously on the press.

The roll stand is usually positioned in a direct line with the printing units. Where space limitations prevent an in-line arrangement, side infeeds can be used, which will decrease the overall length of the press. In such a setup, the roll stand is mounted off to one side, and the web is turned into the press. Most newspaper and some commercial presses are fed from a level below that of the printing units. This keeps roll handling out of the pressroom.

Operation principles. The paper roll usually turns on a shaft inserted through its core. The shaft is expanded—either mechanically or pneumatically (through the use of an air bladder)—to hold the roll fast.

Figure 4-1. *An expandable shaft, which holds the roll in exact lateral position. (Courtesy Double E Co., Inc.)*

The roll stand also incorporates a brake. The need for the brake can be understood by considering how much force it would take to pull the paper from a new roll weighing half a ton. Compare that to the force it would take to pull the paper from a roll that is almost expired and weighs 25 lbs. The force required constantly decreases

Figure 4-2. *A roll mounted on a shaft that has been placed on the roll stand.*

as the roll is expiring. The roll stand brake compensates for this change in resistance to maintain constant tension on the web.

Paper feed rate is established by the press operator. The dancer roller and roll brake working together maintain consistent tension on the unwinding roll. For the dancer to function, the paper feed rate off the roll must be controlled by a braking mechanism that acts on the unwinding roll of paper. Roll stands may employ either electromechanical, hydraulic, pneumatic, or magnetic brakes. The roll brake is automatically triggered by the position of a dancer roller mounted between the roll stand and the press. Keep in mind that these brake changes can occur within a fraction of a second.

The dancer roller is mounted in the infeed frame so that it rides in a loop of the web. Consider the following example to conceptualize the function of the dancer roller. If the dancer is at its normal position, moderate braking is applied to the roll shaft. However, when the roll feeds paper too fast, the web loop increases in size and the dancer shifts with it. This motion will signal an increase in braking to slow the roll, increasing tension. The dancer then returns to its normal position and the brake is decreased. If the roll feeds paper too slowly, the web increases in tension between the roll and the dancer. This makes the web loop smaller, signaling a decrease in braking to increase roll speed. The result is that the web loop lengthens and the dancer returns to its normal position.

The infeed must hold the paper taut from the roll to the infeed tension control, and from the infeed tension control to the first printing unit. Theoretically, the dancer eliminates all tension variations emanating from the unwinding roll. Actually, the conventional dancer reduces, but does not eliminate, tension variations. As explained earlier, constant tension is maintained as long as the effective weight of the dancer remains unchanged. The dancer adjusts to changes in paper feed rate off the roll by moving in the web loop; however, the effective weight of the dancer changes during acceleration and deceleration. Thus, the dancer does not maintain consistent tension into the first printing unit.

The total amount of movement allowed in the dancer varies from one infeed design to the next and depends largely on the means used to link the dancer with the brake. With a mechanical linkage, the dancer may move as much as 6–8 in. (150–200 mm), but only in much older presses. With electrical linkages, the dancer will move only a small fraction of that amount and may not be perceptible.

The Conventional Dancer Roller

The dancer roller in the infeed section of the web press has two primary functions: to regulate the rate at which paper unwinds from the roll and to establish and maintain web tension. The conventional dancer roller is capable of doing a satisfactory job of controlling the unwind rate of the paper but is not capable of establishing and maintaining a constant tension.

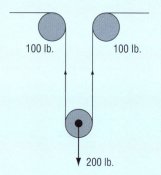

In looking at the drawing above, it can be seen that a roller weighing 200 lb. hanging in a loop of paper is going to impart to the web 100 lb. of tension. That is, half of the weight of the roller is supported by the other side of the paper loop in which it is hanging. Looking at the drawing below, it becomes evident that tension remains the same as long as the roller weighs 200 lb. whether it is in the high, normal, or low position. Carried further, this then means that the amount of brake applied to the shaft of the unwinding rolls does not establish or control tension because tension is a function of the effective weight of the dancer roller.

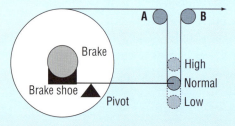

The function of the brake is to control the rate at which paper unwinds at whatever tension is established by the weight of the dancer roller. In the diagram, it can be seen that if the same amount of paper passes point A (coming off the roll) as passes point B (going into the press) the position of the dancer roller will remain at normal and no correction or change is made in the brake setting. If, however, less paper passes A than passes B, the loop in which the dancer roller is hanging will shorten and the dancer roller will move toward the high position. The dancer roller is connected to the brake in such a manner that movement toward the high position will relax the amount of brake being applied to the unwind roll. This, in turn, will allow the roll to unwind more rapidly at the established tension and will return the dancer roller to the normal position. Should more paper be passing point A than is passing point B, the loop will lengthen and the dancer roller will move toward the low position. This movement will apply more brake to the roll and will slow the rate at which the roll is unwinding and restore the dancer roller to the normal position.

It should be remembered that tension variations create problems with register, doubling, and slurring. A system such as is illustrated here does nothing to dampen out tension variations that originate in the web prior to it. For example, any tension variations at point A (which can be caused by an uneven roll, the action of the brake, etc.) will be passed through the loop and to point B and into the press. In a system like the one illustrated, not only the action of the brake but also movement of the dancer roller will create tension variations. Any acceleration or deceleration of the dancer roller, either up or down, will change its effective weight, and thus produces a change in tension. Unless a constant infeed is situated between this point and the press, these tension changes will produce printing problems.

On the simplest web presses, the force that draws the paper off the roll comes from the blanket-to-plate nip of the first unit of the press. In such a simple system the infeed would contain only a roll stand, a dancer roller, and a brake and would feed paper directly into the first printing unit. This system results in extreme paper tension variations. These variations in tension will cause the paper to vary in width as it is stretched to varying degrees. This may result in web breaks and certainly severe color-to-color registration problems.

Metering Rollers and Tension Control

For quality web printing, the infeed must be consistently tensioned and be taut and flat when entering the first printing unit. With a conventional infeed, three (sometimes two) metering rollers located between the dancer and the first unit of the press perform this tensioning function. Two of the infeed metering rollers are driven steel rollers and the third is a nondriven rubber roller. All three are set with a squeeze between them to grip the web.

The second dancer roller activates the variable-speed drive of the metering rollers. Since the first dancer roller already controls the unwinding tension of the web, the second dancer roller compensates for a tighter range of tension variation.

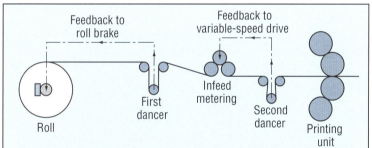

Figure 4-3. Constant tension infeed.

Feedback to roll brake — Feedback to variable-speed drive — First dancer — Infeed metering — Second dancer — Roll — Printing unit

Constant tension infeeds possess a number of distinct advantages: (1) the second dancer corrects for a narrower range of variation than the first dancer, greatly reducing dancer-related variation; (2) by its placement in the infeed, the second dancer corrects for variations arising after the infeed metering system; (3) because the masses involved are much smaller and because the variable-speed drive offers inherently finer control than the roll stand brake, control by the second dancer is more precise, and (4) becasue the second dancer can hold tension variations to a minimum, tension levels in the infeed can be higher without risking a web break. Higher tension levels in the infeed can lessen blanket wrap and elastic recovery in the printing units, which means better control over register.

An inertially compensated dancer roller is a device that operates at a constant effective weight during acceleration and deceleration. The roller is equipped with a flywheel at each end, which maintains the effective weight during a variable rate of movement by the dancer.

A speed control device varies metering roller speed, keeping it relative to press speed. This control will operate over a very short range of speed change. One such speed control, the PIV (positive infinitely variable) drive, is often linked with a planetary gear box to narrow the variable speed range to within 2% of press speed. This slightly slower speed maintains paper tension. In many advanced-design presses, tolerances for the speed of the infeed metering rollers may be as little as ±0.3% of press speed—or even less.

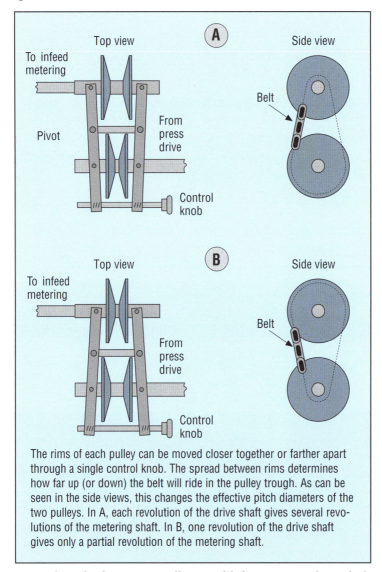

Figure 4-4. *A variable-speed drive that varies speed through two pulleys and a metal link belt.*

The rims of each pulley can be moved closer together or farther apart through a single control knob. The spread between rims determines how far up (or down) the belt will ride in the pulley trough. As can be seen in the side views, this changes the effective pitch diameters of the two pulleys. In A, each revolution of the drive shaft gives several revolutions of the metering shaft. In B, one revolution of the drive shaft gives only a partial revolution of the metering shaft.

The infeed metering rollers establish tension in the web; however, they do not always adequately control it. The tension of a given paper under a consistent drawing rate can still change with the unwinding of the roll: this is due to variations in the structure of the paper from one point in the roll to another. The drawing rate established by the infeed metering rollers will be constant. Thus, if the physical structure of the paper changes, the draw applied by the infeed metering rollers will not change with it, and this will affect paper stretch. These stretch variations can lead to register problems in the printing units. Thus, paper quality and atmospheric consistency (temperature and humidity will affect paper) is essential to quality printing, regardless of a good infeed system.

The conventional infeed creates tension by running the metering rollers slower than the blanket cylinders. With a given paper, the tension established in this span directly relates to the speed difference between the printing unit and the metering rollers—assuming no slippage is occurring at either point, and assuming that the paper's reaction to stress has not changed. Unfortunately, the web commonly slips a small amount over the metering rollers. To help control slip, many infeed metering systems carry two rollers—one steel driven roller and one rubber idler. The web is led between and wrapped around nearly the entire circumference of the steel roller. The high amount of wrap on the steel roller provides the considerable traction needed for control. The rubber roller compresses against the steel to form the nip; therefore, the surface speed differs slightly between the two rollers. As a result, paper speed through a conventional infeed is somewhat unpredictable because the paper runs at the surface speed of the rubber roller, not the surface speed of the driven steel rollers. More importantly, and as previously discussed, no conventional infeed metering system controls tension variations arising from the paper.

Because varying tension adversely affects register and print quality, a constant tension infeed is a crucial part of the web press. Inconsistent tension causes unit-to-unit misregister, doubling, and slurring.

Splicers

The basic function of any splicer is to adhere (with two-sided tape) a new roll to an expired roll on the fly. Continuous roll-feeding devices (automatic splicers) are standard on almost all web presses. The savings in time and waste reduction can be substantial. Automatic splicers are classified according to the running speed of the rolls when they are attached—flying and zero-speed. A flying splicer or paster splices rolls while the paper is running at operating speed (on the fly). The zero-speed splicer makes the splice while the paper is stationary.

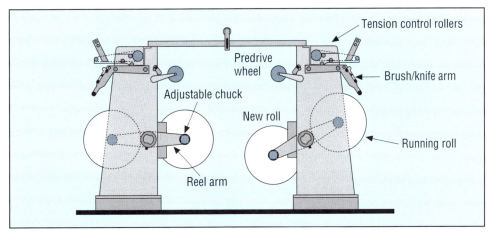

Figure 4-5. *Typical two-arm flying pasters mounted in tandem, with rolls in running position. With this paster, drive wheels accelerate the new roll up to press speed.*

Figure 4-6. CD 13 S splicer. (Courtesy MAN Roland Druckmaschinen AG)

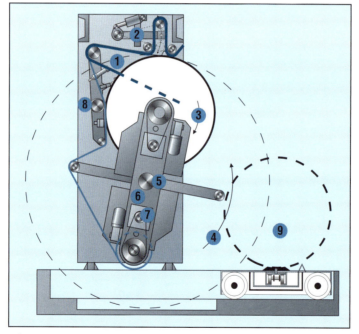

Figure 4-7. Components of the CD 13 S splicer: (1) draw-in device, (2) oscillating roller, (3) sensing of reel rotation, (4) sensing of arm rotation for gluing, (5) support axle, (6) support arm, (7) core drive, (8) gluing unit, (9) loading system. (Courtesy MAN Roland Druckmaschinen AG)

Flying paster principles. Flying pasters may have two or three arms, each with a shaft to hold the roll. These arms are constructed around a central shaft that rotates the rolls. In this way, one roll can be loaded on one arm, while another roll is being fed into the press.

Sensing devices monitor the status and position of the expiring roll. Some pasters monitor the roll with one or two butt switches. A butt switch is a spring-loaded metal finger pointing toward the roll core. It rides against the side of the roll with its tip set some predetermined distance from the roll core. As the roll diameter

---

Tech Tip—Label Flying Paster Arms

The arms of flying pasters should be numbered. In the event of a missed paste or a web break, the number of the feeding arm should be recorded. This system quickly identifies a malfunctioning arm if missed splices are attributable to it.

---

decreases to approximately 8 in. (200 mm), the first butt switch signals the paster cycle. At this time, the paster rotates from the running position (which is also the loading position) to the splicing position. As this happens, the splicing arm and accelerating belts move into position. The belts accelerate the roll to press speed.

The actual splice is initiated by a signal from the second butt switch when about 0.25 in. (6 mm) remains on the expiring roll. Anything left on the expired roll after splicing is core waste, which should always be held to a minimum.

During the splicing sequence, the splicer employs sensors to track the position of the *roll nose* (the lead edge of the splice bearing the adhesive). Some splicers employ a reflector tab on the side of the roll below the roll nose. When the second butt switch kicks over, a photoelectric eye is activated. The first time the reflector tab passes the activated eye, a signal is sent to execute the splice. Other splicers accomplish the same thing electrically with two metal fingers riding over a metallic tab. In this case, once the fingers are activated and make contact with the tab, a circuit is closed.

Both sensing systems perform the same function. They precisely locate the roll nose before signaling a pressure roller to drop, pressing the expiring web to the face of the new roll. Exact timing of the drop of the pressure roller is essential to assure that the entire adhesive-covered area of the roll nose is fused to the expiring web. With exact timing, there is low risk of a loose, sticky lead edge wrapping up in the press.

A second function of the roll nose sensing device is to time the action of the severing knife, which is synchronized with the drop of the pressure roller. This ensures a sufficient paper tail to overlap, completely covering the adhesive area on the new web, which again prevents the splice from wrapping up in the rollers of the press.

A high-speed automatic splicer requires several minutes to execute a splice. This includes the time of the signal that the roll is approaching the core until the completion of the splice. However, the actual splice—from the second butt switch signal until the splice is completed—occurs in a fraction of a second.

On all high-speed flying splicers, the splice roll must be in a preestablished position when the splicing begins. To make this possible, the rolls are mounted at the ends of arms, which rotate the rolls. A two-arm splicer can hold two rolls at once, and a three-arm splicer can hold three. With either, the newly spliced roll moves into the running position shortly after the splice. This action rotates the expired core into the load position where it can be removed and a new roll can be mounted in its place and prepared for the next splice. To operate at high speed, a flying splicer has to be both rugged and precisely engineered. (See sidebar on next page.)

Splicing Sequence on a Flying Paster

With this typical three-arm flying paster, the slots on the roll core are inserted over keys on the paster arm. The butt switches ride against the side of the roll and kick over as the roll surface approaches the core.

The signal from the first butt switch occurs when the roll diameter is down to 8 in. (203.2 mm), as shown in figure 1.

This signal causes (2) the three-arm assembly to rotate into splice position, the acceleration belts to drop on the roll, and the splicing arm to press against the surface of the running web.

In figure 3, the second butt switch kicks over when the roll is 4 in. (101.6 mm) in diameter, activating the photocell. Note that the tab is located on the roll end just behind the splice area.

Once the photocell is activated the next time the reflector tab passes the cell, the pressure rollers press the expiring web against the face of the splice roll (4). The roll is allowed to rotate some fixed distance—in this case, 270°—and then the knife severs the expiring roll (5). In the last phase (6), the acceleration belts and splice arm rise and the roll rotates into running position.

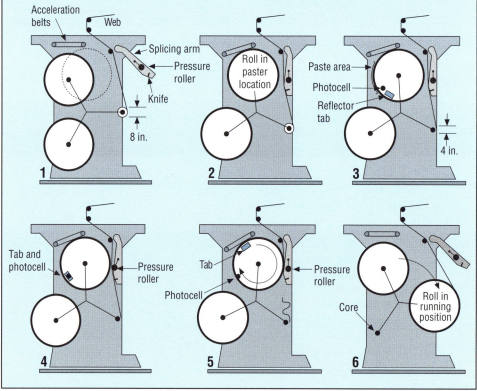

Zero-speed splicer principles. Zero-speed splicers operate on a completely different principle than flying pasters. At the moment the rolls are spliced, neither is moving. The press continues to print at operating speed, feeding from a reserve store of paper. A collapsible festoon stores enough paper to supply the press during the actual splicing sequence. The movement of the festoon is controlled through a pneumatic piston. After the splice, the new roll is accelerated beyond press speed. This allows the festoon rollers to open up, returning to their original position and replenishing their paper supply. Subsequently, the roll is slowed to press speed.

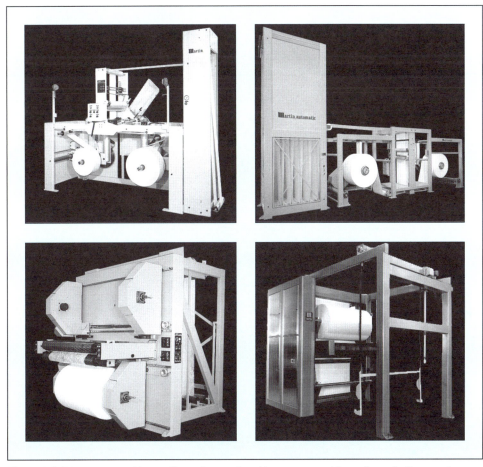

Figure 4-8. *Two zero-speed butt splicers (top row) and two zero-speed lap splicers (bottom row). (Courtesy Martin Automatic, Inc.)*

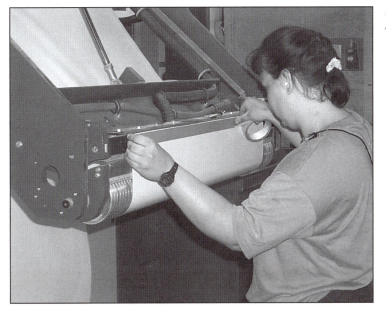

Figure 4-9. *Preparing a splice.*

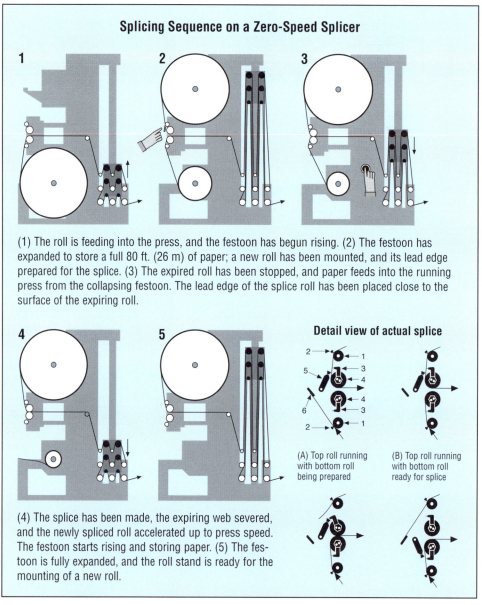

Splicing Sequence on a Zero-Speed Splicer

(1) The roll is feeding into the press, and the festoon has begun rising. (2) The festoon has expanded to store a full 80 ft. (26 m) of paper; a new roll has been mounted, and its lead edge prepared for the splice. (3) The expired roll has been stopped, and paper feeds into the running press from the collapsing festoon. The lead edge of the splice roll has been placed close to the surface of the expiring roll.

Detail view of actual splice

(A) Top roll running with bottom roll being prepared

(B) Top roll running with bottom roll ready for splice

(4) The splice has been made, the expiring web severed, and the newly spliced roll accelerated up to press speed. The festoon starts rising and storing paper. (5) The festoon is fully expanded, and the roll stand is ready for the mounting of a new roll.

The time and cost savings are substantial on any properly operated automatic splicer. The automatic splicer permits continuous operation, thus saving signatures that would be spoiled while ink and water are being balanced after a press shutdown. Ideally, the only waste generated by an automatic splicer should be one signature—the one containing the splice.

Infeed Preparation

Roll inspection. The first step in preparing the infeed is to inspect the roll before mounting it. This is the time to prevent press problems that arise from running a substandard roll. The roll tender should look for any of several defects. The first and eas-

---

Preparing a Roll for the Web Offset Press

Note: Keep the outer wrap on the roll until the splice is to be prepared to (1) minimize moisture pickup of the roll and (2) minimize running waste.

- Record the necessary information from roll wrappers for plant job production files into a company roll report.
- Use roll turner for proper roll unwinding direction.
- Remove roll wrappers and place in proper disposal container.
- Properly record information and fill out roll manufacturer's card; save the card and file for future use.
- Strip away surface cuts and damaged edges from new roll.
- Check for crushed cores, and straighten.
- Weigh and record all strip waste.
- Properly align roll shaft into core. Never force crushed cores onto the shaft. Do not use a hammer to beat the shaft into a roll.
- Assure necessary margin adjustments in roll shaft.
- Hoist roll into roll stand. Caution: Do not stand under hoisted roll.
- Pull outer layer of the roll smooth before preparing the roll lead.
- Use a splicing template to guarantee consistent, rapid preparation of splices.
- Place timing mark on roll, as necessary.
- Remove excess paper from around roll lead.
- Make sure that the paster material is applied evenly.
- Signal changeover of roll to press crews. (Signal also when mill splice goes through press.)
- Remove old butt roll from press and record the amount of paper remaining on roll.
- Make sure that all roll stand parts are properly lubricated and functioning.

---

iest to find is torn or damaged wrappings. The wrappings protect the roll from physical and moisture damage. Torn wrapping exposes the roll, which can significantly increase or decrease the amount of moisture in the outside layers of the roll. These moisture changes can affect web performance on press.

Figure 4-10. Roll with end wrapper removed to facilitate roll inspection.

Figure 4-11. *Roll tender removing paper from the outside of the roll.*

There should be no mill splices in the first 1–1½ in. (25–38 mm) of the outside of the roll. Paper manufacturers have greatly improved the quality of their product by reducing the number of splices per roll and by identifying each splice on the side of the roll. Defects appearing in a new roll should be recorded on a roll record form. This information should then be relayed to the manufacturer.

Next to damaged wrappings, probably the most common paper defect is out-of-round rolls. An out-of-round roll unwinds unevenly on the press and causes variations in tension. Rolls become out-of-round when dropped, picked up with excessive pressure by the roll clamp truck, or stored horizontally. Rolls should always be stored on end so that their own weight does not collapse the roll into an out-of-round shape.

Roll preparation. After removing the wrapper from the roll, the ends of the roll are brushed to remove dirt or slitter dust. Align the roll on the shaft exactly in line with the roll that is feeding. This procedure ensures exact side-to-side splice alignment of

Figure 4-12. *The AUROload loading and unloading system. (Courtesy MAN Roland Druckmaschinen AG)*

Preparing the Infeed System

- Inspect the roll before mounting it in order to (1) prevent press problems that may arise from running a substandard roll, (2) find torn or damaged wrappings, and (3) spot out-of-round rolls, which unwind unevenly on the press and cause variations in tension.
- Remove the wrapper from the roll carefully.
- Brush the ends of the roll to remove dirt or slitter dust accumulated there. A damp sponge may also be used to remove dust from the ends of the roll.
- Position the roll on the shaft and align it exactly (from side to side) with the roll that is feeding in order to (1) ensure exact superimposition of the new and old webs when the splice is made, (2) decrease the likelihood of the new web sticking to the ink build-up on the edges of the printing unit blankets, (3) avoid a misaligned splice which results in paper running through this tacky area, and perhaps breaking the web, and (4) lessen the web's tendency to stick to the ink on the edge of the blanket by brushing both edges of the splice roll with glycerine.
- Leave the protective backing on the exposed side of the two-sided tape applied to the roll until just before the rolls are spliced. Keep the exposed tape from picking up dirt and dust and preventing a secure splice. Keep adhesive away from the web edges because adhesive can cause the web to wrap around an idler roller.
- Use a splicing pattern recommended by the manufacturer to prepare the lead edge of the new web. Use a metal template to guarantee consistent, rapid preparation of splices. Separate templates are constructed for each of the roll widths run on the press.
- Thread the new web through the dancer and metering rollers.
- Adjust the tension of the metering rollers to correspond to the speed of the press.
- Always follow the manufacturer's instructions for splicing procedures because infeed systems vary.

---

the new and old webs. This is particularly important because ink tends to build up on the printing unit blankets just beyond the edges of the running web. Sidelay controls maintain the side-to-side position of the web; therefore, the distance between the buildup of tacky ink on each edge of the blanket often exactly equals the web width. A misaligned splice will result in paper running through this tacky area, which can break the web. If the web's edge consistently catches on sticky blankets during splices, the press operator should brush both edges of the splice roll with glycerin. This action will lessen the web's tendency to stick to the ink on the edge of the blanket and allow time for the sidelay controls to align the new web.

Dirt or dust can settle on the two-sided tape that is applied to the roll for splicing; therefore, the protective backing should be left on the exposed side until just before the rolls are spliced. Also keep the adhesive away from the web edges. Adhesive can cause the web to wrap around an idler roller. For best results, consult the manufacturer's splicing instructions on how to apply the tape.

Using a template for splice preparation. For splicers that require the lead edge of the new web to be cut to a certain shape, make a template to ensure accurate and consistent preparation of the splice edge. Splice patterns are not interchangeable from one paster to another. Most paster manufacturers give specifications on preparing the roll for splicing. These should be followed exactly. A template constructed from sheet metal according to the paster manufacturer's specifications is recommended. A template will guarantee consistent, rapid preparation of splices (figure 4-13).

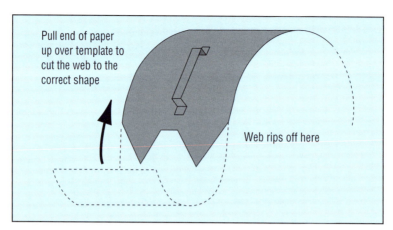

Figure 4-13. *Splice template.*

Properly spacing the idler rollers. Longer web leads maintain flatness and help to control the web during printing. Extra rollers in the infeed lengthen web leads. Specific webbing arrangements depend on the number of webs being run and the particular printing unit to which each web is being led. For this purpose, press infeeds carry a number of nondriven, nonadjustable idler rollers. While longer web leads help maintain flatness, paper should not traverse very long unsupported spans, which easily wrinkle and reduce infeed tensions.

Web Preconditioners

Moisture variations in the web may result in tension variations that might lead to register problems. Web preconditioners or preheaters are located before the infeed metering rollers. Their principal job is to moisture-condition the web, balancing any moisture variations that might exist. Web preconditioners also burn off paper lint and slitter dust. Paper lint tends to cling to the edges of the web, a result of the slitting process at the paper mill. This dust has a tendency to build up on the blanket, causing press problems. Preconditioners may also help to reduce paper stretch and blistering problems.

Preconditioners consist of two sections; a heating section and a chilling section. In the first section, the web is heated to 175–200°F (80–90°C). Because of these high temperatures, chill rolls are needed following the preconditioners to cool the web so that its temperature does not affect the operation of the first printing unit. One

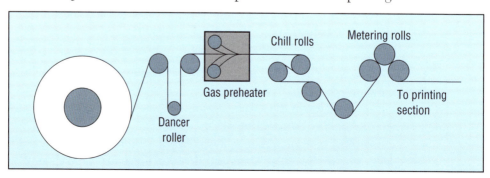

Figure 4-14. *Web preconditioner, or preheater.*

potential problem with preheating is that the paper's exposure to high-temperatures both at the preheater and in the dryer may make paper more brittle, aggravating problems of paper cracking in the folder.

Sheet Cleaners

Sheet cleaners are usually mounted on the press in front of or in place of preconditioners. These cleaners remove loose paper, lint, and dust from the surfaces and edges of the web before it reaches the first printing unit. A sheet cleaner usually consists of a brush that rides the web jarring loose any debris, and a nozzle that vacuums

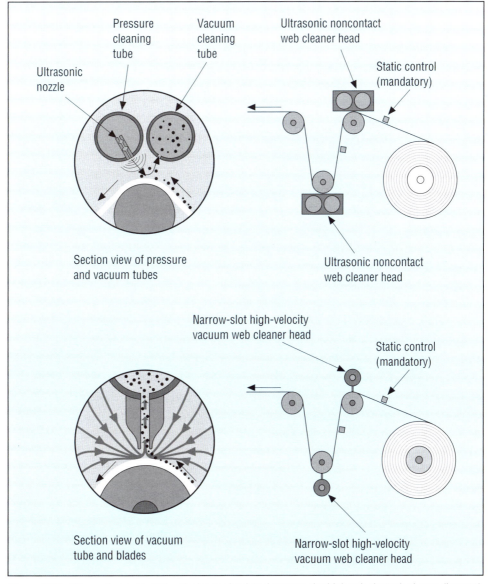

Figure 4-15. *Ultrasonic noncontact web cleaner (top) and narrow-slot high-velocity web cleaner (bottom). (Courtesy Web Systems, Inc.)*

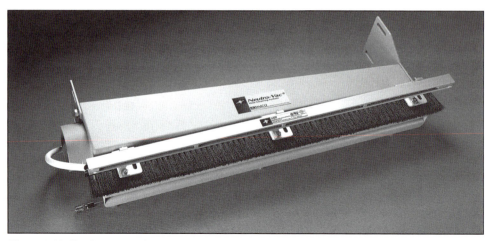

Figure 4-16. *The Neutro-Vac® web cleaning system. (Courtesy SIMCO, an Illinois Tool Works Co.)*

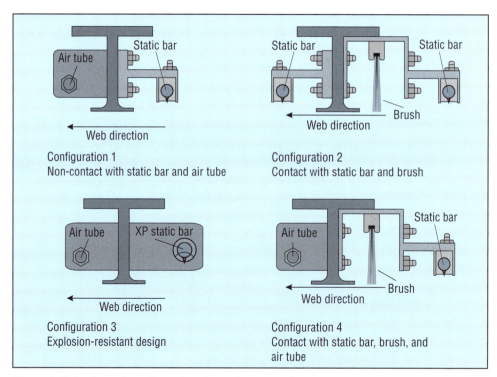

Figure 4-17. *Intake configuration options for the Neutro-Vac® web cleaning system. (Courtesy SIMCO, an Illinois Tool Works Co.)*

the web clean. Other models employ an ultrasonic blast to loosen particles, followed by a vacuum tube to draw the loosened particles from the paper surface.

It is especially useful in an operation running high-lint paper like newsprint. Sheet cleaners, however, do not remove paper fibers that are partly bonded to the web surface. Unfortunately, forces at the printing nip can pick these off the paper surface, creating lint in the press even though a sheet cleaner is in operation. While preheaters remove lint and dust, it is sheet cleaners that have the advantage of being able to pick up loose particles of coating from coated stock that can't be burned off.

Figure 4-18. *ION-O-VAC MK IV system for sheet and web cleaning. (Courtesy SIMCO, an Illinois Tool Works Co.)*

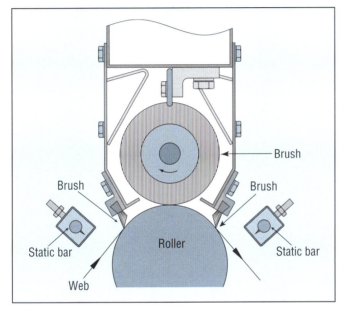

Figure 4-19. *ION-O-VAC MK IV system for sheet and web cleaning, which uses a rotary brush, high-velocity air flow, and static elimination to clean the material. (Courtesy SIMCO, an Illinois Tool Works Co.)*

5 Paper

Paper mills around the world must operate 24 hours a day 365 days years to keep pace with the demand for paper, much of which goes to web offset printers. Web offset presses can print on a wide variety of paper types, including newsprint, offset, coated free sheet or groundwood, and super-calendered. However, the quality of the paper will affect the appearance of the printed images and the runnability of the paper on press. Consequently, it is imperative that the press operator has a sound knowledge of paper. This chapter will present information on paper composition and manufacture, as well as key paper properties that impact web press operation.

The surface of web offset papers should resist picking and be relatively free of dust and lint. The paper should resist moisture, thus preventing fountain solution from loosening the surface fibers or coating pigment. It must also exhibit good ink receptivity and high holdout.

Composition

Cellulose fiber. The principal raw material for producing paper is cellulose fibers, which are short, threadlike structures. Cellulose fiber is the basic building block of plant matter, and large amounts of it can be extracted from wood. There are four main sources of cellulose fiber used in the manufacture of paper: softwood trees, hardwood tress, recycled fiber, and rag (usually composed of textile cuttings and cotton). As a fifth option, synthetic fibers are sometimes used for specialty papers. Softwood and hardwood trees are the most commonly used sources of fiber for web offset papers. Each source produces fiber with slightly different characteristics. Hardwood trees like poplar, birch, and maples produce shorter fibers, about 1 mm in length. Softwood trees like spruce, pine, and fir produce longer fibers, about 3 mm in length. The longer softwood fibers tend to give paper more strength due to better interlocking of the fibers. The shorter hardwood fibers provide paper with bulk and better surface smoothness.

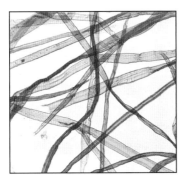 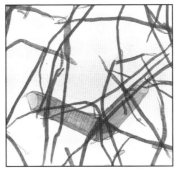

Figure 5-1. Softwood (left) and hardwood kraft fibers before beating, magnified 90×. (Courtesy Institute of Paper Science and Technology)

Paper is made up of a variety of ingredients in addition to fiber, including sizing materials, mineral fillers, and coloring matter.

Sizing. Sizing materials include starch and rosin. These ingredients may be added internally, externally, or both. Internal sizing is intended to give the paper water resistance, a key factor in papers used for lithography. Surface sizing controls the absorption of printing ink, allowing for crisper images on the paper surface. Surface sizing also reduces the release of surface fibers onto blankets, a problem called picking. In addition, sizing may also serve as a preliminary treatment for subsequent coating of the paper.

Fillers. Mineral fillers are added to the fiber before the sheet is formed to improve smoothness, opacity, and color. They also reduce strike-through, a condition whereby ink penetrates the paper and shows up on the other side. Fillers also improve the ink receptivity of offset papers. Paper that has been sized but not filled may not accept printing ink quickly enough for good initial setting, especially at high press speeds. Fillers also reduce dot distortion due to its enhanced surface smoothness. Show-through is also reduced, which occurs when an image printed on one side of the sheet can be seen from the other side, due to a lack of opacity. Fillers improve the brightness (whiteness) of paper, which gives printed images more "pop."

Pigments. Colored paper requires the addition of pigments. Though colored papers are rare in web offset lithography, some specialized jobs call for these papers. The print designer should understand the adverse effect that colored papers will have on colored inks and images.

Paper Manufacture

There are two fundamental steps in papermaking. First, the fibrous raw material, or cellulose, from pulpwood, nonwood fibers, or recycled papers is converted into *pulp,* a mass of fibers suitable for papermaking. Second, the pulp, or fibrous material, is interwoven and bonded into a structure known as paper.

The three primary types of pulp are mechanical pulp, thermo-mechanical pulp (TMP), and chemical pulp. Each type of pulp results in varying degrees of paper quality.

Wood itself is made up of both cellulose fiber and lignin, a glue-like substance that holds the fibers together. Generally speaking, the more lignin that is removed from the wood, the better the quality of the paper. Lignin tends to discolor paper over time. Chemical pulp removes lignin from pulp, while mechanical and thermo-mechanical pulping do not.

Mechanical pulp. Mechanical pulps are made by grinding wooden logs against a rotating stone, which releases fibers into water. Alternatively, the wood may be put into a refiner, which has powerful rotating grinding blades. Papers made from mechanical pulps are very opaque and can be quite strong. These papers will discolor easily and generally do not have good whiteness, which is so important to the quality of printed images.

Thermo-mechanical pulp. Thermo-mechanical pulp (TMP) is made by cooking wood chips or sawdust under pressure, softening the lignin. The softened chips are then ground into pulp and bleached for whiteness. These papers tend to be whiter and brighter than mechanical pulp papers and have good opacity for two-sided printing. Many high-quality newspapers are printed on TMP.

Chemical pulp. Chemical pulp is the most expensive pulping method, resulting in the highest quality papers. In this process wood chips are cooked under pressure in a cooking liquor designed to break down and remove the lignin from the pulp. The pulp is cleaned and bleached, resulting in a bright white paper that has great permanence.

Paper forming. Most printing papers are made on fourdrinier machines, named after the Fourdrinier brothers, Henry and Sealy. The pulp is mixed with large amounts of water before the paper forming stage, resulting in a 99% water ratio. The mixture of pulp and water, called a furnish, flows onto a continuously cycling fine screen through which water drains—leaving the fibers in a thin sheet or web on its surface. The fibers

Figure 5-2. A fourdrinier machine, viewed from the dry end. (Courtesy Glatfelter)

align with the moving wire, which forms the paper's grain direction. The web transfers from the wire to a continuous felt blanket that carries it to steam-heated dryer cylinders. Each side of the paper has slightly different characteristics, designated as either the wire side or the felt side. Print quality may vary from one side to the other. Twin-wire machines form the paper between two wires; thus, the resulting web has two wire sides.

Surface-sized paper is made by passing the partially dried web between a roller-based coating system that applies the sizing material. The size press is located in the dryer section of the machine. After being dried, the paper web is calendered between polished steel rollers to give it the desired smoothness, making it machine-finish (MF) paper.

Supercalendering. Supercalendered paper is made by running machine-finish paper through the supercalender, a machine consisting of alternate polished steel and compressed paper or cotton rollers running together under high pressure. It compresses the paper and increases surface smoothness and gloss.

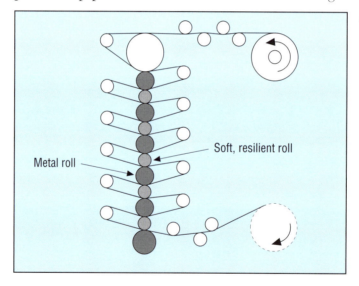

Figure 5-3. A supercalender.

Soft, resilient roll

Metal roll

Coated papers. Coated papers are stocks that have received a mineral coating, usually clay, to improve their printability and appearance. They are available in a variety of finishes. Coatings improve pick resistance, ink holdout, and overall print quality. Usually, paper is coated on two sides (C2S). Labels and some specialty products are only coated on one side (C1S).

Offset newsprint. Offset newsprint is a low-grade paper used mainly to economically produce a large quantity of newspapers by the offset method. Most newspapers were formerly printed by letterpress; however, offset lithography prints much finer and better halftones, thus lending itself to the wider use of quality pictures and color in news and advertising. The web offset presses designed for newspaper printing generally do not have dryers. The inks dry entirely by absorption and oxidation.

Paper Finishing

Paper machines vary considerably in width, potentially producing webs over 300 in. (7,620 mm) wide. However most printing papers are made on 100- to 200-in.-wide (2,540- to 5,080-mm-wide) machines. The full-width manufactured roll is called a parent or log roll. This roll is slit to the needed width either in-line at the dry end of the paper machine, or in a separate off-line slitting operation. Rolls of the slit paper are re-wound on cores or reels under tension for use directly on web presses.

Figure 5-4. *Optireel, which is used in the reeling process on paper machines. (Courtesy Valmet Paper Machinery, Inc.)*

It is during the paper finishing operations that two important aspects of quality control—namely, elimination of physical defects and maintenance of moisture content—can either be emphasized or neglected. It has been stated that 50% of the mechanical troubles in paper are either caused in the finishing room or can be reasonably eliminated there. Automatic optical and electronic systems are used to detect and eliminate flaws such as holes and dirt.

Paper Properties

Paper is a material that has many measurable physical properties. Many of these properties are a direct result of how the paper is manufactured. The printer should have knowledge of paper properties and how these characteristics affect the pressrun as well as the finished product.

Grain. Grain is a characteristic of all machine-made papers. It results primarily from fiber alignment during the formation of the sheet. The fibers tend to align themselves parallel to the direction traveled by the wire of the paper machine. On all roll papers, grain direction is lengthwise on the web, parallel to the direction of web travel. The effects of grain on paper properties are as follows:

- Paper tears and folds more easily in the grain direction than across the grain.
- Paper exhibits greater stiffness and higher tensile strength in the grain direction. Tensile strength is the ability of paper to resist tearing or breaking when pulled.

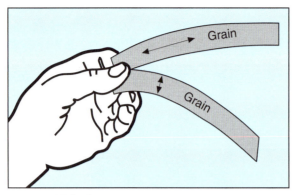

Figure 5-5. *Paper cut in the direction of the paper grain (top strip of paper) and across it. Paper is stiffer in the direction of the grain than across the grain, and folds are cleanest when made parallel to the grain direction.*

- With changes in moisture content, paper expands or contracts more in the cross direction than in the grain direction. This is important to know because paper absorbs or gives off moisture when in contact with a wet blanket, when passing through a dryer, and when running in contact with chill rolls.

Two-sidedness. Because of the nature of the fourdrinier wire, the two sides of paper are different in structure. The addition of coating makes the two sides more consistent. The wire side of uncoated paper has an open structure, contains less sizing and filler, has fewer short fibers, and has a more pronounced grain. The felt side has a closer structure and less grain, because the fibers are more completely interwoven. The sophistication of modern papermaking machines has greatly reduced two-sidedness. Both sides of the stock manufactured on conventional machines print equally well.

Density. Density is defined as the weight of paper relative to its volume. Dense papers are compact with their fibers more tightly bonded together. Surface sizing and supercalendering tend to increase density. In softer, bulkier, more porous papers, fiber clusters can swell or shrink without much change in the overall dimensions of the sheet. Denser sheets will have less latitude, changing more dramatically in dimension with moisture variations. The dimensional stability of paper, therefore, closely relates to density. Nonuniform density in a web can cause tension variations. With uncoated stocks, uneven density can also lead to nonuniform ink absorption, which appears most noticeably in solids and halftones (as mottle).

Color. Paper can be made in almost any color. Process color reproduction, however, should employ white paper, because any color in paper affects the colors in the reproduction. The colors affected most are those complementary to the paper color. For example, bluer sheets make yellow darker, and reddish sheets cause green to appear gray. Slight variations from white (blue-white, cream-white, or pink-white) may still produce acceptable results; however, they may be visibly discernible. Bright white sheets will produce the best color with the maximum color gamut from any given set of process inks.

Brightness. Brightness of paper is measured with a blue wavelength (457 nm), which provides a value indicative of the degree of bleaching. Optical brighteners that increase blue-light reflectance contribute to better contrast in the printed image (in blue and black areas), resulting in more brilliance, snap, and sparkle. Brightness will actually reduce the color gamut of yellows, reds, and greens. Any variation in brightness will detract from print quality, most noticeably in large areas of halftone tints.

Whiteness. Whiteness is the degree of reflectance, in uniform amounts, of red, green, and blue light. White objects are highly reflective. Conversely, black objects reflect little or no red, green, and blue light, even though the reflectance—or the lack thereof—may be uniform.

Reflectiveness. Paper reflects light superficially and internally. Some of the light that falls on the white areas in printed halftones penetrates the paper, where it is scattered. Thus, part of the scattered light is trapped behind the halftone dots. As a result, the white areas between dots appear lower in brightness than large white areas under the same illumination. Deeper light penetration increases light scattering. The result is increased contrast in the highlights and decreased contrast in the shadows. In addition, all tone values are darkened. Low contrast will result in lack of brilliance, snap, sparkle, and clarity of detail.

Mineral filler and coating pigment reflect more light than cellulose fiber, preventing the deep penetration of light, and yielding higher halftone contrast.

Opacity. Opacity is defined as the extent to which light transmission through paper is obstructed. This property affects show-through of printed matter. Show-through is a lack of opacity that allows the printing on one side of the sheet to be seen from the opposite side. Excessive show-through reduces image contrast and makes two-sided printing quality suffer. Show-through is different from strike-through. Strike-through is excessive penetration of ink through the sheet.

Smoothness. Smoothness describes the continuous evenness of a paper's surface. Smoother paper surfaces allow thinner ink films to provide coverage. A thin ink film results in less ink usage, less dot gain and sharper dots, and better clarity of detail.

Gloss. Gloss may be a property of either a paper surface or a printed ink film. Gloss is determined by the degree to which specular reflection exceeds diffuse reflection. A mirror or glass surface has nearly perfect specular reflection, with light waves reflecting in straight lines. Diffuse reflection results from the scattering of light rays, and results from a rougher surface.

High-gloss papers are desirable for some applications but inappropriate for others. They enhance the brilliance and intensity of colors but are objectionable for reading matter because of glare. Paper gloss has an important effect on the gloss or finish of printed ink films. When identical ink films are printed on papers of equal ink absorbency, the ink gloss is always higher on the glossier paper.

Performance Requirements of Web Papers

Papers used in web offset come in a wide variety of classes and finishes. Their basis weights range from about 17 lb. to more than 100 lb. (25–148 g/m^2) at a basic size of 25 × 38 in. Heavier weights can be run but do not handle well in folders and so are usually delivered in sheets. Heavier-weight coated papers are also more prone to blistering. Lighter-weight papers are difficult to run because of their lower tensile strength and lower stress resistance. For satisfactory performance in web offset, papers should meet the following basic requirements.

Flatness. Webs should be flat enough to pass through the printing nips without wrinkling or excessively distorting. Flatness is only in part achieved through proper press settings. Webs with wavy edges and baggy centers sometimes show up on the press regardless of settings. The problems are usually the result of varying moisture content or basis weight within the web. The flatness of these webs can be substantially improved on press by increasing the length of travel and the tension exerted between the roll and the first printing unit.

Dimensional stability. In printing, the paper web remains under tension all the way from the roll to the cutoff. Uniform mechanical stretch is necessary for consistently good register and requires uniformity of fiber density and paper moisture content.

Moisture content. Web stress remains constant on the press once tension settings have been set. As the web picks up moisture from unit to unit during the pressrun, sheet dimensions will change, potentially causing misregister. Bustle wheels are used to counteract the web growth.

Nonuniformity in moisture content is more frequent across the web width; for example, the center may be moister than the outer edges. Variation in this direction can result in one or both edges of the web running tight while the center runs baggy or vice versa. This can lead to localized misregister, wrinkling, and doubling in the loose parts of the web. Modern infeed tension control devices improve the runnability of such rolls by subjecting them to maximum pre-stressing before they reach the first printing unit. Alternatively, the web may be run through the longest possible lead in the infeed under high tension.

Paper Printability

Printability is the characteristic of paper that is directly related to the quality of the images printed on it. Physical and optical properties of paper and its surface affect tone and color reproduction, smoothness of print, and therefore the appearance of the printed reproduction. Diverse papers provide varying degrees of fidelity in tone and color, contrast, smoothness of halftones and solids, and clarity of detail.

The attractiveness of a printed reproduction depends on the nature of the original and the end use of the printed piece. Some subjects such as portraits, greeting cards, or abstract art appear more attractive when reproduced with the soft effects obtainable on a rough vellum stock. Other subjects such as machinery, furniture, or food illustrations, requiring sharpness and clear detail, are best reproduced on a high-finished or coated stock.

Moisture content should be as uniform as possible throughout the roll. It should also be as high as possible—but within web paper specifications. The dryer drives much of the moisture out of the web. Dry paper cracks easily in the folder, resulting in extremely dry, brittle signatures. If a bound book has low moisture, it will pick up atmospheric moisture, turning the edges wavy. Remoisturizing restores adequate moisture content to paper, ensuring that the delivered signatures remain supple and dimensionally stable. This process also reduces static electricity in the web.

Many of the sheetfed printer's concerns about conditioning paper to pressroom atmosphere do not apply to the web printer. If wrappings are left on until the roll is to be run, atmospheric humidity has little effect on roll moisture content. The edges of the roll are tightly wound and relatively impervious to atmospheric moisture for a reasonable amount of time.

Surface durability. Fountain solution weakens the coatings of some papers, especially in four-color process work. Material from a weakened coating can be picked off and cause print defects or blanket piling.

Because web papers that run on blanket-to-blanket presses are printed on both sides simultaneously with thin films of high-tack inks, they are subjected to forces that can cause internal rupture or ***delamination.*** Delamination only occurs in the printing units of blanket-to-blanket presses when the web adheres to both blankets past the point of impression in an S-wrap. Unlike blistering, delamination occurs on one side of the paper only. It is long in the press direction and ragged in appearance. The effect can be cumulative in multicolor printing, worsening from unit to unit.

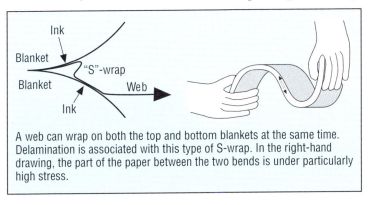

Figure 5-6. S-wrap and delamination.

A web can wrap on both the top and bottom blankets at the same time. Delamination is associated with this type of S-wrap. In the right-hand drawing, the part of the paper between the two bends is under particularly high stress.

Surface strength. Surface strength is the characteristic of paper that enables it to resist external forces—chemical and mechanical—as it passes through the printing press. Diverse papers exhibit varying degrees of surface strength.

The thin film of ink on the blanket exerts a very strong pulling force on the moving web. In a condition called ***picking,*** weak paper will rupture at the surface, releasing paper fibers onto the blanket. When the surface strength of paper is a problem, some adjustments can be made to decrease the tack of the ink. For example, reducing press speed will decrease the effective tack of ink, and thus eliminate picking.

Paper condition. The condition of the paper rolls is important. Paper rolls should be round and uniformly wound under proper tension. Many problems with paper condition are possible including soft spots, welts, corrugations, and water streaks. Rolls may also arrive at the press starred. It is important to fully protect rolls from the atmosphere, maintaining constant moisture content throughout by proper wrapping. The wrappings should be undamaged and kept on the roll until just before it is run on the press. Several layers of paper that must be slabbed off the outside of a damaged roll represent a substantial loss.

Paper Problems

There are a host of potential paper problems. These may be caused during paper manufacture, by improper storage, or by the press conditions themselves. It is important for the web press operator to be able to determine the cause of paper problems as they arise.

Piling. Piling is the accumulation of material on the blanket or plate. This buildup usually contains paper coatings, ink, and/or substances from the fountain solution. Coated and uncoated papers can generate piling. With uncoated papers, mineral filler can accumulate on the blanket, forming an abrasive layer. With long pressruns, this may wear the desensitizing film off of the nonimage areas of the plate, leading to scumming. Coated papers cause piling if the coating is too water-soluble. To reduce piling, coated papers must have enough moisture resistance to prevent separation of their coatings during printing.

Linting. Linting is the transfer of loosely bonded paper surface fibers to the blanket. It differs from picking in that the sheet may have adequate pick resistance, but loosely attached individual fibers are picked up by the blanket. Each impression may carry only a few of these fibers. After a few thousand impressions, the blanket, plate, and inking system can become so contaminated with fibers that the press has to be stopped and washed up. Linting is a prevalent problem with newsprint and uncoated groundwood papers. Linting is usually worse on the wire side than on the felt side of the paper. Sometimes linting occurs on the second, third, or fourth units of a multicolor press with no sign of it on the first unit. This happens because the starch surface size holds the fibers on the web during one or more impressions. After the first unit, the sizing becomes softened and the surface fibers are released and picked up by the tacky ink.

Picking. Picking is the separation of a paper's surface fibers by adhesion to inked blankets. Picking occurs when the ink tack (pulling force) exceeds the paper's surface strength. To counteract picking, press operators can reduce the ink tack, the press speed, or both. Reducing ink tack is often an unsatisfactory remedy, because it may adversely affect print quality.

Picking appears in various forms, but each involves the rupturing of the paper, and it occurs principally in solid print areas. Uncoated papers may yield small clumps

A Gallery of Paper Defects

Burst:
An irregular separation or rupture in the wound roll, usually shortest in the cross-machine direction.

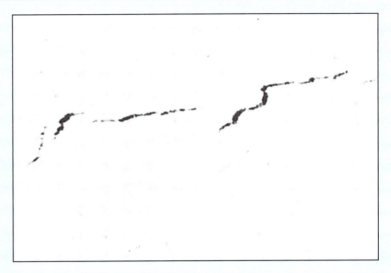

Calender cut:
A straight, sharp cut with a glazed edge, running for a relatively short distance at an angle to the direction of web travel. It is caused by a wrinkle in the web going through and being hard pressed by the calender.

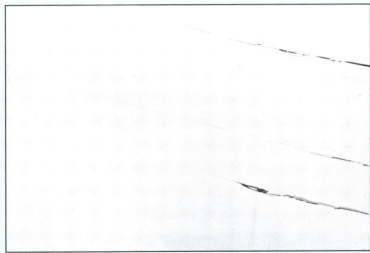

Calender spot:
A defect or imperfection in paper appearing in the form of a glazed or indented spot, often transparent. It is caused by a small flake or piece of paper that adheres to the calender rolls or is carried through the roll on the paper sheet.

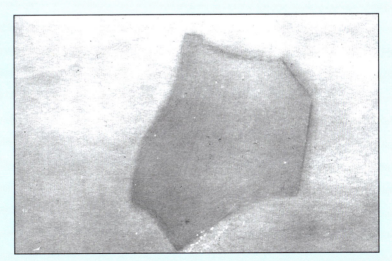

A Gallery of Paper Defects
(continued)

Corrugated roll:
A roll with bands of relatively uniform width that extend around a roll parallel to the machine direction and within which are diagonal markings resembling a rope or tire-like pattern.

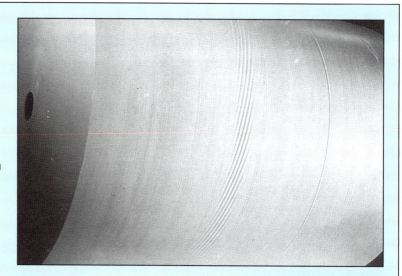

Damaged edge:
The outer portion of the roll that has been rendered useless due to mishandling. The damaged layers must be slabbed off the roll so that this paper does not enter the press.

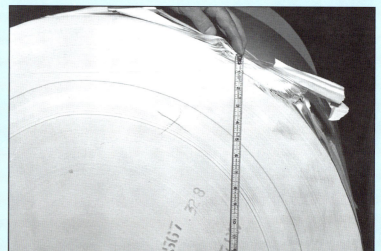

Fiber cut:
A short, straight, fairly smooth, randomly located cut caused by passage through the calender of a fiber or shive embedded in the web of paper.

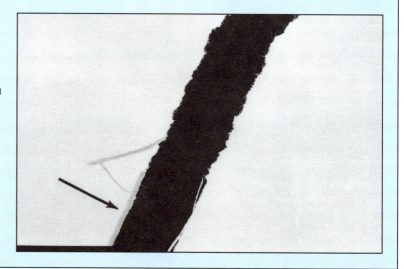

A Gallery of Paper Defects
(continued)

Hair cut:
A sharp, smooth curved cut having no characteristic length or direction, which is caused by hair or some synthetic fiber getting into the web.

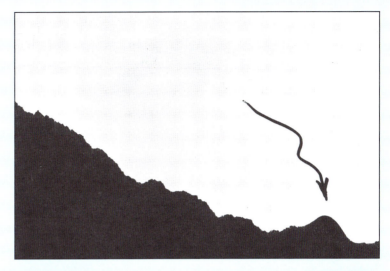

Out-of-round roll:
A roll with an irregular shape caused by storage on its side, excessive roll clamp pressure, or dropping or bumping the roll.

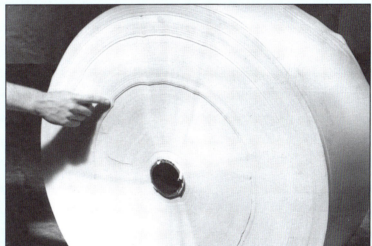

Telescoped roll:
A roll with progressive edge misalignment, concave or convex, due to slippage of its inner layers in the direction of its axis as a result of a thrust force on or within its body after being wound.

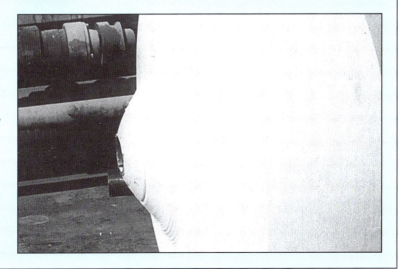

of fibers that stick to the blanket, marring printing on succeeding impressions until they are removed. The fibers quickly soak up water and reject ink, thus producing white spots in the solid print areas.

Similar spots are caused by slitter dust, which may be misinterpreted as picking. To find the specific cause, select a typical spot and trace its origin to the signature where it first appeared. Examine the spot with a magnifying glass or microscope. A ruptured paper surface indicates picking. Otherwise, a loose paper particle probably generated the defect. Though uncoated papers pick most often, coated papers pick when improperly bonded coating separates from the base stock.

Offset papers must resist the force of thin ink films, which have relatively high tack. Furthermore, the resilient surfaces of offset blankets conform to paper and pull on printing and nonprinting areas.

Blistering. The combination of (1) a heavy, tightly sealed coating, (2) high basis weight and density, and (3) high moisture content can be especially troublesome in web offset. When such a stock is printed with heavy ink coverage on both sides, internal moisture that is vaporized by the dryer cannot escape through surface pores. As a result, small bubble-like formations called blisters appear on the web surface. Blisters may be round or oval shaped, and will occur on both sides of the web. About the only remedy, aside from avoiding such stocks, is to reduce the press speed and minimize dryer temperature.

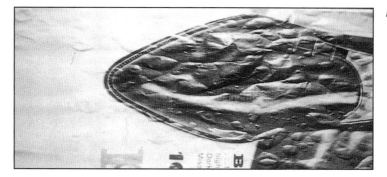

Figure 5-7. *Blistering.*

Selecting and Ordering Rolls

The selection of paper should be influenced by production factors such as the speed of the press, whether process color is required, area of coverage, whether printing one or two sides of the sheet, and the image quality desired. The job to be printed is therefore an integral part of paper selection. Because paper affects print quality and press performance, it makes sense to limit the number of stocks run on a given press. This will help the press operator reduce paper/ink performance variations.

Ordering and handling. The following tips provide guidelines to increase productivity, increase quality, and reduce waste:

1. Order rolls made to the maximum diameter, allowing the production planner to maximize image space per cutoff. Fewer roll changes during a run reduce the number of press adjustments required as well as cost due to a shorter run length.

2. Limit the number of splices in rolls. This also reduces the chance of tension-related problems.

3. Mark all splices so that the signatures containing them may be removed from the delivery.

4. Check basis weight for accuracy. Mills sell paper by weight, not length. For example, when ordering 17×22 in.—20-lb. (75 g/m^2) bond paper, buyers may receive an actual weight of 21-lb. or even 23-lb. paper. The heavier the basis weight, the shorter the web length, yielding less product.

5. When planning production, allow for about 2–4% spoilage for wrapper and core waste. (The weight of the roll includes the outside wrapper and the core.) The last few feet of paper nearest the core are unusable.

6. Specify that the manufacturer send paper free of slitter dust. Slitter dust often piles up on the blanket of offset presses, resulting in spoilage and lost time due to excessive washups.

7. Provide the proper core information. Although there is no standard size for cores, the 3.5-in. (88.9-mm) size is most common. Roll chucks dictate the proper core size. Specify returnable or nonreturnable core. Request slotted cores if the press requires them. Large, heavy rolls may be shipped on returnable iron cores.

8. Stipulate "rewound" when rolls are a narrow width. Narrow gauge equipment is high-precision equipment. Slitting must be exact, and winding must be hard, firm, and consistently tensioned.

9. Specify wire side out or felt side out. Some splicers allow the roll to be unwound in one direction only. If it is desirable to run one side or the other up, winding must be specified.

10. Specify directional arrows to appear on wrappers. These are an aid in proper orientation of the roll when it is delivered to the pressroom.

11. Give handling instructions to shippers. Shipping and storing rolls on their ends is recommended.

12. Ask the paper manufacturers for shipping papers that identify all rolls by number and list their weights. Your shipping, stock, and accounting departments need this information.

13. Keep roll cards with their respective rolls. The lot number, roll number, roll weight, and manufacturer's code are listed on these cards.

14. Inspect paper carefully before the shipper unloads it from the car or truck. Photograph any damage using an instant camera, and report it to the common carrier at once. Otherwise, the printer pays for the damage. Some printers equip their paper handlers with instant cameras and instructions to

Checklist for Ordering Paper in Rolls

Grade _____   Type of press _____

Quantity _____ pounds   Speed of press _____

Maximum basis weight _____ pounds   Maximum ink drying temperature _____

Maximum thickness ____ pages-to-inch _____   Type of delivery _____

Maximum roll width _____ inches   Maximum roll diameter _____

Core inside diameter _____ inches

Type of core _____

- [ ] returnable   [ ] nonreturnable
- [ ] slotted   [ ] nonslotted
- [ ] slots in juxtaposition [ ] dimensions of keyway ____ × ____

Roll winding   [ ] felt side OUT   [ ] felt side IN

Note: Mark Directional Arrows on Wrappers

Splicing   Maximum acceptable to roll _____
   flag   [ ] one side   [ ] two sides
           [ ] diagonally across roll
           [ ] use splicing material for heatset reproduction

Wrapping   [ ] moisture proof   [ ] nonmoisture proof

Delivery   [ ] on side   [ ] on end

Plant Humidity Requirements _____

Special Instructions
- [ ] Rolls must be wound uniformly firm
- [ ] Indicate weight of each roll on wrapper
- [ ] Surface of paper must be free from lint and other extraneous materials
- [ ] Indicate roll number and winding direction on ends of rolls
- [ ] Provide roll cards in core of each roll
- [ ] Provide packing slips complete with roll number, weight per roll, and number of splices per roll
- [ ] Number each roll on wrapper

Shipping   [ ] siding on _____ RR
               [ ] plant can accommodate trailers up to _____ feet long
               [ ] sidewalk delivery by winch truck

Most satisfactory delivery hours _____ AM to _____ AM   _____ PM to _____ PM
Receiving platform closed _____ to _____

Special Markings on Roll Wrappers _____

_____

Special Instructions Not Identified Above _____
- [ ] Sample for Matching Enclosed

Note: Be sure to send out-turned samples in advance of shipment.

routinely take pictures of the condition of all shipments. This practice helps greatly in documenting shipping problems. If concealed damage is discovered while the shipment is being unloaded, the paper handlers should take another picture showing the extent and nature of the problem.

Once rolls are received in good condition, they should be handled carefully to prevent damage. Each roll should be inspected before being placed in storage. If a wrapper is torn when a roll is handled, it should be repaired immediately to protect the paper from dirt or moisture damage.

All roll handling equipment must be in good condition. Roll clamp trucks should be equipped with pressure-regulated jaw pads.

In the pressroom, where rolls are handled manually, all areas should be kept immaculately clean. A great deal of damage can be avoided if all areas of the floor where paper is rolled are kept free of any debris that could punch holes in the surface of the roll.

Do not remove the outer wrap from a roll until the splice is to be prepared; this minimizes the chances for moisture pickup or loss on the roll surface and keeps running waste to a minimum.

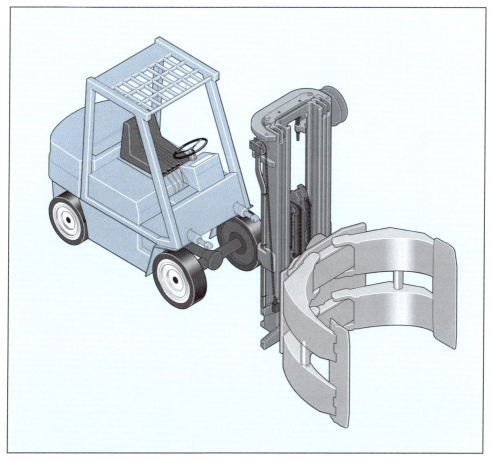

Figure 5-8. Roll lift truck. (Courtesy Cascade Corporation)

Roll Retrieval

- Follow proper safety guidelines when operating a roll-clamp truck.
- Equip all roll-clamp trucks with a yellow flashing light and an alert signal to indicate reversing.
- Slow down at aisle intersections and traffic areas such as doorways and stairs.
- Place rubber pads on the clamp surfaces of the roll-clamp truck to help protect the roll from damage.
- Lift the clamp and place it gently around the roll. Avoid damaging other rolls next to the roll being retrieved.
- Apply a minimum amount of pressure to the roll. Keep the roll over the stack and lift it up about 1 ft. (305 mm). Make sure the roll is secure before lifting it away from the stack.
- Slowly tilt the clamp slightly to one side to help prevent the roll from falling.
- Lower the roll to the ground.
- Check the aisles behind the truck before reversing.
- Again, watch for pedestrians and carry the roll to the pressroom.

Important: Carry only one roll at a time, unless the roll-clamp truck is specifically designed to carry two rolls. Stacked rolls cause damage and excessive pressure to the bottom roll.

Figure 5-9. The AUROSYS, an automated guided vehicle (AGV), transporting a roll of paper to the automatic splicing preparation station. (Courtesy MAN Roland Druckmaschinen AG)

6 Controlling Paper Waste

A press must run continuously at high speed to achieve maximum productivity, and shutdowns during the pressrun should be avoided. Every shutdown leaves surplus water and/or ink in the press, throwing off the ink/water balance. With conventional dampening, a long stop may result in a dried-out system. When starting up, any press will generate some spoilage as ink/water balance is being restored.

Stoppages in the middle of the pressrun are sometimes unavoidable; however, they are always costly. Studies have indicated that the average range for run intervals is between 0.8 and 1.8 hr., depending on the plant and the type of work being done. Stop intervals average between 0.4 and 0.7 hr., also depending on the plant. These general figures indicate that between the start of production and the end of the job the press is running only about 70% of the time. Every plant should have a program aimed at extending the "length of the run" interval and reducing the "length of the stop" interval. Improving either figure increases the amount of salable signatures produced per running hour.

Problems Requiring Press Shutdowns

Web tension variations. Most paper-induced tension variations can be completely controlled by a constant-tension infeed. However, occasionally web tension will suddenly change for no apparent reason, throwing off color register or folding accuracy. Possible causes are mill splices in the middle of the roll and excessive variations in mill winding tension that surfaces as the web unwinds on the press. Waste due to mill splices can be minimized if the roll tender watches for splices and warns the press operator when they are about to go through the press.

Piling problems. Blankets must transfer a clean sharp image during the entire pressrun. Several factors affect blanket performance. Some paper coatings readily build up on the blanket (this is called piling), which reduces print quality. Often, the only recourse is to wash the blankets frequently during the run, requiring shutdowns. Some presses are equipped with automatic blanket washers that retard the rate of buildup. A press without automatic blanket washers should never be washed on the fly. The savings in time and paper that might be realized from washing during running are not worth the risk of personal injury or mechanical damage.

Ink problems. Heatset inks can begin to set on the inking system rollers during the pressrun. This is due to solvent evaporation caused by excessive heat generated in the inking system. Water-cooled ink oscillators help to control this by limiting temperature increases. If the press is not so equipped, it is best to use inks with solvents that have relatively high boiling points. Such inks yield their solvents less quickly and remain stable on the press; however, they dry more slowly. Any ink that sets on the rollers is not properly formulated for existing conditions, and the ink manufacturer should be consulted.

Minimizing Downtime

Reduce downtime by anticipating the stop, assigning specific duties to crew members, and bringing the required materials to the press in advance.

When the press is stopped, immediately gum the plates if it appears that the shutdown will exceed 30 min. (depending on the plate). The plates must be gummed properly to prevent gum streaks in the image areas, which would require additional stops to clean the plate. When restarting the press, some press operators increase speed incrementally while adjusting register and color. If plates blind during the run, rub them with the solution that is recommended by the manufacturer.

Wash blankets as needed to maintain image quality. Several successive jobs may be printed without washing the blankets in between pressruns. The duration of downtime also dictates whether or not blankets should be washed. Manually washing blankets between every pressrun consumes valuable time and reduces productivity. Ink may set on the rollers during makeready. Spray solvents on the rollers to overcome this problem.

Accurate Run Lengths

Web offset printers often encounter discrepancies between run length and final signature count in the bindery. Jobs that come up short in the bindery require rerunning to make up a shortage, an unnecessary expense.

For this reason, some pressrooms normally run 2–3% more signatures than the job order calls for just to prevent such a shortage. If 1% of press production per year costs $70,000, a standard overrun of 3% represents an added cost of $210,000 each year. The math shows that it is critical to determine the minimum amount of overage required to get the job done and then to run just that amount. If bindery spoilage allowances have been figured into the count, there is no need to overrun. The bindery needs only the signatures called for on the print order.

Causes of inaccurate counts. In some cases, neither the pressroom nor the bindery counts are accurate, making it difficult to locate the cause of the waste. With some presses, two counters are attached to the press drive. One will track the total count including waste and the other salable signatures only. With this system, the salable signature counter should be turned on and off with the change from makeready to salable signatures. However, because of the pressures of production in the press-room, this is usually not accurately done.

There are other common practices that cause errors in counting. First, signatures totaled on the good counter are used as inspection sheets. These signatures are usually discarded. With three or four people on the average crew inspecting signatures, this can become a significant source of inaccurate counts. A second source of shortage—perhaps the most significant source—arises when bad signatures are printed and thrown out while the salable counter is running. Catch-ups, off-register signatures, and/or poorly folded signatures account for numerous signatures that are wasted. Any of these items can create inaccurate counts. Each by itself is probably small enough that it could be ignored; collectively, they may represent a significant cause of shortages.

Keeping accurate count. Counters that track individual signatures in a delivery stream should be mounted far into the delivery system so that they are less likely to count waste signatures or signatures that have been removed for inspection or splices.

In addition to a counter, scales may be used to weigh stacked and bundled waste signatures. Simple math will determine the total number of signatures in a stack in the following way:

$$\frac{\text{Weight of the stack}}{\text{Weight of a single signature}} = \text{Number of pieces}$$

Salable signatures can be put on the scale and accurately counted. This will allow a double-check of the accuracy of the counter.

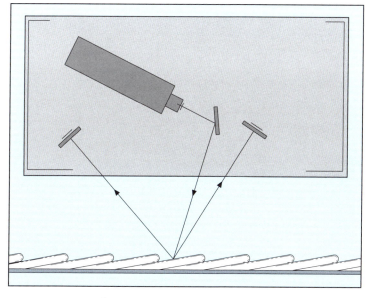

Figure 6-1. Denex laser counter.

Keeping seconds. Regardless of the accuracy of the counting procedure used, unforeseen bindery problems can still cause a job to come up short. For this reason, seconds—signatures produced at the end of running makeready before the counter is turned on—are saved in many shops. Usually, seconds are only slightly off-register or

off-color. Seconds, sometimes called salvages or substandards, can be especially useful if the job has several sections to be bound together. Simply by saving a few hundred sections, the bindery may finish a job that otherwise may have been short in
only one of the sections. It is also good practice for the bindery to use seconds in the
bindery setup.

Record Keeping

Records are necessary for the analysis and control of any business. This is especially
true for a business that is as complicated and difficult to analyze as printing. Since
waste, makeready, and downtime costs vary from one job to the next, press records
should be kept for every job. Computer-controlled data recording systems are used to
accurately monitor production on modern web presses.

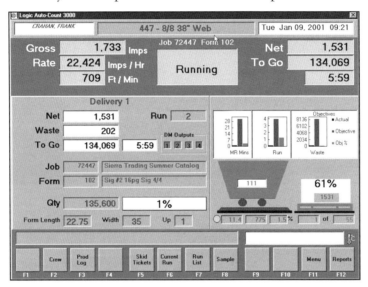

Figure 6-2. Two screen
captures from Printcafe's
Auto-Count software,
which delivers production
counts and also collects
real-time press and pro-
duction statistics for
plant-wide management
systems. (Courtesy
Printcafe Software Inc.)

If it is necessary to keep manual records, the printer should use data forms that are simple to complete. The data sheet below was developed by GATF for gathering information in the pressroom. Wherever possible, the data is entered by merely making a check mark. All of the required information is printed on the form. The exception is quality level, which could be printed on the back of the data sheet for easy reference.

JOB DATA SHEET

Plant _____   Year 19 [ ] Month [ ] Day [ ]
Job Number [   ]   Press Number [ ]

Makeready Hours (to nearest tenth)   Total paper   [   ] Pounds
Makeready Signatures   Transit Damage   Pounds
Running Hours (to nearest tenth)   White Waste   Pounds
Good Signatures   Paper Costs (cents/lb.)   (nearest tenth)
Gross Count (incl. makeready)   Basis Weight (pounds)   25 × 38—500 sh.

Paper Finish
1 Offset (Book)
2 Newsprint
3 Supercalendered
4 High Bulk
5 Embossed
6 Other Uncoated
7 Dull Coated
8 Gloss Coated
9 Coated—One Side
10 Other—Coated

Type of Plate
1 Photopolymer—Neg
2 Photopolymer—Pos
3 Bimetal—Neg
4 Bimetal—Pos
5 Trimetal—Pos
6 Wipe-on Al.—Neg
7 Presensitized—Neg
8 Presensitized—Pos
9 Other

Type of Blanket
Conventional
1 3-Ply
2 3-Ply—Quick rel
3 4-Ply
4 4-Ply—Quick rel
Compressible
5 3-Ply
6 3-Ply—Quick rel
7 4-Ply
8 4-Ply—Quick rel
9 Other

Number of:
Webs
Press Units Used
Plates (originals, do not include makeovers)
Split Fountains (0, 1, 2, etc.)

Job Frequency
1 Plate Change
2 Multiple Form
3 Weekly
4 Monthly
5 Every 2 Months
6 Every 3 Months
7 Every 4 Months
8 Every 6 Months
9 Annually
10 One-Time Job

Job Folding
1 Double Former
2 Combination Folder
3 Ribbon
4 Ribbon—Interleaved
5 Ribbon + Former
6 Sheeter
7 Two Folders

Blankets Changed (during makeready)

Paper Changed (from last job)
1 No Change
2 Change Paper (type, brand, or basis weight)
3 Changed Width (or number of webs)
4 Changed Paper and Width (or number of webs)

Perforation
1 No Perforation
2 Running Perforation
3 Cross Perforation
4 Running and Cross Perforation

Paste
1 No Paste
2 Job Pasted on Press

Folder Change from Last Job
1 No Change
2 Minimum Change (tabloid to chop)
3 More Difficult Change
4 Change from Folder (or sheeter) to Folder
5 Change from Folder to Sheeter
6 Make Change That Requires Machinist

Accuracy of Fold
1 Not Important
2 Average
3 Important
4 Critical

Sheets Okayed By:
1 Production Dept.
2 Quality Control
3 Art Department
4 Salesperson
5 Customer

Total Ink Coverage
1 0–50%
2 100% (±50%)
3 200% (±50%)
4 300% (±50%)

Spot Color
Number of Plates with Spot Color (0, 1, 2, 3)

Quality Level
1 See Separate Sheet for Definitions
2
3
4
5

No. of Products Delivered

No. of People Used on Crew

	1st Shift	2nd Shift	3rd Shift
1st Operator			
2nd Operator			
3rd Operator			

Figure 6-3. Job data sheet.

Figures 6-4 and 6-5 show two approaches to record keeping. To accurately narrow down the cause of waste, many other items (in addition to those on these forms) may have to be recorded. For example, it may be important to record (1) the frequency that blankets had to be washed during the run of a given paper, (2) the mechanical troubles encountered on the press, and (3) the length of downtime.

A job data sheet can be used to determine the following information:

- Press production capabilities and limitations.

- Costs associated with running poor paper compared to high-quality paper. When this information is relayed to the sales force, appropriate papers may appear more frequently.

- Overall crew efficiency. A good reporting system indicates specific areas in which the crew needs help.

WEB PRESS RECORD FORM

Press no. _____ Press operator _____ Date _____

Customer _____ Issue _____ Job no. _____

Time: a.m.
   At start of makeready _____ p.m.   Form no. _____

                          a.m.
   At end of run _____ p.m.   Imposition style _____

Impressions required _____   Press speed _____ IPH

Counter Reading: Strip waste _____

   Total counter at start of run _____   Printed waste _____

   Total counter at end of run _____

   Job counter at end of run _____

Paper: Code no. _____ Roll no. _____   Record all stops

   Roll damage: _____   _____

_____   _____

_____   _____

_____   _____

Performance of paper on press (circle most applicable number)

(worst) 1 2 3 4 5 6 7 8 9 10 (best)

Web breaks during run _____

    Location of break   Reason for break

_____   _____

_____   _____

_____   _____

_____   _____

Plates: No. of plates required for normal job ____   Excess plates needed _____

   Reason for excess plates _____

Quality: Of pressrun (circle most applicable number)

   (unacceptable) 1 2 3 4 5 6 7 8 9 10 (excellent)

Comments and Problem Areas: (continue on other side if necessary)

_____

_____

Figure 6-4. A form for recording both numerical and verbal information, with emphasis on the identification of paper and press problems.

PAR-WEB OFFSET REPORT FORM*

Press Mfgr. _____ Rated Speed _____ IPH _____ Auto Pasters _____ Roll Stands _____ Perfecting Units _____ Type of Folder _____ Cutoff _____ Width _____

Job. No.	Type of Makeready	Type of Paper	Chargeable Hours				Standard Hours	Non-chargeable Hours	Percent of Productivity	Makeready Impressions	Net Good Impressions	Run Average Per Hour	Gross Run Impressions	Run Spoilage Percent	Paper Consumed Pounds	White Waste	
			MR	WU	Run	Total										Pounds	%

*All data will be treated confidentially.

Courtesy Web Offset Association, Printing Industries of America

Figure 6-5. A record keeping form that emphasizes quantitative information regarding production.

- Actual production figures that may be used to determine the accuracy of estimates (job costing).
- The amount of excessive "white waste." This unprinted paper is pure spoilage (wrapper, strip, and butt).
- Press problems including mechanical or electrical weaknesses.
- Amount of waste and the reasons for it during makeready and during the pressrun.
- Damaged paper rolls. This information can be relayed to the manufacturer.
- Bad splices and reasons for problems.

Computerized Monitoring Devices

Computerized monitoring devices attached to the press record most of the information required to analyze the printing operation. These systems provide real-time press performance information, allowing the crew to make adjustments to maintain print quality. Speed of the press, color register, paper temperature at the dryer and chill roll, web tension, ink film density, trapping, print contrast, and other factors can all be measured and displayed at the console. It is important for the press crew to realize that computer monitoring devices only provide data; interpretation of the information, evaluation of press performance, and adjustments require highly skilled operators.

7 Controlling Web Tension

Paper is drawn through a web offset press under tension. The press operator is responsible for controlling and understanding the conditions affecting web tension. Uniform tension must be maintained to achieve and hold register. Some degree of consistent tension control is possible, but there are paper-related factors that make exacting control impossible. Applying even a minimal pulling force to paper changes it dimensionally, and the change is not constant for a given type of paper or even within a specific roll. The erratic behavior of paper under stress often results in inconsistencies in the finished product—inconsistencies due to variation in stretch leading to unit-to-unit register, side-lay, cutoff, and folding accuracy problems. Press-related factors like dampening levels and even operations such as drying and chilling can affect web tension. Thus, one fundamental problem in web printing is controlling paper behavior through the press. This chapter provides the underlying factors that affect web tension and describes devices and techniques for controlling tension in the web.

Tension Spans and Control

The distances between successive points of tension are referred to as *tension spans,* as in the unsupported span of paper between the first and second units of a press. Ideally, paper tension would be uniform throughout each span of the press. There are a variety of devices and mechanisms that help maintain tension through these spans. Tension-control devices act on the web in one of two ways:

- **Constant control.** Constant-torque devices directly apply an unchanging force to the web, as in the torque created by the rotation of the blanket cylinder.
- **Variable control.** Variable-speed devices control the flow rate of the web and can be adjusted to increase or decrease tension. The variable-speed metering roll is an example here.

Web Control Aspects

Web control involves three major aspects: (1) tension, (2) length, and (3) the modulus of elasticity of the web. All three aspects are fundamentally interrelated. A change in any one will effect changes in at least one of the remaining two.

Tension. Tension is defined as the force that the web span is placed under. Tension is required to keep the paper flat and taut as it enters the printing units. Papers used for printing vary widely in strength, influencing the amount of tension required. Generally, heavier stocks can be placed under greater tension, while lighter stocks are best kept at lower tensions to avoid web breaks.

Length. The longer the ribbon of paper, the more force that is required to keep the paper taut. For example, long spans in the dryer require more force to be exerted on the paper so that the paper does not sag.

Modulus of elasticity. Paper stretches relative to the amount of force acting upon it. The relationship between force and (elastic) stretch of paper is known as the ***modulus of elasticity.*** The modulus of elasticity depends on the paper's composition. When a paper's composition changes, the modulus changes, and so does the paper's reaction to tension. Bulk weight, fiber length, fiber composition, filler material, and moisture content, among other factors, all affect the modulus of elasticity. (See sidebar on the next page.)

Because most of these factors can vary slightly during the papermaking process, some variation in modulus can be expected within the roll. Devices that compensate for these variations in the roll must therefore be built into the press.

Variations in modulus due to on-press processing stem from changes in the web's moisture content. The dryer of the heatset press radically changes the web's moisture content. Moisture picked up from the fountain solution similarly affects the web.

Varying web tension or varying flow rate are the two means by which variations in modulus are controlled. When the modulus changes and tension is to be held constant, the flow rate must be varied. This is the function of the dancer roll and brake system. When modulus changes and the flow rate must be held constant, tension will change as the modulus changes. This occurs in the printing units.

Tension Variables

To understand and control tension, several variables must be considered. Surface speed, paper type, drying, draw, and slip are factors that can influence tension. Each of these are described below.

Surface speed. When used in connection with web control, *speed* always refers to *surface speed*, as opposed to a speed measurement like RPMs (revolutions per minute). Surface speeds of the driven rollers and cylinders on the press determine web speed, which will vary only slightly due to slip (nondriven idlers and angle bars only guide the web).

An increase (or decrease) in the circumference of driven rollers and cylinders will change their effective surface speed. This can be understood by comparing a small wheel turning at a constant RPM (revolutions per minute) with a larger wheel

Modulus of Elasticity

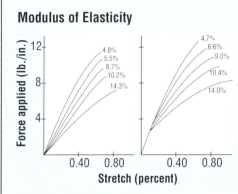

The slopes in the above diagrams represent the modulus of elasticity for different newsprint stocks. The percentage figures indicate moisture content. Both diagrams show that higher moisture content gives a lower modulus of elasticity (greater stretchability).

The diagrams below and at right show the results of two tests, each involving two different web offset papers (A and B). Each paper was manufactured at three moisture contents (4%, 5%, and 6%). The tests were conducted in two plants, on two presses, by two different crews.

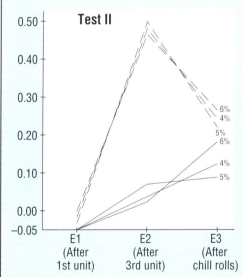

This conclusion, in turn, suggests that, overall, on-press influences such as press speed, variable speed settings, press water, and dryer temperature are more important in determining stretch behavior than is paper moisture content. A measure of the effect of press conditions on web behavior can be seen by comparing the readings at E1 for both tests. In Test I, all six webs gained in length between 0.10% and 0.13%. In Test II, five of the six webs actually shortened in length, from 0.02% to 0.04%. This latter effect is probably caused by the relatively high infeed tension on the presses used in Test II.

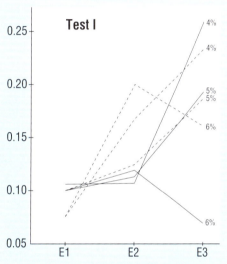

Although moisture content varied only 2% (and not the 10% shown in the newsprint diagrams), one might expect relatively consistent stretch behavior. Actually, the tests showed the paper behaved differently on the two presses. Both paper A and paper B reacted about the same to the conditions that existed during Test I. Under the different conditions of Test II, however, the two papers show no similarity in behavior. The two tests suggest that modulus of elasticity determines stretch behavior only in some cases, and, furthermore, that these cases cannot be predicted beforehand.

rotating at the same RPM. While the RPMs are equal, the speeds at the surface of the wheels are quite different, with the larger wheel moving much faster. Adding packing to a blanket cylinder increases its circumference and hence its effective surface speed. Taping a chill roll to prevent wrinkling may increase its effective surface speed in the same way.

Nips between driven rollers and cylinders can greatly affect surface speeds. Hard roller nips, as in steel-to-steel contact, have a negligible effect on surface speed. Rubber, however, compresses under nip pressures and squeezes through a nip at a faster rate than the surface speed of a steel roller. The effective surface speed of a rubber roller also varies slightly as it passes through a nip, unpredictably affecting web surface speed. Surface speed variations are encountered in the rubber-to-steel nips of some infeed metering devices and in blanket-to-blanket printing nips.

Draw. Tension can be varied across a span of the web with nips at each end by changing the surface speed of the web at either the beginning or ending nip. The resulting difference in surface speed is called draw and is positive when the surface speed at the second nip is higher than at the first. Positive draw applies force to the web and increases tension. An increase at the infeed nip or a decrease at the output nip reduces draw.

If effective surface speed of the paper through both nips is the same, and the elastic properties of the paper do not change, tension in the span remains the same as it was in the previous span.

Slip. Web speed through a nip may differ from the surface speed of the rotating rollers. Slip between the web and roller surfaces slightly slows the paper. Slip commonly occurs at many points on the press making web control more difficult. At the chill rolls, for instance, the traction between the web and the roll surface controls web speed.

This traction is the result of friction. ***Static friction*** occurs when there is no slip between the web and the driven roller or cylinder. As long as there is no slip, the web will move at the speed of the chill rolls. ***Kinetic friction*** occurs when there is a slip between the moving web and the driven roller or cylinder. When slip occurs at the chill roll, a reduction in traction and, hence, tension will occur. For this reason, chill rolls are sometimes run at a slightly higher speed than the last printing nip, creating a positive draw to counteract slip.

Paper. Surface speed, draw, and slip are the mechanical, press-related factors that influence tension. Paper changes also effect tension changes. As a viscoelastic material, paper reacts to pull (tension) by elongating in two different ways:

- **Elastic recovery.** Like a rubber band, paper stretches under applied force and recovers most of its original dimensions when the force is removed. The recoverable elongation is referred to as elastic recovery.
- **Plastic flow.** Paper that is stretched rarely fully regains its original dimension. The nonrecoverable change is referred to as plastic flow.

This physical property reduces tension control, especially in the earlier spans in the press, before paper is conditioned. A series of nip-to-nip spans in the infeed create a relatively long distance between the roll and the first printing unit, more adequately conditioning the paper for printing. (See figure 7-1.)

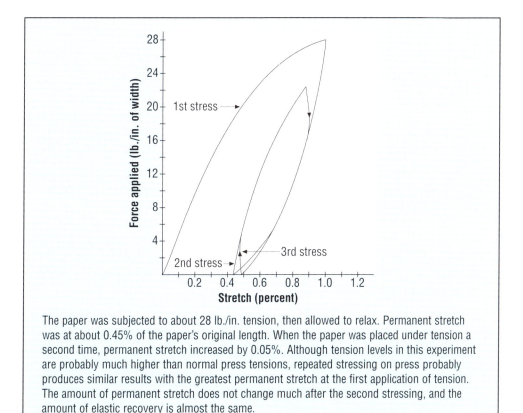

The paper was subjected to about 28 lb./in. tension, then allowed to relax. Permanent stretch was at about 0.45% of the paper's original length. When the paper was placed under tension a second time, permanent stretch increased by 0.05%. Although tension levels in this experiment are probably much higher than normal press tensions, repeated stressing on press probably produces similar results with the greatest permanent stretch at the first application of tension. The amount of permanent stretch does not change much after the second stressing, and the amount of elastic recovery is almost the same.

Figure 7-1. *The effect of repeated stretching on paper. (Adapted from TAGA Proceedings)*

Dryer. The dryer is one point on the press where the modulus of paper can be expected to change significantly. Moisture is driven out of the web, and as a result the paper becomes less elastic. The same amount of tension produces less stretch in the paper after it leaves the dryer.

Paper also shrinks in the dryer. As water molecules leave the web, the fibers draw closer together and increase tension. It is not unusual for the chill rolls to run at the same speed as the last printing unit, and for all tension in this span to be supplied by paper shrinkage in the dryer.

Tension Dynamics

Testing done on a four-unit blanket-to-blanket press shows that tension variations propagate only in a forward direction. For example, small tension variations in the infeed carry all the way through to the folder. On the other hand, fairly large tension changes in the span between the fourth unit and the chill rolls are rarely detectable between the second and third unit and never between the first and second unit.

The slip that occurs at the printing nips and at the chill rolls allows tension variations to have some upstream effect, but mainly the effects are felt in the direction of web travel.

Implications of tension dynamics. Because the effects of tension variations move forward and not back, the needs of the infeed and press sections should be met first. Then the dryer-chill section should be balanced, and finally the delivery section.

If there is too much or too little tension for the folder (which moves from the press units), it is common practice to go back to the infeed and increase or decrease tension there, which then moves forward to the folder. However, if the initial infeed settings were correct, the readjusted settings could cause tension problems in other areas of the press.

If folder tension problems are commonly experienced, the folder may have to be adjusted or modified to match proper press tension settings. This can be done by changing the diameter of the nip rollers, which controls web draw into the folder. For example, an increase in the nip roller diameter will increase the surface speed and thus increase draw. An alternative method is to change press output to match folder needs by changing the blanket cylinder packing. As previously discussed, an increase in blanket packing will increase surface speed, thus increasing draw. A decrease in blanket packing will have the opposite effect—decreasing surface speed and decreasing draw. To maintain adequate squeeze, changes in blanket packing should be accompanied by equal and opposite changes in plate packing.

Paper Control Factors

Rolls of paper are wound at the mill under tension several times greater than is generated by the press. Thus, paper usually undergoes considerable elastic recovery in the infeed and perhaps beyond, in the printing units. This process necessitates long web leads (as in festooning) in the infeed. A long lead gives the paper more time to recover fully before it reaches the printing units, where significant elastic recovery may cause register problems.

Dancer roll factors. Variations in winding tension at the mill, variations in the paper's basis weight, and rolls that are out of round or tapered produce tension variations as the roll unwinds. These variations can travel downstream and affect tension throughout the press, all the way to cutoff. It is the job of the dancer roller to offset these variations in the infeed; however, no dancer maintains perfectly constant tension. The conventional dancer roll has limitations. A tension change can pass by the dancer before it has time to react.

Infeed metering roller factors. The infeed metering rollers must run slower than the blanket cylinders of the press to produce a positive draw. The web runs from the infeed metering to the first printing unit under high tension, the object being to produce a taut, flat surface as the web enters the first unit.

The infeed metering rollers comprise one of the most important tension control points on the web offset press, but some slip occurs here. A rubber idler at the metering rollers offers some slip resistance, rubber giving better traction than steel. (See figure 7-2.)

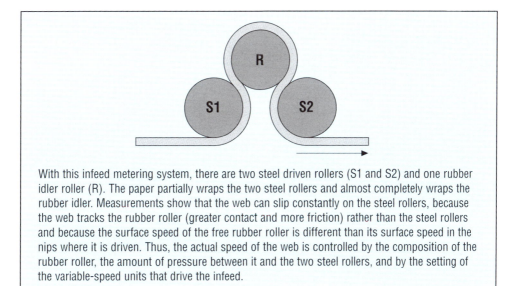

With this infeed metering system, there are two steel driven rollers (S1 and S2) and one rubber idler roller (R). The paper partially wraps the two steel rollers and almost completely wraps the rubber idler. Measurements show that the web can slip constantly on the steel rollers, because the web tracks the rubber roller (greater contact and more friction) rather than the steel rollers and because the surface speed of the free rubber roller is different than its surface speed in the nips where it is driven. Thus, the actual speed of the web is controlled by the composition of the rubber roller, the amount of pressure between it and the two steel rollers, and by the setting of the variable-speed units that drive the infeed.

Figure 7-2. The common "pyramid" infeed metering system.

Printing unit factors. The printing units exert the strongest, most positive control on the web. The rubber blankets provide good traction, which is assisted by pressure in the printing nips. The printing units can be thought of as establishing the nominal web speed from which tension is set throughout the press.

More slippage will occur between the blanket and impression cylinder on an in-line nonperfecting web press than on a blanket-to-blanket web press. This is due to the steel-to-rubber nip formed in the in-line press versus the superior rubber-to-rubber nip traction in the blanket-to-blanket nip.

The high tension established between infeed metering and the first unit usually decreases through succeeding units for several reasons:

- **Moisture pickup.** Dampening solution picked up by the web expands paper, which reduces tension unpredictably. (See figure 7-3.)
- **Blanket pull.** Blanket pull is the action of the sticky, inked blankets, which causes the web to adhere to one or both blankets on the unit past the point of impression. Blanket pull tends to increase tension. The amount of pull needs to be held constant. High tension reduces adhesion, and hence blanket pull.

Progressive packing—increasing the packing on each of the blankets of the last one or two units by 0.001 in. (0.25 mm)—builds a slight positive draw into the printing units. Progressive packing can increase web stability between the units and will decrease blanket pull.

Dryer and chill roll factors. Normally, the longest unsupported web span on the press extends from the last printing unit, through the dryer, to the chill rolls. Because the span is so long, the web may corrugate (form longitudinal wrinkles) as it goes through the dryer. In extreme cases, this will produce wrinkles on the chill rolls (see figure 7-4). As discussed in the chapter on dryers and chill rolls, corrugation can be

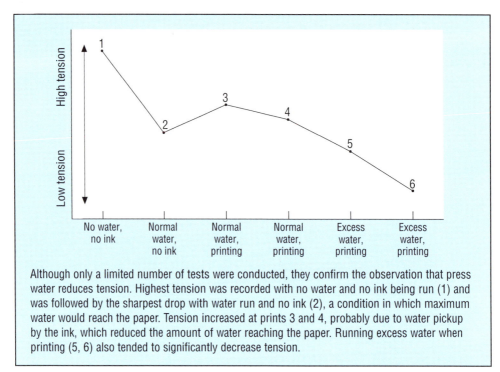

Although only a limited number of tests were conducted, they confirm the observation that press water reduces tension. Highest tension was recorded with no water and no ink being run (1) and was followed by the sharpest drop with water run and no ink (2), a condition in which maximum water would reach the paper. Tension increased at prints 3 and 4, probably due to water pickup by the ink, which reduced the amount of water reaching the paper. Running excess water when printing (5, 6) also tended to significantly decrease tension.

Figure 7-3. The effects on web tension of various ink/water conditions set up on a four-unit press.

controlled by staggering air blasts on the top and bottom of the drying web, producing a sine wave in the web. To produce the sine wave, some slack in tension is necessary.

Another major factor affecting tension in the dryer span is web shrinkage, which will effectively increase web tension in the dryer. Because the web tends to slip on the chill rolls, consistent tension is maintained. However, slip at the chill rolls can cause other problems. The friction created from slip can cause ink smearing on the chill rolls. Therefore, ideally chill roll speed should be set to produce zero slip. However, the shrinkage caused by the dryer can cause slippage on the chill rolls. For maximum control, set the nip to pinch the web against the last chill roll to maintain constant chill roll speed with no slip. A no-slip nip on the chill rolls can be achieved through the use of air-loaded nip rollers. Air-loading makes high nip pressure attainable (400 lb.—1,780N—total force or higher) and permits precise equalization of pressure at each roller.

Folder factors. Nipping rollers in the folder are set to a higher surface speed than the web units, creating a positive draw. A controlled amount of slip is designed into the folder and varies from press to press. Because the setting is sensitive to the caliper and finish of the web, the press operator has to acquire a feel for nipping roller adjustments. These settings must be precise, as they are critical to web control.

Alternatively, a no-slip nip may be created between the rollers below the former board by air-loading the nipping rollers. The surface speed of the nipping rollers is made equal to blanket speed by machining their surfaces to the required diameter. This makes the setting of these rollers easier and less critical than on conventional

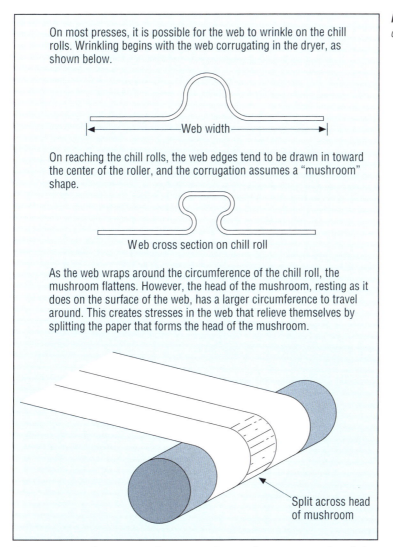

On most presses, it is possible for the web to wrinkle on the chill rolls. Wrinkling begins with the web corrugating in the dryer, as shown below.

|◄──────────────Web width──────────────►|

On reaching the chill rolls, the web edges tend to be drawn in toward the center of the roller, and the corrugation assumes a "mushroom" shape.

Web cross section on chill roll

As the web wraps around the circumference of the chill roll, the mushroom flattens. However, the head of the mushroom, resting as it does on the surface of the web, has a larger circumference to travel around. This creates stresses in the web that relieve themselves by splitting the paper that forms the head of the mushroom.

Split across head of mushroom

Figure 7-4. Wrinkling on the chill rolls.

slip systems. The major advantage of a no-slip system is the ability to change from a relatively thin coated stock to a thick uncoated stock without changing folder settings.

The last tension span in the press is between the nip rollers and the press cutoff cylinder. Folder pins on the cutoff cylinder pierce the lead edge of the web, maintaining control of the uncut portion. The size and shape of the holes left by these pins indicate folder gain. If the pins tear through the lead edge, too much slip has been set into the nip rollers, and gain between these rollers and the pins is too high. Circular holes indicate that web tension is too low. Proper tension leaves holes slightly larger than the pin itself and not perfectly circular, but rather oval.

8 Web Control

As press speeds have risen, so has the necessity for reliable web guides, cutoff controls, and web break detectors. It is now unlikely that any new press would be specified without web guides. Cutoff controls are less critical than web guides for productive press operation but are still regarded as essential for modern heatset web presses. Higher operating speeds in recent years have meant that a wrap could prove very costly, thereby increasing the need for web break detection equipment.

Web Guiding

Side-lay control. *Side-lay* is the lateral placement of the web through the printing units and folder. Side-lay affects the printed image margins on the signature pages and the position of the former fold line. For this reason, web side-lay should be controlled at two points along the press: before the printing units and before the folder.

The primary function of a web guide is to maintain the alignment of paper as it moves into the printing unit, especially after splicing has occurred. This is achieved by matching the centerline of the paper to the centerline of the plate. Web guides, however, are not fast enough to cope with large side-to-side, cyclical variations caused by bad alignment of the paster.

Web guides generally have a sensitivity of no more than 0.005 in. (0.13 mm) and an overall guiding accuracy in the range 0.005–0.01 in. (0.13–0.25 mm). Correction rates of 0.5 in. (13 mm) per second are typical.

One common web guide, or web-steering device, is the ***box tilt,*** which has four rollers arranged in a box shape (figure 8-2). The first and fourth rollers in contact with the web are in fixed position, while rollers two and three can be moved in parallel and skewed relative to the other two. When the web crosses the skewed rollers, it shifts. In this way, side-to-side position can be precisely controlled before the web enters the printing units and the folder. Other side-lay devices have three (sometimes only two) rollers. These devices should be set to produce minimum stress in the web during correction. Wrinkles will form in the web if the action of the side-lay generates excessive stress.

Normally webs are center-guided. Center guiding positions the web so it is centered with the press and folder. This is accomplished by mounting side-lay sensors in pairs across from each other on each side of the web. The side-lay pairs can be mounted at each station on the press where guiding is necessary. Center guiding is best for work where folder operations are needed. This assures that the web is consistently folded at the same position across the web. Edge guiding, on the other hand, holds only one edge of the web fixed relative to the press and folder. Edge guiding is

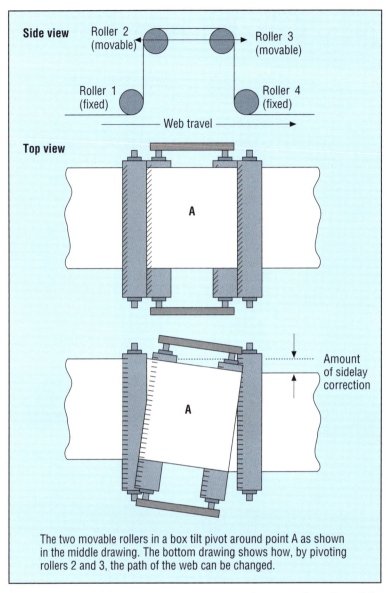

Figure 8-1. Box tilt.

The two movable rollers in a box tilt pivot around point A as shown in the middle drawing. The bottom drawing shows how, by pivoting rollers 2 and 3, the path of the web can be changed.

fine for sheeting, where cutting occurs in relation to the edge of the web. The guiding device monitors the edge of the web that will be used to register later operations. A web guide should be located in the infeed in order to center the printing on the web. A second web guide between the press and folder keeps the fold in the center of the web.

Side-lay sensors. Web side-lay sensors (figure 8-3) detect minute lateral web movement and signal steering devices to adjust the paper's position. One automatic side-lay device is a U-shaped apparatus with the web moving through the space between the two arms. One arm houses a light source and the other a photoelectric sensor cell, or eye. As the web passes through, the light to the cell is partly blocked. The strength of the signal from the photocell unit varies directly with the amount of

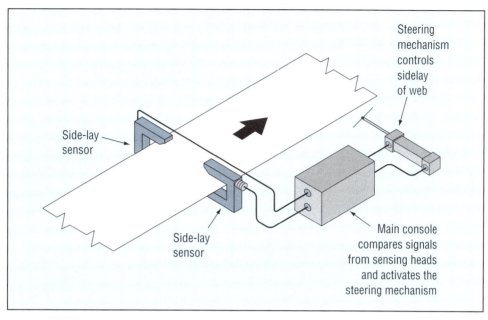

Figure 8-2. Center guiding.

light reaching the cell, which in turn is determined by the position of the web edge across the face of the cell. The strength of the signal determines the amount that the web steering device (e.g., box tilt, cocking roller, etc.) corrects the web's position.

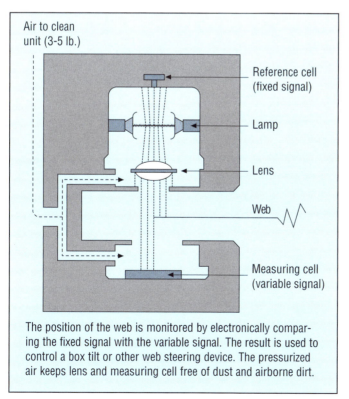

Figure 8-3. A typical sensing device for automatic side-lay control.

Cutoff Control

Cutoff determines the image position on the signature along the length of the web. The parallel or jaw folds are always made at a fixed distance back from the leading edge of the cutoff section. If the printed image in this section is improperly positioned, the signatures will be folded inaccurately.

The function of cutoff controls is to provide continuous compensation for paper shrinkage, shifts, and similar occurrences. They cannot compensate for such errors as incorrect cutting. There are two basic approaches to cutoff controls: the conventional system and the constant-stretch system. A conventional cutoff control system works by correcting errors. To do so, this system utilizes (1) a scanner, although this scanner will not have the precision needed for register control, (2) an encoder or pulse generator on the folder or sheeter, (3) a compensator to advance or retard the web, and (4) an electronic control system. The system compares the signals generated by the scanner and the encoder and then adjusts the compensator until the two signals coincide. The constant-stretch system aims to maintain constant web stretch between the final printing unit and the cutoff point. Constant stretch is regarded as a prerequisite for good print-to-cut register. Cutoff controls have the ability to (1) read both color and black-and-white and (2) automatically select and find an appropriate mark on the web.

Controlling cutoff using a compensator roller. Cutoff is controlled through a compensator roller or rollers mounted in advance of the folder (figure 8-4). The compensator is a steel nondriven roller that can be advanced or retarded while the press is running. The compensator is set in accordance with signature size and creates a fixed distance between the last printing unit and the cutoff knives in the folder. This fixed distance is evenly divisible into a predetermined number of impressions—for example, ten. The web is looped around the compensator. By moving the compensator ahead or back, the size of the loop can be increased or decreased, stretching the web lead so that there are always exactly ten cutoff lengths in the span.

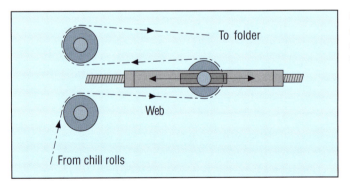

Figure 8-4. *Controlling cutoff using an adjustable compensator roller. (Courtesy Heidelberg USA, Inc.)*

No-slip cutoff system. A no-slip cutoff system requires no compensators. To understand the system, remember the three factors that affect register in the web supported by tension: (1) the length of the web; (2) the tension in the web; and (3) the modulus of elasticity of the paper. With the no-slip cutoff system, the surface speed of the chill rolls is set to match exactly the surface speed of the web coming out of the press. Nip rollers are added at the point where the web leaves the final chill roll to ensure that the paper does not slip on the chill rolls. The nip rollers in the folders are sized so that the surface speed also exactly matches the speed of the web coming out of the press. Considerable pressure is applied at this nip to ensure that the web does not slip. Once register is established in a system such as this, and as long as the no-slip provisions are maintained, misregister does not occur.

A key advantage to this type of system is that no adjustment is required for a change in either the surface properties of the paper or the thickness of the paper, which reduces makeready time for jobs that require paper changes. In addition, variable-speed drives on the chill rolls are eliminated.

Automatic cutoff controls. Automatic cutoff controls precisely divide the web into signatures so that the printed images are consistently positioned. Without cutoff controls, colors may fit exactly; however, the entire image may move around the page from signature to signature because of variation in cutoff. Conventional systems can maintain cutoff within a tolerance of $\pm\frac{1}{32}$–$\frac{1}{16}$ in. (0.8–1.6 mm). Automatic controls have reduced the variation to $\pm\frac{1}{64}$ in. (0.4 mm) or less.

In an automatic cutoff system, an electric eye is set to detect a specific area of the printed form. This eye is mounted somewhere between the cutoff compensator and the cutoff knives. The control accurately measures the interval between the signal from the eye, as detected on each succeeding impression, and a signal generated by the press.

The control is connected to the compensator. If the signal from the web is received before or after the signal from the press, the web lead is increased or decreased (by moving the compensator) until the signals coincide and the web and folder are in time with each other. Proportional control should be built into these systems. Monitoring a large area of the printed form permits precise adjustment to maintain image position.

Web Break Detectors

Web break detectors are devices that mechanically, optically, or pneumatically detect a break in the web of paper, subsequently stopping the press. Web wraps can prove extremely costly—in time and money. When a web break occurs, the detector triggers an electric circuit pinpointing the location of the break.

The original design of web break detector consists of a mechanical roller or finger that rides against the moving web. In the absence of paper, the roller or finger changes position and triggers a switch to shut down the press.

Another system uses an infrared light source on one side of the web and a photoelectric eye on the other. The web prevents light from reaching the eye; otherwise, the press would be stopped. Note that if the photocell is dirty or the light source burns out, this detector will cease to detect web breaks. Thus, this system is a non-fail-safe system. To make the system fail-safe, the light and the photocell can be mounted on the same side of the web, with the light angled off the web to the photocell. If the

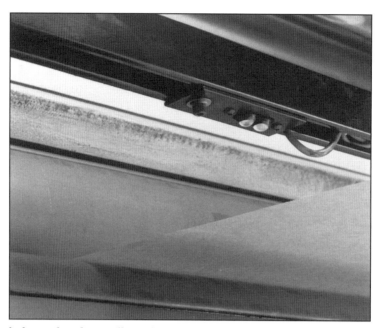

Figure 8-5. Electronic web break detector.

light or the photocell malfunction, the press will not start. Thus, this system design is fail-safe. The higher speeds of newer web presses have ensured the dominance of infrared break detection systems since these systems can stop the press much more quickly and, therefore, save money by reducing waste.

Another web break detector operates pneumatically. A jet of air is blown at the web. On the other side of the web is a pressure-sensitive switch that is activated when the air jet strikes it.

Strip rewind hazard. Web break detectors should always be mounted so that they protect or monitor both edges of the web. A strip rewind, although uncommon, is a particularly dangerous form of web break that is possible on webs with break detectors on only one edge of the web. With a strip rewind, only a portion of the full width of the web is moving properly through the press. The missing portion of the web is being wrapped up on one of the blankets. A strip rewind often starts as a tear on one side of the web but does not cause the entire web to break. If web break detectors are not monitoring the edge of the web that is missing, the break detection system will not sense the problem. Unless the press is stopped manually, the paper will accumulate on the blanket and damage the blanket and printing unit.

Web break detectors should be mounted in pairs at each appropriate point along the press, one detector at each side of the web. They should be set no more than 3–4 in. (75–100 mm) inside the edges of the web, and when web widths change, their proper adjustment should be a routine part of makeready.

Web severing. Many detector systems also incorporate web severing devices that immediately cut off the web on signal of a web break. These prevent excessive paper from wrapping up in a printing unit—a particularly useful feature on large, fast presses with a relatively long stopping time.

Figure 8-6. Web anti-wrap system, which is installed between the last printing unit and the dryer to prevent web wrap-ups on the blankets of the last printing unit in the event of a web break. (Courtesy Baldwin Technology Co., Inc.)

Mechanical detector advantages. Unlike optical and pneumatic detectors, mechanical web break detectors sense not only breaks but also any decrease in tension. This is because the tail end of the break need not run under the web break detector before it senses the break. Rather, it just needs to lose contact when the web becomes slack. For this reason mechanical detectors are effective when mounted in long spans of web, such as those that occur when a second web is run through the press to the third unit of the press. If mechanical detectors are not present, the press may run for some time after the break occurs.

Relative Print Size

Center register marks are often placed down the length and across the width of an image. In maintaining register, the press operator strives to overprint these marks precisely as each successive color is printed. During the pressrun, register marks along the image length may align perfectly, while the marks across the width may be out of alignment, and vice versa. Infrequently, this results from inaccurate placement by the platemaker. Usually, this situation results from the expansion or contraction of color images independently of one another.

Relative print size refers to the size of one color image compared to the size of another color image. This size change may occur in two directions, along the length of the web or across the width of the web. Relative print length (RPL) and relative print width (RPW) describe the specific direction of variation. RPW is commonly known as *fan-out.*

To clarify, the concern is not that the overall full-color image has enlarged or contracted, but rather the individual colors have changed in size compared to one another. Most jobs have margins of error that allow the overall image length to increase by approximately 0.012 in. (0.30 mm) without adversely affecting the image integrity.

With relative print size, a length difference of 0.012 in. (0.30 mm) between the magenta and cyan images prints visibly out of register. Again, this misregistration is caused by the colors making up the images not increasing in size proportionally.

RPW or fan-out is the result of the web width increasing as it passes through the units and picks up moisture. Image width changes in proportion to web width changes after the ink is laid down. One of the characteristics of fan-out is that the amount of overall variation from signature to signature is small. GATF research indicates that ±0.001 in. (±0.025 mm) on a 35-in. (889-mm) form is the normal range of variation for as many as fifty consecutive signatures. Also, fan-out usually decreases at higher press speeds.

A small amount of fan-out may not distort the image beyond acceptable limits, and this may be the case with coarser line screens. For example, newspaper halftone images are commonly reproduced at 85 lpi, a fairly coarse line screen requiring less stringent registration control. The press operator can minimize the fan-out effect by moving the plate cylinders one way or the other to split the difference in misregister on both sides of the image.

Controlling fan-out. Fan-out can be controlled in prepress by slightly distorting the aspect ratio of images on the plate. The aspect ratio of an image indicates the relative height to width ratio of the image. A 1:1 aspect ratio indicates no distortion in the image. By changing the aspect ratio to 0.99:1 the image becomes distorted by stretching slightly in one direction. In fully digital prepress workflows, the prepress output technician imposes the pages on computer as they will appear on the plate. If the amount of fan-out is known, the output technician can change the aspect ratio of each individual color plate to counteract the expansion that will take place on press. The key is to determine the amount of fan-out that will occur.

Determining fan-out values. Fan-out will vary from paper type to paper type and from press to press. Fan-out values will be somewhat approximate, rather than an exact value that can always be used. To determine fan-out follow these procedures:

1. Prepare four plates, each having a fine line grid pattern image. Each plate should have a bold centerline running the length of the plate both horizontally and vertically. A suggested grid pattern increment may be $\frac{1}{32}$ in. or perhaps 1 mm.

2. Mount the plates on a press already inked with the process colors and prepare the roll for which fan-out will be measured.

3. Register all four colors along the centerline on the grid-imaged plates, bringing the press up to normal press speed.

4. Examine the full cutoff press sheet. There will be four color grids on the sheet and fan-out will have caused the grid lines to misregister on the outside edges of the cutoff sheet.

5. Use a linen tester or a high-power reflecting microscope to measure the amount of fan-out for each color. Because these values are very small, some approximation may be necessary.

These fan-out values can now be given to the prepress department to change the aspect ratio of the plates. If various paper types and thicknesses are used, the fan-out test should be done for each paper. Also, different presses may produce different amounts of fan-out. Each press should be tested separately.

Bustle wheels. Fan-out may be controlled on press with the use of devices called ***bustle wheels*** (figures 8-7 and 8-8). They are wheels mounted on the press just before the printing nip. The wheels run along the underside of the web. When the wheel is forced up into the web, the resulting tent-like deformation of the web draws the edges together. The wheels shown in the illustration below can be finely adjusted by thumb screws on the bottom. Bustle wheels should be mounted so that they always run true in the direction of web travel. It is important that the wheels do not wobble; they should be mounted on ball bearings. Bustle wheels are normally 3–3.5 in. (76–90 mm) in diameter and less than 0.25 in. (6 mm) wide. Larger, wider wheels can be forced further into the web, allowing for greater correction from a single wheel. Two or more bustle wheels are commonly employed on the third and fourth units, where fan-out reaches its greatest proportions. Multiple wheels correct fan-out more effectively than a single wheel. Air bustles can replace the mechanical bustle wheel if ink marking becomes a common problem. Air bustles blow a stream of high-pressure air upward to the web surface, drawing the outside edges of the web inward, serving the same purpose as the bustle wheel.

Figure 8-7. Bustle wheels.

Controlling relative print length. RPW and RPL are two distinct factors. RPW (width) is paper-related, while RPL (length) seems to be more press-related. RPW changes occur between units. A change on the top of the web is matched by the same change on the bottom of the web. RPL changes, on the other hand, appear to occur within the printing couple. For example, the top magenta can print long (relative to the top black image) on one impression, while the bottom magenta is printing

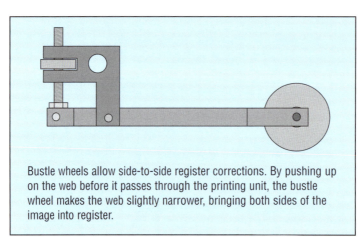

Figure 8-8. Bustle wheel. (Courtesy Network Industrial Services, Inc.)

Bustle wheels allow side-to-side register corrections. By pushing up on the web before it passes through the printing unit, the bustle wheel makes the web slightly narrower, bringing both sides of the image into register.

short—and the relationship can reverse on the very next impression. The range of variation encountered from signature to signature in RPL is much greater than that associated with RPW. Fan-out (RPW) can be controlled fairly easily, while the same cannot be said of RPL.

RPL refers to variations in image size measured in the lengthwise direction of the web, which do not seem to occur due to stretching. A stretch variation would cause the first-down color to print shorter than succeeding colors, with the last-down color printing the longest. However, actual tests show that the first- and second-down images typically contract, while the third-down image expands and that there are differences between the top and bottom of the web.

Figure 8-9. Ribbon control systems help reduce makeready time and start up waste. (Courtesy QTI)

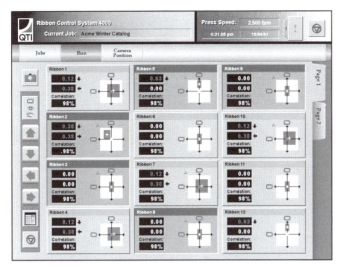

Figure 8-10. The run screen of QTI's Ribbon Control System shows the status of all ribbons in one glance. (Courtesy QTI)

The source of RPL variation seems to lie within the printing couples. Variations in RPL on the top of the web are usually significantly greater than those occurring on the web bottom. The cause of this is not clearly understood. It is known that the tension setting at infeed metering sometimes affects RPL. Press speed also seems to have an effect, but both effects are unpredictable. Further, RPL can vary radically from web top to bottom, from one impression to the next, or from one unit to the next. Thus, even if the causes of the problem were well understood, the means for controlling it would not be simple.

Web-to-web and ribbon-to-ribbon. Multiple webs or ribbons run together over a former board and must be registered if folding accuracy is to be maintained. Because of the potential variations in each web (or ribbon), side-lay and cutoff must be independently maintained in each web (or ribbon).

Most ribbon folders have a full-web compensator to maintain cutoff register before the web is slit. They also carry smaller compensators to register each ribbon after the slitting operation. If the press has automatic cutoff controls, the full-web compensators are usually automatic and the ribbon compensators are manual.

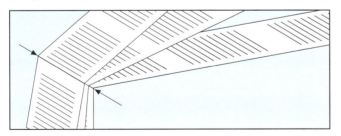

Figure 8-11. An example of web-to-web register.

Section IV
The Blanket and Plate Cylinders

9 Blanket and Plate Cylinder Principles

Each printing couple on a blanket-to-blanket (BTB) web offset lithographic press consists of four basic elements:

1. A *dampening system,* consisting of a fountain pan holding a supply of dampening solution, and a series of rollers that apply the solution to the plate.

2. An *inking system,* consisting of a fountain trough holding a supply of ink, and a series of rollers that carry the ink from the fountain to the plate.

3. A *plate cylinder,* on which the plate is mounted. The plate is a thin metal sheet that wraps around the cylinder surface and carries the image.

Safety Precautions

- Do not operate equipment unless authorized. Heed all verbal and written instructions before operating the press.
- Wear protective gear for eyes, ears, head, hands, and feet where necessary to protect against injury.
- Wear clothing that will not become entangled in any part of the press equipment.
- Make sure the press is completely stopped before touching any of its operating parts.
- Check all safety devices on the press every day to ensure that they are reliable and working. Never switch off safety devices or remove or otherwise bypass guards.
- Before working on the press, check to make sure it has been put on "safe" by using the "stop" (security) push button.
- Check that all guards, covers, and swiveling steps are securely fastened or completely locked in place before operating the press.
- Do not work on moving rollers with rags, tools, etc., because of the high risk of accident and damage.
- Replace lattice guards immediately after removing the press washup devices.
- Use the "reverse" button on the press only for plate and blanket removal—not for cleaning or gumming cylinders.
- Do not start a machine that has stopped without an apparent reason.
- Check for persons, tools, or pieces of equipment between the units and around the press before starting it.
- Remove all used plates, tools, and equipment from the press area and alert your coworkers before starting the press.
- To avoid the possibility of dropping tools into the press or other hazardous locations, do not carry them in pockets.
- Use only tools specifically recommended and provided by the manufacturer to remove and mount blankets and plates.

4. A ***blanket cylinder,*** on which the blanket is mounted. The blanket is a sheet of fabric that is covered with synthetic rubber. It picks up the inked image by contacting the plate and transfers it to the paper.

The couples in a blanket-to-blanket press are organized into pairs, each pair representing one unit. The blanket from one couple provides the impression pressure to transfer ink to the paper required for the opposite couple. No separate impression cylinder is required because of this design. A BTB press may consist of one or more units, each unit carrying a different color of ink to the paper. Full-color printing, also called process color printing, is possible on both sides of the web when four units are grouped together. In this particular instance, colors printed on both sides of the web would include yellow, cyan, magenta, and black. With properly made plates, water from the dampening system adheres to the nonprinting, or nonimage, areas, while ink from the inking system will adhere to image areas.

A variety of conditions must be controlled for proper transfer of ink from plate to blanket, and finally to paper. Proper pressure between plate and blanket cylinders is one of the most crucial. Accurately measured pressure between the blanket and the paper must also be maintained for optimal image transfer. (This topic is discussed in detail in chapter 12, "Cylinder Packing.) All cylinders must be timed relative to each other on multi-unit presses as well. Cylinder timing and bearer pressures will be presented in detail after a discussion of plate cylinder and blanket cylinder anatomy.

The Plate Cylinder

The plate cylinder has four primary functions:

- To hold the lithographic printing plate tightly in position
- To hold the plate while the dampening rollers are contacting it and wetting the nonimage area
- To hold the plate while the inking rollers are contacting it and applying ink to the image area
- To help transfer the inked image to the blanket

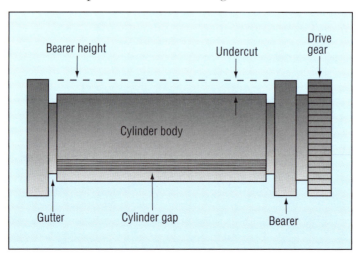

Figure 9-1. *A printing unit cylinder with the major elements identified.*

The basic features of all plate cylinders are the same. Almost all have **bearers:** smooth, flat metal rings at the extreme ends of the cylinder. Just inside each bearer (between bearer and cylinder body) is a narrow groove, called the **gutter.** Between the two gutters is the **body**—the main portion of the cylinder—on which the plate and packing are mounted. The body of the cylinder is always lower than the surface of the bearers; the exact difference in height—called the cylinder undercut—varies from press to press. Often, the amount of undercut is specified by the plant ordering the press. The exact amount of undercut on the plate cylinder must be known in order to set proper pressures in the printing unit.

Figure 9-2. Closeup of printing unit cylinders and bearers.

The surface of the plate cylinder body does not extend all the way around the cylinder circumference. On nearly all presses, a gap runs from gutter to gutter across the cylinder. This gap contains clamping devices that hold the plate tightly onto the cylinder.

The **leading edge** of the plate cylinder is the edge along the gap that is followed by the cylinder body as the cylinder rotates in the running direction. It is this edge that grips the lead edge of the plate when it is mounted onto the cylinder. The **trailing edge** is followed by the cylinder gap. The leading edge of the gap is machined at an acute angle to the surface of the cylinder body, and the leading edge of the plate is pre-bent to exactly this angle before mounting on the press. The plate is inserted in the slot at the lead edge, then the plate is rolled around the cylinder. Next, the tail of the plate is placed in the lockup at the trailing edge which, when tightened, will provide the gripping force necessary to hold the plate tightly and smoothly against the cylinder. (See figure 9-3.)

The cylinder gap represents a nonprinting area where no ink can be carried and transferred to paper. The white space left on the web by the cylinder gap represents the web cutoff area, where cutting will result in sheets and subsequently folded signatures. It is important to note that press manufacturers try to keep plate gaps to a minimum, because a wider nonprinted cutoff area on the paper represents more waste over the pressrun. For example, consider a cutoff gap of 0.5 in. (12 mm) versus one of 0.25 in. (6 mm). The 0.25-in. difference between these cutoff gaps may not

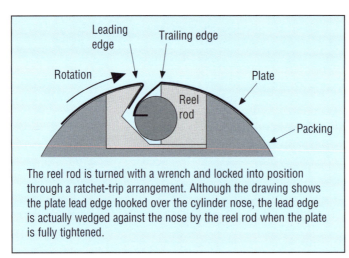

Figure 9-3. *Cross section of a typical plate lockup mechanism, showing the printing plate fully tightened. (Courtesy Heidelberg USA, Inc.)*

seem like much on a single sheet, but if one considers 1,000,000 impressions, the difference becomes very significant:

0.25 in. × 1,000,000 impressions equals 250,000 in., or 20,833 ft. of paper

The gap on the plate cylinder is usually about 0.125 in. (3 mm) narrower than the one on the blanket cylinder. The reason for the wider blanket gap is that the blanket and its mounting bars are much thicker than the plate and thus require a wider lockup area. If the cylinders are properly timed, the plate cylinder gap should fall centered between the blanket cylinder gap. (See figure 9-4.)

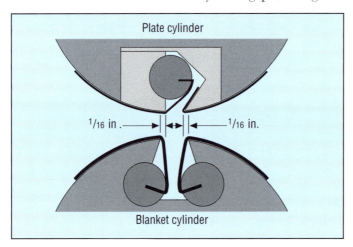

Figure 9-4. *The plate cylinder gap falling inside the blanket cylinder gap, indicating proper timing. On this press, the blanket cylinder gap is about ⅛ in. (3 mm) wider than the plate cylinder gap. The cylinders are shown spread apart for easier visualization. (Courtesy Heidelberg USA, Inc.)*

Because plates are thinner than blankets, the undercut on a plate cylinder is usually much less than the undercut on a blanket cylinder. Plate thickness, or gauge depends on the size of the plate, and usually increases with plate size. Gauges normally vary from about 0.012 in. (0.30 mm) for a 17×22-in. plate up to 0.015 in. (0.38 mm) on large plates.

On many presses, the bearers of the plate cylinder run in contact with the bearers of the blanket cylinder during printing. The diameter of the bearer is the effective diameter of the cylinder and is the same as the pitch diameter (i.e., the working diameter) of the gear attached to the journal. The plate cylinder is driven by this

---

Care and Maintenance of Offset Press Cylinders

To keep the printing unit running properly, a regular maintenance routine must be followed. Lubrication of the cylinders is of extreme importance.

- Place the press in the safety operating mode.
- Clean and lubricate the moving parts of the plate and blanket clamps or reel rods.
- Check and replenish oil reservoirs for automatic and self-oiling systems.
- Cylinder gears that do not run in oil baths must be cleaned before lubrication.
- Use a small stiff brush to clean the gear teeth to remove any particles of ink, paper, lint, or gum that have accumulated. Clean to the bottom of each tooth.
- Lubricate the cylinder gears with a lubricant recommended by the manufacturer.
- Bearers should be clean and dry. Remove all traces of gum and ink from the bearer surfaces.
- Although most cylinders are chromium-plated to prevent rusting, some cylinders do rust. Rust causes high spots on the cylinder which add pressure. Remove any rust with a nonabrasive scouring pad. (Never use files, razor blades, or coarse abrasives on the cylinder surface.)
- Wipe the cylinder surface with a thin film of lightweight oil.

---

gear, which is, in turn, driven by a similar gear on the blanket cylinder. The cylinder gears may be spur (on older presses) or helical (on newer presses). A **spur gear** has teeth cut straight across it, and a **helical gear** has teeth cut at an angle. A spur gear used as a plate cylinder gear nearly always has a **backlash gear,** a thin second gear bolted to it to reduce **play** (free or unimpeded movement) between gears. Presses that print with the plate and blanket cylinder bearers out of contact always have helical gears to reduce gear play and provide a smooth drive.

Figure 9-5. Helical gears.

The Blanket Cylinder

The **blanket cylinder** carries the printing blanket and has two primary functions:

- To carry the offset rubber blanket into contact with the inked image on the plate cylinder
- To transfer, or offset, the ink film image to the paper (or other substrate)

The basic design of the blanket cylinder is the same as that of the plate cylinder. Almost all blanket cylinders have bearers that run in firm continuous contact with the plate cylinder bearers. The blanket cylinder gutters help to prevent chemicals from working in under the blanket, and they also keep foreign matter picked up by the bearers from moving onto the blanket surface. The body of the cylinder is the area around which the blanket and its packing are mounted.

Usually, the blanket cylinder body also has a gap containing the lockup that holds the blanket at the leading and trailing edges. The gap on the blanket cylinder is wider than the one on the plate cylinder, usually by 0.125 in. (3 mm), in order to accommodate the thicker blanket and mounting bars. (See figure 9-6.)

Three-ply blankets are generally used on blanket cylinders with undercuts of 0.075 in. (1.9 mm) or less. Four-ply blankets are used on presses with undercuts of more than 0.075 in. A "ply" was originally intended to refer to the fabric layers within

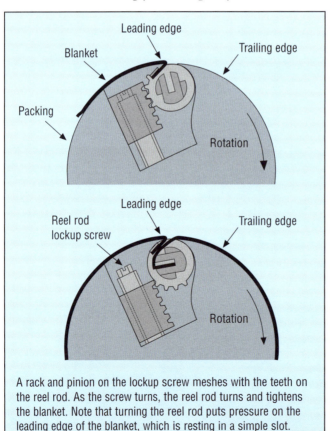

Figure 9-6. *A blanket lockup.*

A rack and pinion on the lockup screw meshes with the teeth on the reel rod. As the screw turns, the reel rod turns and tightens the blanket. Note that turning the reel rod puts pressure on the leading edge of the blanket, which is resting in a simple slot.

the blanket design. While blankets manufactured today contain only one true "ply," the "three-ply" and "four-ply" designations have been retained to reference the thickness of the blanket.

Cylinder Pressures

As previously explained, the printing unit of a blanket-to-blanket press has two printing couples that run with their blankets in contact. During operation, problems encountered may involve only one couple, but any adjustments will affect the entire unit. Because of this, the printing unit should be viewed as a single, dynamic system. Changes in one part of the system affect every other part. This is most true when cylinder pressure and cylinder timing are involved.

Setting Bearer Pressure

Caution: The press operator must follow the proper safety precautions when setting bearer pressures. It is very important to have the proper pressure settings between bearers. If not enough pressure exists, packing pressure will force the bearers away from one another, creating a gap. When the bearers are not running in contact at fast speeds, the plate and blanket cylinders can bounce as the gaps come in contact. This bounce may cause slurring, and eventually streaks will begin to appear.

Checking for run-out. Prior to setting the bearer pressure, the press operator or a maintenance worker must check the cylinders and bearers for ***run-out,*** a condition in which the cylinders or bearers are out of round. To check for run-out, a magnetic-base dial indicator is attached to a frame member, the press is inched a couple of inches, and a reading is taken. The press is again inched, and another reading is taken. This procedure is repeated until the cylinder has made one complete revolution. The readings on the dial indicator are recorded, and the difference between the highest and lowest readings is compared to the manufacturer's specifications for maximum allowable run-out. If run-out is within specifications, the bearer setting procedure can proceed. If run-out exceeds specifications, the cylinder may have to be repaired or replaced.

The light method. One of the more popular bearer pressure setting procedures is known as the "light method." This method employs a lamp to illuminate the space

---

Importance of Maintenance

The blanket-to-blanket unit is capable of providing excellent printing results. However, without proper maintenance, quality printing and timely makereadies are bound to suffer. It is highly recommended that a regular maintenance schedule be established that includes measuring plate, blanket, and packing caliper and verifying proper squeeze values with a packing gauge. In addition, determine the effective cylinder undercut on a regular basis to track and correct for variations. Finally, the proper timing of the cylinders in the blanket-to-blanket unit should be verified and corrected on a regular basis.

between bearers, providing a visual check. General procedures for the "light method" of checking bearer pressures is as follows: (Note that the exact values for packing and caliper of materials used may vary from one press to another. Check with the manufacturer for exact caliper values.)

1. Clean bearers of all dried ink or other residue.

2. Pack the plate and blanket cylinders according to manufacturer's specifications.

3. Prepare 0.003-, 0.004-, and 0.005-in.-thick (0.075-, 0.10- and 0.125-mm-thick) sheets of polyester (e.g., Mylar). Each sheet should be the width of the plate and blanket.

4. Inch a 0.003-in. (0.075-mm) Mylar sheet between the plate and blanket.

5. Place a pair of strong bright lights behind the nips of the bearers on both the operator and gear side.

6. Turn the impression on. At this point, if the bearers are properly set, no light should be visible between the contact point of the bearers. If light is visible, the bearers are too loose and need to be reset.

7. Next, place a 0.006-in. (0.15-mm) Mylar sheet (two 0.003-in. sheets) between each plate and blanket. With the light on, a crack of light should be visible. If no light is visible, the bearers are set too tight and must be reset.

8. Repeat this procedure for each plate-to-blanket bearer setting.

9. Place a 0.008-in. (0.2 mm) Mylar sheet (two 0.004-in. sheets) between the blankets. With the impression on, a thin crack of light should appear between the blanket cylinder bearers. If no light appears, the blanket cylinder bearers are set too tight and should be readjusted.

10. Place a 0.005-in. (0.125 mm) Mylar sheet between the blankets. With the impression on, no light should be visible. If light is visible, the bearer pressure is set too lightly and should be readjusted.

Bearer pressure imprint. To obtain a "picture" of the proper bearer setting, use a thin strip of aluminum foil less than 0.001 in. thick (0.025 mm) and wider than the bearers. Place the aluminum foil between the bearers and put the impression on and off, and then measure the width of the stripe using a 20–30× magnifier with built-in measuring reticle. This measurement is recorded and is then used as a standard against which future measurements are compared. It is recommended that this bearer pressure stripe procedure be performed monthly to determine if any setting changes have occurred. If the width of the stripe on subsequent measurements is narrower than the width of the original stripe, the bearer pressure has decreased and needs to be reset.

If changes occur repeatably over a predictable length of time, this time frame can be a gauge for the plant's preventive maintenance program. For example, if the stripe width changes every three months, the light method of bearer setting should be repeated every three months as a part of preventive maintenance.

Cylinder Timing

Cylinder timing involves accurately registering multiple plate cylinders circumferentially to one another, and then timing the blanket cylinders to the plate cylinders. The result is that all cylinders will be synchronized. There are a variety of methods used to time web presses in the printing industry. Many press operators are not responsible for the timing of their press, but they should know how timing works to troubleshoot timing problems. The following method will present the principle of cylinder timing using the GATF Register Test Grid.

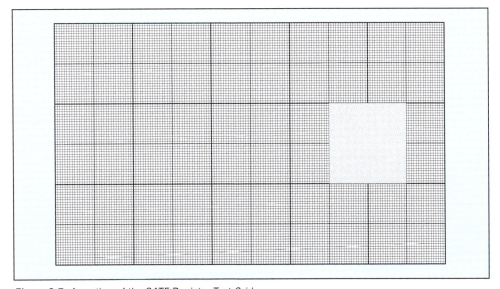

Figure 9-7. *A portion of the GATF Register Test Grid.*

The GATF Register Test Grid Method. The GATF Register Test Grid can be used for accurate cylinder timing on a web press by following these procedures. The first set of procedures is for accurately timing the top to bottom cylinders for accurate image backup:

1. Make plates for all the top and bottom units of the press from the same GATF Register Test Grid.

2. Mount the plates on the plate cylinders according to the press manufacturer's recommendations.

3. Center all the plate cylinders circumferentially from the press register console.

4. Ink the printing units and run the press until the ink laydown is satisfactory for all the colors on the top and bottom of the sheet.

5. Stop the press and evaluate the placement of the GATF Register Test Grid on the press sheet.

6. Identify at least one set of cylinders for the same color of ink that are backing themselves up on the top and bottom side of the press. Keep these cylinders that are backing up in the locked position.

7. Identify cylinder pairs that need to be adjusted (those not backing up properly).

8. Using the printed GATF Register Test Grid image, accurately measure and record the amount of misregister for each color that is not backing up top-to-bottom.

9. Put the press on safe and open the press housing on the gear side of the press.

10. Unlock (loosen) the bolts on the gears of the plate cylinders that need to be adjusted, and then circumferentially forward or reverse each plate cylinder by the distance required to get the units to register and back up top and bottom.

11. Lock the plate cylinder gears, close the frame housing, and start printing again.

12. Evaluate the newly printed Register Test Grid to determine how close the units come to backing up circumferentially.

13. If circumferential register is still off, repeat the above procedure on the offending plate cylinder(s) until circumferential register is accurate.

Aligning the blanket cylinder to the plate cylinder: Mylar strip method. The next set of procedures will assure that the blanket and plate cylinders are timed properly by aligning all of the blanket cylinder gaps with their respective plate cylinder gaps:

1. Pack the plate and blanket cylinders according to manufacturer's specifications.

2. Prepare several 2×18-in. (51×457-mm) clear Mylar strips.

3. Tape the narrow edge of a Mylar strip to the plate (in-running side) of the first unit being timed with the long dimension of the Mylar parallel to the direction of web travel.

4. Inch the press until the plate cylinder gap of the unit is easily accessible.

5. With the press on safe, manually rub some ink on both sides of the plate cylinder gap.

6. With the impression and safe both on, press the taped Mylar strip against the plate cylinder gap to transfer an inked imprint onto it. The resulting imprint represents the leading and trailing edges of the plate gap.

7. Next, flip the loose edge of the taped Mylar against the blanket. The inked imprint on the Mylar will show the location of the blanket cylinder gap with respect to the plate cylinder gap.

8. If the blanket cylinder is not properly aligned and must be adjusted, loosen the bolts on the blanket cylinder gear, take the press off impression, and forward or reverse the blanket cylinder so that its gap will align with the inked imprint on the piece of Mylar.

9. This procedure should be repeated once or twice to make sure that the gaps are perfectly aligned.

10. Finally, with the impression on, the press operator locks up the bolts on the blanket cylinder.

11. Move to the next printing unit to repeat the procedure.

After all units have been adjusted, the plate cylinders should all be in circumferential register. The blanket gaps will be timed to match up with the plate cylinder gaps. At the end of a printing job, the press operator will now be able to put the plate cylinders back to their circumferentially centered "zero" position, with all cylinders falling into alignment.

10 Lithographic Plates

There are a variety of plates used in the web offset industry. Each type of plate may have characteristics that make it suitable for a given type of press or type of production workflow. Even so, any plate used in web offset will fall into one of two general categories that describe how the plate is imaged: (1) contact or (2) digitally. Contact plates are made by exposing the plate to imaged film in a contact frame. This type of plate assumes that the printer is using a film-based workflow. Presensitized surface plates are the most common type of contact plate, but some use of multimetal plates also exists. Digitally imaged plates are quickly gaining a share of production work in the web offset industry. These plates are made in some type of platesetting device, which is an output device within a complete digital workflow, requiring no film. There are several types of digitally imaged plates including silver halide and photopolymer.

Lithography demonstrates the principle that oil and water, generally, do not mix. A conventional lithographic plate consists of image areas, which accept ink and repel water; and the nonimage areas, which accept water and, when wet, repel ink. Image and nonimage areas are distinguished by differences in surface chemistry. These image and nonimage areas exist on essentially the same plane; thus, lithography is a planographic process.

Types of Plates

Contact plates. A high percentage of today's commercial printing plates are made photographically using either negative or positive film. These plates are coated with a light-sensitive material that changes its solubility when exposed to light in a contact frame. The contact frame is a device that holds the film and plate together in firm contact under vacuum pressure. A light source that is appropriate for exposing the plate is activated with a timer or light meter for an exacting exposure. Negative-working plates are exposed with high-contrast film with the images in negative form. Positive-working plates are exposed with films in a positive format. In either case, the light that penetrates the transparent areas changes the solubility of the plate coating. After exposure, the plate is developed or processed to remove the coating from nonimage areas.

Negative-working contact plates are the most popular. They have soluble coatings that become insoluble upon exposure to light. The image (clear) areas of the negative allow light to penetrate and form the image areas of the plate. The unexposed coating remains soluble and is removed during development.

Positive-working plates are completely covered with an ink-receptive, light-sensitive coating. They are exposed through positives. The unexposed areas (image areas) are insoluble in the developer. The exposed areas are solubilized by exposure to light and then removed during development, leaving the water-receptive coating as the nonimage areas.

Surface plates. Whether negative or positive, there are two general categories of contact plates including (1) surface and (2) multimetal. Surface plates may be either presensitized or wipe-on. Wipe-on plates are still sometimes used in the newspaper industry but are becoming more rare in recent years. Presensitized plates are coated with a sensitizing agent by the manufacturer; wipe-on plates are coated by the platemaker before imaging the plates.

Presensitized plates consist of a thin film of light-sensitive material, usually a diazo compound or a photopolymer that is coated on the sheet metal. Photopolymer coatings consist of polymers and photosensitizers that react (crosslink) during exposure to light to produce a tough, long-wearing image area. Diazo coatings also react with light to produce a tough, long-wearing image area. After exposure, the plates require special organic or aqueous solvents for processing. Both negative-working and positive-working plates are available with diazo or photopolymer coatings.

Presensitized plates are sometimes designated as being either *additive* or *subtractive.* These terms describe the differences in processing procedures. A presensitized plate is additive when the platemaker rubs on (adds) an image-reinforcing material, called lacquer, to the image areas during processing. The sensitized coating over the nonimage areas must be removed during development. Subtractive presensitized plates are by far the most common types of presensitized plate. These plates are manufactured with a lacquer coating by the manufacturer. During processing, the developer removes (subtracts) the unexposed lacquer coating along with the sensitizing agent.

Wipe-on plates are chemically similar to presensitized plates, but they are coated with aqueous diazo coatings in the plateroom, in many cases with a simple roller coater. A specially treated aluminum or anodized aluminum base is used. Wipe-on coatings are thin and lack durability on press, so special developers are required that contain lacquer or plastic to build up on the image to greatly increase durability, much like additive presensitized plates.

Multimetal plates. Multimetal plates include both bimetal and trimetal types. Some types of metal are naturally oleophilic, most notably copper. Bimetal plates consist of a base metal, usually aluminum or stainless steel, on which a thin layer of copper has been electroplated. Copper forms the ink-receptive image areas and is removed from nonimage areas. The bare base metal, the nonimage area, is easily desensitized to become water-receptive. Bimetal plates are supplied in smooth and grained form.

A special type of bimetal plate is made by electroplating a thin layer of chromium over a base metal of copper or brass. The chromium is etched away during processing, exposing the base metal in the image areas. Chromium remains on the plate in the nonimage areas.

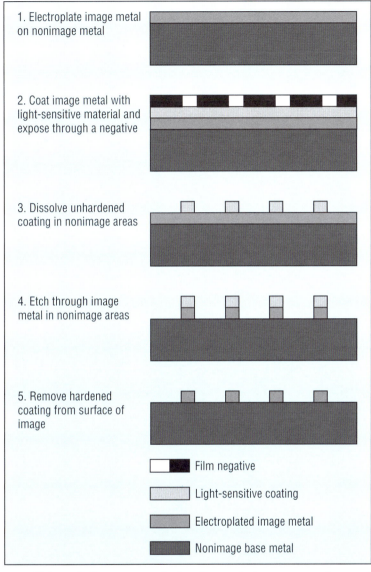

Figure 10-1. General processing sequence for a bimetal plate.

1. Electroplate image metal on nonimage metal

2. Coat image metal with light-sensitive material and expose through a negative

3. Dissolve unhardened coating in nonimage areas

4. Etch through image metal in nonimage areas

5. Remove hardened coating from surface of image

Film negative

Light-sensitive coating

Electroplated image metal

Nonimage base metal

Trimetal plates are made by electroplating two metals on a third base metal. The base metal is usually aluminum, mild steel, or stainless steel. A layer of copper is first electrodeposited on the base metal, followed by a thin layer of chromium. Chromium forms the nonimage areas and is etched away to bare the copper in the image areas. If grained plates are desired, the base metal sheet is grained before the other metals are plated on it. The films of these metals are so thin that they have little effect on the grain.

Digital-direct plates. The digital-direct plate workflow is usually referred to as *direct-to-plate* or *computer-to-plate (CTP)* technology. The prepress workflow requires that imposition and proofing are done digitally, eliminating the need for film, stripping, and film-based proofing.

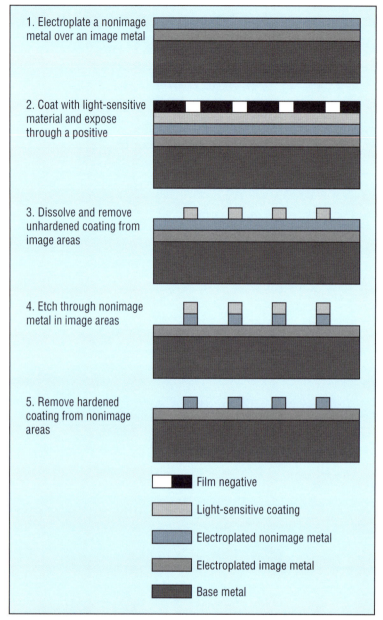

Figure 10-2. *General processing sequence for a trimetal plate.*

1. Electroplate a nonimage metal over an image metal

2. Coat with light-sensitive material and expose through a positive

3. Dissolve and remove unhardened coating from image areas

4. Etch through nonimage metal in image areas

5. Remove hardened coating from nonimage areas

Film negative

Light-sensitive coating

Electroplated nonimage metal

Electroplated image metal

Base metal

   Digital-direct plates are imaged in a platesetter (figure 10-3), directly from digital data sent from the computer workstation. Lasers, UV light, or light-emitting diodes (LEDs) may be used to make the exposure. Inkjet technology may also be used to image the plate. With this technology, an inkjet forms the image areas on the plate to form a mask. The plate is then exposed, and the inkjet mask is dissolved and removed during processing.

   There are three basic types of platesetter: internal drum, external drum, and flatbed. Internal drum platesetters require the plate to be inserted within a drum. The drum stays stationary while a rotating mirror moving through the center of the drum on a worm gear guides the laser beam to the plate. With external drum plate-

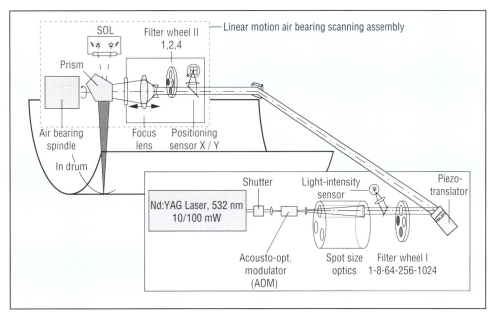

Figure 10-3. *Internal drum platesetter. (Courtesy Heidelberg Prepress)*

setters, plates are loaded around the outside of the drum. While the drum spins, a head with multiple lasers moves from one side of the plate to the other, exposing the plate. Plates are positioned flat on a bed with flatbed platesetters. One type of flatbed platesetter employs the use of two LCD panels mounted on separate carriages. A UV light exposes the plate through the LCD panels, which can move over just the image areas of the plate, reducing time required to expose plates that have light areas of coverage.

Plates for use in a platesetter may be coated with various materials including a silver-halide emulsion, diazo, and photopolymer. Light-sensitive plates are exposed by the light of the exposing device (laser, UV light, or LED) while thermal plates are exposed by the heat created by the laser.

Some types of digital-direct plates are preheated (prebaked) before exposure and/or post-heated (baked) after exposure. This step in processing serves to harden the ink-receptive layer of the plate, extending the run length.

Waterless plates. Plates for waterless lithography are constructed of an aluminum base, a primer, a photopolymer layer, an ink-repellent silicone rubber layer, and a transparent protective film on top. Plates are manufactured for either contact or digital-direct exposure and processing. Care must be taken handling the plates; any scratches in the silicone rubber layer will attract ink. Testing at GATF has proven that the waterless process produces a sharper image than conventional web offset does. The waterless plate, as of the present, has a shorter run life than comparable conventional plates.

A ***positive-working waterless contact plate*** is processed by first exposing it to UV light through a film positive in a vacuum frame. The exposure causes the silicone rubber layer to bind to the light-sensitive layer in the nonimage area. The top

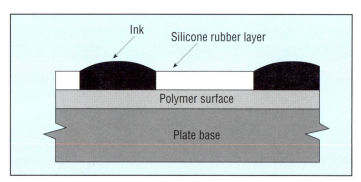

protective layer is then peeled off, and a developer that removes the silicone rubber layer from the light-sensitive layer in the image areas is applied to the plate surface.

Processing of **negative-working waterless contact plates** begins in the same way. However, with a negative-working plate, exposure to UV light through a film negative weakens the bonds between the light-sensitive layer and the silicone rubber layer in the exposed image areas. After the exposure, the protective cover film is peeled off and a pretreatment solution is applied. This solution strengthens the binding between the silicone rubber and light-sensitive layers in the plate's unexposed nonimage areas. The silicone rubber layer is then removed from the light-sensitive image layer in the plate's exposed areas.

Digital-direct waterless plates are imaged by thermal ablative technology, which is processless and also may be performed directly on press. The energy from the laser removes (ablates) the ink-repelling silicone layer of the plate, unveiling the oleophilic photopolymer layer. Because the ablation leaves burrs (specks of the removed silicone layer) on the plate surface, a brush/vacuum must be incorporated into the platesetter to clean the surface before use.

Plate Composition

A conventional lithographic plate consists of two distinct surface areas. The image areas are oleophilic, meaning that the surface will readily attract oily ink. Nonimage areas are hydrophilic, or water attracting. Because oil-based ink and water tend to repel each other, these two areas remain separate on the plate, even though water and ink are distributed over the entire plate by the dampening and inking systems.

Plate base. Today, most lithographic plates have a thin metal base. Aluminum is by far the most common metal used, although plates can also be made of stainless steel, mild steel, or brass. Aluminum has the advantage of being relatively light, flexible (for wrapping around the cylinder), and hard enough to withstand reasonable surface compression forces.

- **Gauge tolerance.** It is critical that sheet metal used in the manufacture of plates meets strict gauge tolerances and is free of surface defects. Plates up to 22×34 in. (559×864 mm) should not vary more than 0.001 in. (0.025 mm) in thickness. For example, a 0.012-in. (0.305-mm) plate should not be more than 0.0125 in. (0.318 mm) nor less than 0.0115 in. (0.292 mm) thick in any area. In this example, the tolerance is expressed as ±0.0005 in. (±0.013 mm). An excessive variation in tolerance will create pressure variations in different areas of the plate, resulting in uneven tints and solids.

- *Plate gauge.* The plate gauge varies depending upon the cylinder undercut and press size. Standard thicknesses range from 0.0055 in. (0.14 mm) to 0.020 in. (0.51 mm), while sizes of the plates may be as large as 59×78 in. (1,499×1,981 mm). Generally, a large-diameter press cylinder requires thicker plates to minimize stretching and plate cracking. Plate thickness should be matched to the cylinder undercut so that minimal packing is required. Several layers of packing are more likely to creep and compress than a single sheet.

Plate graining. Before a metal can be used as a base for a lithographic plate, its surface must be properly prepared. The surface of a sheet of aluminum is very smooth. When water is coated on the smooth surface it will have a tendency to bead. Because a lithographic plate must accept a film of water without beading, the surface of the plate is given a grain. This is accomplished either by roughening the surface mechanically or treating it chemically or electrolytically. This mechanical or chemical graining process significantly improves press latitude, reduces drawdown time in the vacuum frame, and helps to eliminate halation.

Most plates in the United States are grained on a machine in which a continuous web of aluminum is passed under a series of rotating nylon brushes and grained with a mixture of abrasives and water, a process referred to as *slurry brush graining.* Introducing a uniform grain produces a surface that is dark in color. This brush grain is very fine and is satisfactory for presensitized and wipe-on plates. Using a chemical etch after brush graining produces a much lighter, cleaner, and slightly rougher grain.

As an option to mechanical brush graining, several methods of cleaning and slightly roughening plates chemically are currently in commercial use. They are used primarily for treating relatively smooth, short-run plates prior to coating in the manufacture of presensitized plates. These plates are usually double-sided plates for use on small presses. Most bimetal aluminum plates are also chemically grained, but they are made much rougher than the presensitized chemically grained plates. Some premium long-run plates are electrochemically grained to produce a uniform, relatively rough grain. As an exception to grained plates, stainless-steel bimetal plates are smooth and without grain. This is because stainless steel is naturally hydrophilic.

Rougher-grained plates have several advantages. They provide better latitude for ink/water balance on press, faster drawdown in a vacuum frame, less trouble with dirt and hickeys, better durability on press, and less tendency for dot slur. However, these plates are not able to hold fine highlight dots as well as smooth grained plates.

Silicating. In addition to roughening the surface, chemical treatments are also needed for some processes, especially wipe-on and negative-working diazo presensitized plates. Diazo compounds, which are ink-receptive when exposed, can react with untreated metals. Therefore, the aluminum is usually treated in a hot sodium silicate

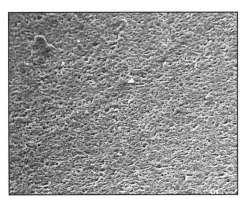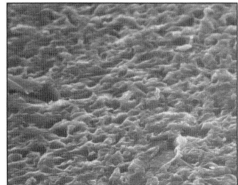

Figure 10-5. *Photomicrographs of a grained, anodized aluminum surface at two magnifications: 200× (left) and 1000× (right). (Courtesy Imation)*

solution to create a barrier layer that prevents a reaction between the diazo and the aluminum. This treatment also desensitizes the plate so that it will be more water-receptive, as well as helping to make the plate surface more receptive to bonding with the diazo.

When positive-working diazo presensitized plates are made, surface treatments may not be necessary; fine graining and/or cleaning usually precede the application of positive-working diazos.

Anodizing. Most high-quality plates are anodized after graining. Aluminum anodizing is a process by which a thin, uniform layer of extremely hard aluminum oxide is produced electrolytically on the grained aluminum. This anodic layer has many extremely small pores, similar to a honeycomb. The anodic layer must be sealed before the photosensitive coating is applied. Usually, hot solutions of sodium silicate are used to treat the anodized layer, making it highly water-receptive. This process also prepares the plate surface to receive the light-sensitive coating. Furthermore, the anodic layer is hard, abrasion-resistant, and highly durable.

Diazo sensitizers. Both water- and solvent-soluble diazos are commonly used as a sensitizing agent for plates. Positive-acting plates are coated with diazo oxides or quinone diazides. Exposure to UV light converts negative-acting diazos directly to insoluble resins that have good ink receptivity and durability for printing. Positive-acting diazos decompose upon exposure to light and become soluble in the developer while the unexposed diazo remains on the plate, forming the image areas.

Photopolymer sensitizers. A thin polymer coating makes an excellent oleophilic surface. A number of different reactive polymers can be sensitized with a suitable photoinitiator for use as a plate coating. On exposure to light, the exposed parts of the coating become insoluble in the same solvents that dissolve the unexposed portions of the coating. Thermal plates often employ photopolymer coatings that harden when exposed to light. The resultant images are tough, and the plates generally withstand long runs. Most photopolymer coatings are aqueous (water-developable).

Silver-halide sensitizers. Silver-halide coatings have been used to sensitize graphic arts films for decades. Manufacturers are now using silver-halides as a sensitizer for plates. The light-sensitive silver halides are suspended in an emulsion layer, which lies on top of what is called the nuclei layer. A barrier separates these two layers. The laser in the platesetter exposes the nonimage areas of the plate, reducing the silver halides. During the processing, the unexposed silver halides diffuse through the barrier layer into the nuclei layer, where it is reduced to molecular silver, which is oleophilic in nature. A second processing step removes the exposed silver and corresponding nuclei layer from the plate surface.

Plate Run Length

Any plate image, if run long enough, will begin to degrade. The first signs of image degradation can usually be seen in the halftone or tinted areas as dot sharpening. That is, the dots begin to decrease in size, resulting in a lighter image. Image areas of plates can also become desensitized, resisting ink adhesion. This is called plate blinding.

A number of on-press factors can affect the number of impressions that a press can handle before the image breaks down in some way. Premature wear can result from excessive pressures between the form rollers and the plate, and between the plate and the blanket. Checking and setting pressures on a consistent basis will help assure maximum plate life.

Inks with a larger pigment particle size will cause faster plate wear. Opaque inks, particularly metallic inks, tend to have larger pigment particles. Also, ink that is not well milled will have poor fineness of grind. These inks will be more abrasive, contributing to premature plate wear.

Checking fountain solution concentration to assure proper acidity and gum concentrations will help prevent plate blinding. Copper-imaged plates, in particular, tend to be sensitive to excessive gum in the fountain solution. However, any plate will blind if acidity is too high. A higher acidity will cause the gum to desensitize rollers as well.

Because on-press factors affect run length, it is difficult to accurately predict the life of a plate. However, some estimation of run length is possible. Presensitized contact plates can last anywhere from 100,000 to over 1,000,000 impressions. The longer-run presensitized plates tend to require post-baking to make the image area harder and more resistant to wear. Shorter-run plates tent to be smoother grained. Presensitized plates print exceptionally high quality and have the advantage of convenience. These plates are relatively susceptible to excessive pressure on the press and the composition of fountain solutions.

Wipe-on plates cost much less than presensitized plates, because they have to be coated by the platemaker prior to exposure. These plates can effectively print 200,000 impressions.

Bimetal and trimetal plates, being electroplated with copper or copper and chromium, are relatively expensive. However, they produce high-quality printing, resist dot sharpening, and are usually good for over 2,000,000 impressions without any noticeable image degradation.

Thermal digital-direct plates have a very long run life when baked—over 2,000,000 impressions. Unbaked thermal plates may last anywhere from 100,000–300,000 impressions, depending upon the manufacturer. If the plate is to be used for a shorter run, it is worth skipping the baking step to save time in plate preparation.

Silver halide plates on aluminum bases are typically rated to last anywhere from 100,000 to 500,000 impressions, depending upon the manufacturer. Silver halide plates with polyester bases, typically not used on web presses, are rated at only about 20,000 impressions.

Photopolymer plates are rated from 100,000 to 2,000,000 impressions, again, depending upon the manufacturer. Some types of photopolymer plates require baking and others are not made for baking. Like thermal plates, baking extends run life considerably with photopolymer coatings.

Waterless plates are rated at about 100,000 impressions. They never receive a preheating or baking step, and they do not require any processing after exposure.

Handling Plates on Press

Nonimage areas sometimes lose their desensitization on press (scumming) and must be cleaned and desensitized to restore their water-receptivity. Some proprietary plate cleaners are designed to be squirted on the plate while the press is running. The cleaner should also desensitize nonimage areas and remove ink and piling from image areas. Most plate cleaners are somewhat abrasive; excessive use can reduce plate life.

Plate cleaners are also used to eliminate scumming. If a plate continues to scum in the same pattern after cleaning, the plate is probably defective and should be replaced.

Aluminum plates sometimes develop an oxidation scum, often referred to as **ink dot scum.** This occurs during shutdowns when residual fountain solution corrodes the aluminum, forming small pits. To avoid ink dot scumming, lift the inking and dampening form rollers for several impressions prior to stopping the press.

Plate Bending

Plates for web offset printing must be bent very precisely before mounting them on press. These plates, because of their bends, do not lend themselves to being twisted on the cylinder. Thus, plate-to-plate registration must be accurate, and the bends must be consistent. Figure 10-6 shows the precision required in maintaining register between the image and the bends.

To ensure proper bending of plates, any plate bending device should have these design features:

- An adjustment mechanism to change the distance between bends.
- A pin register system to ensure accurate register of the image to the plate bend.
- A vacuum hold-down to maintain control of the plate during clamping and bending.

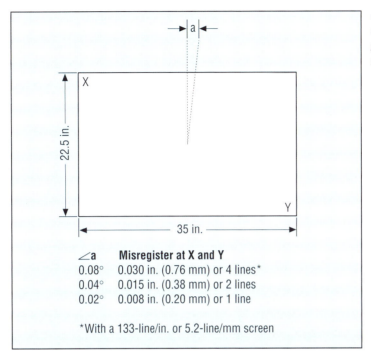

Figure 10-6. Movement required to throw two points out of register a given amount.

∠a	Misregister at X and Y
0.08°	0.030 in. (0.76 mm) or 4 lines*
0.04°	0.015 in. (0.38 mm) or 2 lines
0.02°	0.008 in. (0.20 mm) or 1 line

*With a 133-line/in. or 5.2-line/mm screen

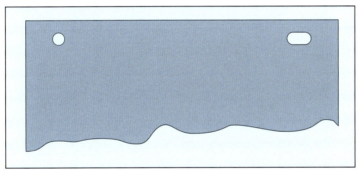

Figure 10-7. Punched plate with a round hole and a slot.

Figure 10-7 shows a punching system that employs a round hole and a slot. The punched plate is fitted to pins that are in fixed positions on the plate bending device. This assures that the plate will not shift when the bend is made.

The quality of the bend in the plate is affected by three key factors. Metals bend differently according to (1) the amount of force applied, (2) the rate at which it is applied, and (3) the radius of the bend. For a plate bending device to ensure uniform and consistent plate bending, these factors must remain consistent for a given plate type.

Having to stop a pressrun to change a cracked plate can create significant costs in downtime. Plate cracking generally occurs when metal fatigue becomes excessive and ruptures the plate metal. This is due to undue stress placed on some area of the plate during the pressrun. A properly bent plate will reduce the chances of plate cracking and should meet the following requirements:

1. The distance between the two bends must correctly match the distance around the plate cylinder body from gap to gap. It is important to recognize

that an increase or decrease in plate packing will change the effective circumference of the plate cylinder, changing bending requirements. If the bends are too close together, the plate will fail to fit snugly against the cylinder and will most likely crack at the tail edge.

2. The bends on the lead and tail edge should be parallel to each other, allowing the plate to sit squarely on the cylinder. Cracking may result in a non-squarely bent plate because the plate must be cocked to square the image with the cylinder. This stresses the plate.

3. If the bend angle at the lead edge is too acute or not acute enough, cracking may occur. The proper angle and shape of this bend are critical for the plate to be held firmly. Plates may also bend differently when the speed of the bend is varied, which can be a problem in manual benders. Power-operated plate bending devices eliminate this variable by bending at the same speed each time.

4. The lip distance put on the plate at the trailing edge should be precise. If the trailing edge lip is too long, the plate will touch the reel rod in the lockup. This will prevent the plate from being pulled tightly against the cylinder. Too short a trailing edge bend may cause the plate to slip out of the lockup.

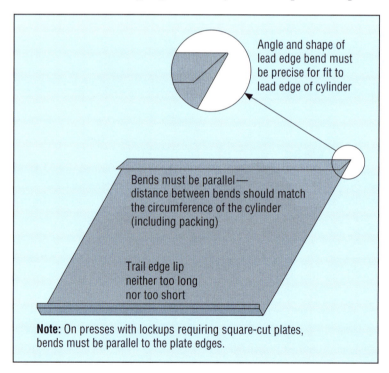

Figure 10-8. Critical dimensions in plate bending.

Plate Mounting

The procedure for mounting a printing plate is simple, but requires care on the part of the press operator. First, the plate is bent on a plate bending device, which bends the leading and trailing edges of the plate so that the bends conform exactly to the leading edge and trailing edge lockup on the plate cylinder. Inaccurate plate bending deforms a plate so that it fails to register with the other plates or fits loosely around the plate cylinder. Improper fit results in plate cracking. This section identifies the proper steps and procedures for mounting an offset lithographic plate:

1. Set the plate cylinder position for mounting the plate.
2. Adjust the lead edge and tail edge of the plate clamps to a centered or zero setting.
3. Lay a thick bead of oil on the outside edges of the plate. Attach the packing material by laying it on the back of the plate and the oil will adhere it to the plate. The oil will also prevent water from seeping in behind the plate and damaging the cylinder. Tape or glue may be used instead of oil to fasten the packing to the plate at the leading edge. The packing is trimmed so that it aligns with the bends in the plate.
4. Put the press on impression. **Note:** Plates should always be mounted on impression.
5. Install the lead edge of the plate and plate packing into the plate gap. Align the center mark on the plate with the scribe mark on the plate cylinder.
6. Hold the tail edge of the plate securely. Plates over 36 in. wide (914 mm) require two operators to hold three positions: the gear side, the center, and the operator side of the plate.
7. Wrap the plate around the cylinder by inching the press.
8. Lock the tail edge of the plate into the plate reelrod, taking up excess slack. The center mark on the tail of the plate should align perfectly with the center mark on the lead edge of the plate. The primary objective is to fit the plate snugly against the cylinder. Do not try to force or shift the plate edge to line up, because this will create a buckle in the plate and the plate will crack during the pressrun. If the marks do not align, check the squareness of the plate bender or the master marks flat.

Figure 10-9. Press operator measuring the combined thickness of the plate and its packing using a deadweight bench micrometer from E. J. Cady & Co. prior to mounting the plate on press.

9. Check the plate-to-bearer height with a packing gauge.

10. Check that the proper above-bearer height or "transfer squeeze" has been achieved between the plate and blanket cylinder.

11. Adjust the blanket impression cylinder according to the press manual instructions or the specific stock to be run.

12. Run the plate and check image transfer for proper reproduction.

13. Twist or cock the plate cylinder (if necessary) to square, align, or register the image.

Automated Plate Mounting

In the last few years, many press manufacturers have added automatic or semiautomatic plate mounting devices to their presses. The exact procedure involved in using these devices varies with the level of the device's automation, as well as from manufacturer to manufacturer. Therefore, consult the operating manual for the specific procedure.

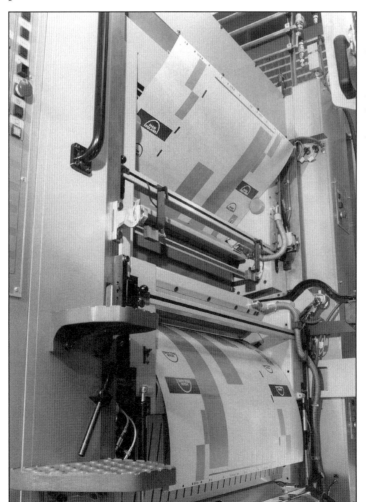

Figure 10-10. A printing unit of a ROTOMAN web press equipped with the PPL plate loading system, showing automatic plate mounting in progress. (Courtesy MAN Roland Druckmaschinen AG)

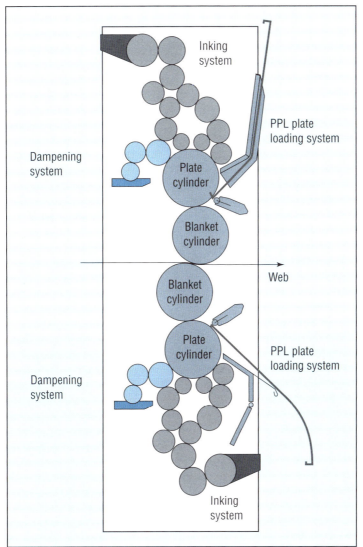

Figure 10-11. *A printing unit of a LITHOMAN web press equipped with the Power Plate Loading (PPL) system, which automates the plate mounting process. (Courtesy MAN Roland Druckmaschinen AG)*

11 Blankets

The use of an intermediate cylinder covered with a rubber blanket is unique to the lithographic printing process. The purpose of the blanket is to transfer the inked image from the plate to the paper with minimum distortion to the original image. Most blankets are capable of receiving and transferring very fine images, up to 600-line/in. halftones. Selecting the most appropriate blanket for the type of work to be printed is essential to good quality as is proper installation and maintenance.

Blanket Types: Compressible and Noncompressible

There are two categories of blanket: compressible and noncompressible. The terms compressible and noncompressible describe how the blanket behaves under the squeezing action of the printing nips—plate/blanket nip and blanket/substrate nip. The noncompressible blanket when squeezed in a nip bulges out on either one or both sides of the nip. The materials that make up the blanket cannot be compressed; therefore, they are displaced. Compressible blankets contain a cellular sponge-like layer that compresses and then recovers, regaining its original thickness. Compressible blankets have several advantages over conventional blankets:

- **Better resistance to smashing.** Blankets are susceptible to damage when compressed too much. A web break or a wrinkled sheet can damage a blanket, creating low spots that will not properly and evenly transfer ink. Compressible blankets are far more likely to recover from this kind of sudden trauma than conventional blankets.

- **More packing latitude.** Excessive packing will create too much transfer pressure at the plate-to-blanket and blanket-to-paper nip. Dot gain, slur, and premature plate wear are just three problems that can result from overpacking. Compressible blankets allow a wider tolerance in packing, making the associated problems less likely.

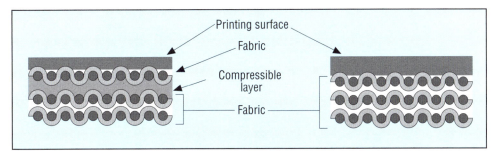

Figure 11-1. Cross section of compressible blanket (left) and conventional blanket (right).

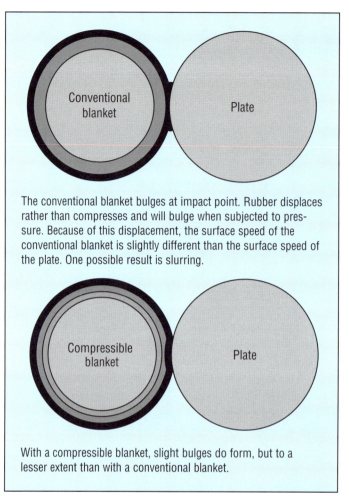

The conventional blanket bulges at impact point. Rubber displaces rather than compresses and will bulge when subjected to pressure. Because of this displacement, the surface speed of the conventional blanket is slightly different than the surface speed of the plate. One possible result is slurring.

With a compressible blanket, slight bulges do form, but to a lesser extent than with a conventional blanket.

Figure 11-2. The displacement of conventional (top) and compressible blankets (bottom) at impact point.

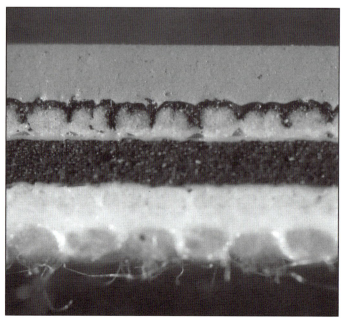

Figure 11-3. A photomicrograph showing the printing surface, compressible layer, and carcass of a compressible blanket. (Courtesy DAY International, Inc.)

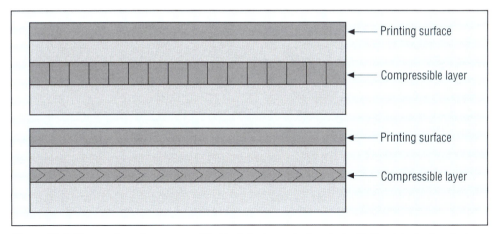

Figure 11-4. *The reaction of a compressible blanket when out of contact with the plate (top) and when in contact with the plate (bottom). The thickness of the rubber surface does not change. Rather, the effects of pressure are absorbed by the compressible layer.*

- **Reduced plate wear.** A conventional blanket has a "harder" surface and will compress less as it contacts the plate. The bulge created at the nip increases friction. Over time, the blanket will wear down the image on the plate.

- **Minimized vibration problems.** As the cylinder gaps meet on press a slight vibration is created. Gears, particularly if worn from use, can also send unwanted vibrations through the press. It is possible for excessive vibrations to affect print quality, distorting large solids or tints with variations in print density. Compressible blankets help to absorb the shock of cylinder gap or gear vibrations.

Noncompressible blankets are often referred to as conventional blankets because the original design of lithographic blankets did not contain a compressible

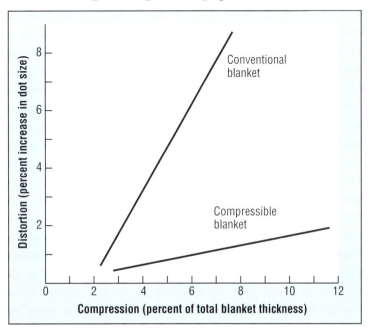

Figure 11-5. *Compressible versus conventional blankets.*

layer. Though not commonly used, noncompressible blankets are still used for applications that require a harder printing surface, particularly when printing textured papers.

Blanket Construction

Carcass. Blankets are constructed of two basic parts: the carcass and the surface layer. The carcass is made up of several fabric plies that are fused together with adhesive rubber cements. This cement used is highly resistant to press chemicals and dampness. Compressible blankets contain a compressible layer, which varies somewhat among manufacturers. This layer may be composed of cork or it may be a synthetic rubber sponge-like material. In either case, this layer is sandwiched between the fabric plies.

The threads in fabric are woven at right angles to one another. One direction is called the warp and the other direction is called the weft. The thread structure is strongest and has the least stretch in the warp direction. For this reason, the blanket is mounted around the circumference of the cylinder in the direction of a fabric's warp, while the weft is run across the cylinder.

Blankets used on most web presses (excluding newspaper presses) are manufactured in two primary thickness ranges. Three-ply blankets usually measure 0.064–0.070 in. (1.63–1.78 mm) thick. Four-ply blankets measure 0.075–0.080 in. (1.91–2.03 mm). The ply designation was initially based on the number of fabric layers that comprised the carcass of a noncompressible blanket. The introduction of better fabrics and the additional space required by the compressible layer of a compressible blanket have led to the manufacture of blankets that do not actually have three- or four-ply carcasses. Currently, ply ratings simply refer to the two thickness ranges.

Three-ply blankets generally are not run on cylinders with undercuts designed to accommodate a four-ply blanket. A three-ply blanket mounted on such a cylinder would require an extra 0.010 in. (0.25 mm) of packing or more. Using a four-ply blanket on the deeper undercut cylinders reduces the amount of required packing, thus minimizing the chance of packing creep.

Thicker and heavier blankets are made for special web applications. For example, high-speed newspaper presses may require blankets that are 0.081–0.086 in. (2.06–2.18 mm) thick.

Surface layer. The surface of the blanket is often called the blanket "face." The type of rubber compound and finish of the blanket surface will affect print quality characteristics. Synthetic elastomers have completely replaced the natural rubbers once used for the blanket surface. A variety of synthetic rubber formulations are available to meet the special requirements of ink formulations and solvents. UV inks are one example. Most UV inks have a tendency to swell rubber.

The texture of the blanket surface can vary considerably. A blanket with a smooth finish will transfer very sharp dots and fine lines, but will create more resistance with

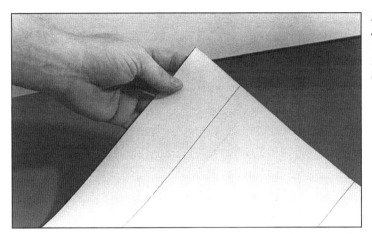

Figure 11-6. *The back of a blanket showing the warp lines, which run in the around-the-cylinder direction.*

ink release. This could cause a higher incidence of paper picking, particularly with papers of weak surface strength. Rougher surfaced blankets produce rougher edged dots and fine lines, however the mechanical release is very high. Newsprint and other stocks with low surface strengths often require rougher textured blankets.

Blanket Characteristics

Release. Release is the ease with which a blanket allows the tacky ink film to break from the blanket surface. Poor release properties can cause a host of printing problems including dot distortion, blanket piling, excessive sheet curl, and uneven solids. If the web flutters between printing units, and this cannot be controlled with web tension, there is probably a release problem.

There are both mechanical and chemical factors that affect the release characteristics of a blanket. As described previously, mechanical release variations are affected when blanket surfaces are manufactured with various finishes. Rougher finishes release tacky ink more easily, while smoother finishes place more pulling force on the paper. Chemical release properties are influenced by the chemical composition of the synthetic rubber face of the blanket. Manufacturers must balance mechanical and chemical release factors to achieve optimal print characteristics for a given application.

Compressibility. Compressibility is defined as the degree to which a blanket reduces in volume under pressure and then rebounds to its original size. Compressibility allows the blanket to exert roughly equal amounts of pressure at the printing nip over a range of packing levels/sheet caliper variations. As described above, most blankets are manufactured with a compressible layer to improve compressibility.

Tensile strength. Tensile strength is a measure of the amount of pulling force the blanket can withstand before tearing or breaking. The fabrics used in the manufacture of the blanket must withstand very high levels of torque when being tensioned on the blanket cylinder. As the number of plies increases, the tensile strength increases as well.

Stretch. Stretch is the amount of elongation that a blanket will undergo under a given load. A blanket must be able to stretch some in order to conform to the blanket cylinder. If the blanket stretches too much, the thickness of the blanket will decrease too much, decreasing nip pressures.

Caliper variability. For equal printing pressures, a blanket must conform to strict caliper tolerances. As a rule of thumb, blankets less than 42 in. (1,067 mm) wide should vary no more than ±0.001. The device most commonly used to measure blanket thickness is the Cady gauge, which is a spring-loaded deadweight micrometer. The model used for measuring blankets should have a wide foot on the bottom of the device. Each blanket should be measured using a standard procedure. First, measure all four corners of each blanket to locate the lowest thickness. This measurement is then used to compute the required packing. The difference between the lowest and highest readings should not exceed the agreed-upon tolerance.

Figure 11-7. Press operator using a deadweight bench micrometer from E. J. Cady & Co. to measure blanket thickness.

Squareness. Blankets that are not cut squarely will not tension properly on the cylinder, possibly resulting in registration problems and web control problems like wrinkling. Each blanket should be measured from opposite corners to check for squareness. As a rule of thumb, variation in the two measurements should be no more than 1/16 in. (1.6 mm) for every 30 in. (762 mm). Also, the blanket should be cut exactly parallel to the warp of the fabric. This will also assure even tension. The press operator should check the cut of each new blanket.

Solvent resistance. The synthetic rubber face of the blanket should resist swelling and distortion when in contact with blanket washes, fountain solutions, and inks. Many manufacturers will apply edge sealants to the blanket to help the edges resist the penetration of solvent and water. Any absorption of chemicals will cause swelling at the edges.

Blanket Mounting Designs

There are several types of blanket mounting designs. Some blankets are prepunched and must be locked into mounting bars by the press operator. Others are premounted, and some do not require mounting bars. There are also blanket sleeves made for gapless presses.

Premounted bars. Many modern web offset presses are equipped to lock blankets without premounted bars into blanket clamps that are built into the blanket reel bars. However a good number of web offset presses require blankets that have some form of a metal bar mounted on each end. These bars are the means by which the blanket is locked into the blanket cylinder gap. There are over two dozen bar styles available for use on different makes and models of presses. The blanket manufacturer should be able to premount the appropriate bars for use on a specific press.

Prepunched blankets. Prepunched blankets are not typically used on larger presses. These blankets have holes that match the bolt pattern of specific mounting bars. Unless otherwise specified, the holes in a prepunched blanket fall on a straight line across both ends of the blanket, at a right angle to the warp lines. The two rows of holes are parallel to each other. Bars for prepunched blankets consist of two metal bands with interlocking surfaces that grip the edge of the blanket from each side. Bolts connect the metal bands through holes in the blanket's edge.

Blankets for gapless presses. Gapless presses are designed without blanket gaps and associated lockup mechanisms. Rather, these presses have cylindrical blanket sleeves that slide onto the cylinders. Gapless presses are manufactured to run at very high speeds and have the advantage of very fast blanket changes. Manufacturers claim that blankets can be changed in 30 seconds or less. In addition, because the blankets are gapless, trim areas between impressions can be reduced to a minimum, allowing for potential savings in paper costs.

Because the blankets are slipped over the cylinders as sleeves, the presses are bearerless. Control of printing pressures is accomplished with adjustable wedge blocks, which control the space between the cylinders. To check the printing pressures, a special set of blanket sleeves with cut-outs at each end are temporarily installed on the

Figure 11-8. Press operator sliding blanket sleeve onto the blanket cylinder for a gapless web press. (Courtesy Heidelberger Druckmaschinen AG)

unit. The press operator can measure the slight gap between the blanket cylinder and the plate surface with a feeler gauge.

The gapless blanket is made of a compressible carcass laminated to a rigid inner sleeve made of seamless nickel. The surface layer is the same as those used on conventional flat blankets. To install the blanket, air pressure is applied under the sleeve to slightly expand the sleeve diameter enough to allow the press operator to slide the sleeve on the cylinder. When the air pressure is released, the blanket sleeve shrinks to tightly hug the cylinder body.

Printing Pressure Considerations

Adequate nip pressures are required to properly transfer ink and ensure quality printing. Inadequate pressure causes uneven ink transfer and may produce broken images. Excessive pressure distorts images, causes dot gain, and prematurely wears plates and blankets. Furthermore, excessive pressure may interfere with web feed and cause web breaks.

How much squeeze? Proper packing for plate and blanket cylinders is typically recommended by the press manufacturer to achieve a specified squeeze pressure, which is required to transfer ink from the plate to the blanket. Squeeze pressure is variable within ±0.002 in. (±0.051 mm); several factors may contribute to such variation. A plate is commonly packed to 0.002 in. (0.051 mm) above bearer height. The blanket cylinder is packed to compensate for the difference between the packed plate and recommended squeeze pressure.

Squeeze pressure for compressible blankets measures 0.004–0.006 in. (0.107–0.152 mm). Noncompressible (conventional) blankets require 0.002–0.004 in. (0.051–0.107 mm) of squeeze. When calculating the packing required to create sufficient squeeze, the press operator should add 0.002 in. (0.051 mm) to compensate for mounting tension and running compaction. All blankets on a press should be packed within ±0.002 in. (0.051 mm) of each other.

Blanket mounting considerations. Blanket mounting tension is more critical than plate mounting tension, due to the elasticity of the blankets. Excessively tightening the lockup can slightly decrease the blanket thickness. Blanket characteristics and varying strength among press operators make it difficult to pack a blanket to the proper torque without a mechanical aid. A micrometer-adjustable torque wrench can be used to apply a prescribed tension to the mounted blanket. The torque wrench can be set to a specific torque value, and the wrench will click when the press operator tightens the blanket to that value.

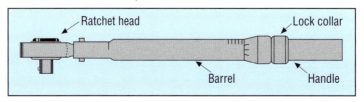

Figure 11-9. Micrometer-adjustable torque-sensing wrench.

Blanket thickness will decrease slightly, due to running compaction; retighten each new blanket with a torque wrench after the first 3,000–4,000 impressions. Consult the blanket manufacturer's recommended torque.

Verifying blanket-to-bearer relationship and torque. After installation, a packing gauge should be used to verify the blanket-to-bearer relationship. This measurement should be taken in three points on the blanket cylinder: (1) at the lead edge, (2) at the trailing edge, and (3) at the center between the lead and trailing edge (180° from the lead edge). The readings should be consistent to within 0.001 in. (0.25 mm). If there is more variability than this, the blanket is likely overtightened.

Bearer compression. Another factor that affects the packing as measured on the press is bearer compression. Hard steel bearers do not remain perfectly circular, but deform under pressure. This changes the amount of undercut on the cylinders and increases squeeze at the nip or point of impression. Normally, web offset presses are run with a considerable amount of pressure between the bearers. This pressure is beneficial in that it provides smooth rolling contact, eliminates gear backlash, and stabilizes the cylinders.

Because of compression, proper packing for the press is not always straightforward. For example, assume that the recommended squeeze between plate and blanket is 0.004 in. (0.10 mm). Assume also that the bearers compress 0.0005 in. (0.013 mm) each. If the plate is packed exactly 0.002 in. (0.05 mm) over bearer and the blanket is packed exactly 0.002 in. (0.05 mm) over bearer, then the actual compression between plate and blanket will be 0.005 in. (0.13 mm) rather than 0.004 in. The extra 0.001 in. (0.025 mm) of squeeze is due to the fact that the packing measurements are made relative to the uncompressed bearers. Bearer compression on newer presses is not excessive and probably runs in the range of 0.001 in. (0.025 mm) total compression for the two bearers involved. On older presses, however, bearer compression can run considerably higher, even as much as 0.002 in. (0.05 mm) compression per bearer. This would mean a total of 0.004 in. (0.10 mm) impression squeeze gained in a printing nip due to bearer compression.

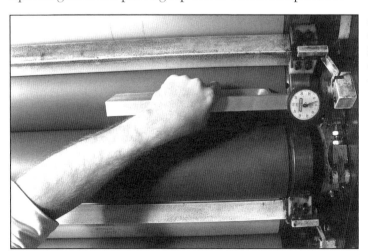

Figure 11-10. A press operator using a packing gauge to check the height of the blanket in relation to its bearers.

Blanket mounting procedures. The proper mounting blankets is important to quality printing, makeready, and production time. Here is a set of procedures for the mounting of blankets on a web offset press:

1. Prepare the blanket for mounting. Clean blanket cylinder with approved solvent. Attach blanket clamps. Clean and inspect the blanket. "Mike" the blanket thickness with a hand-held micrometer or a Cady gauge. Verify the thickness indicated on the back of the blanket. Determine the type of blanket: compressible or conventional.

2. Mount the lead edge of the blanket on the press. Warp lines on the blanket back must go around the cylinder body.

3. Insert the blanket packing underneath and work it slightly into the cylinder gap. The blanket packing should extend all the way around the cylinder, from leading edge to trailing edge. Packing can be cut to web width or slightly less. Cutting the packing materials slightly undersize will prevent ink from building up on the edge of the blanket. Packing may consist of specially manufactured paper or Mylar plastic. Mylar may be glued to the cylinder or the back of the blanket. The grain of packing paper should run across the cylinder width.

4. Inch the blanket forward, slowly, around the cylinder while gently stretching the blanket by the tail edge. It is not necessary to put a blanket in under impression.

5. Use a torque wrench to achieve truly consistent results in blanket tightening.

6. Check the blanket-to-bearer height with a packing gauge. A reading of 0.002–0.004 in. over bearer height usually works best, but always check your manufacturer's specifications.

7. Retorque or check the blanket tension after a short initial run on the press.

Care and Handling

Blanket storage. Handle and store blankets as recommended by the manufacturer. Avoid prolonged exposure to light; continuous direct sunlight will crack blanket rubber, rendering it useless. Over time, fluorescent light can have a similar effect.

Blankets should be stored in a dark, dry, cool place. The tubes in which the blankets are packed is a good place to store them. Never attempt to put too many blankets in a single tube, as this will overtighten the roll, causing damage to the blanket's carcass.

Blankets can be stored flat. If stored in this way, be certain that rubber contacts rubber and fabric contacts fabric. Prolonged contact between a rubber surface and a fabric backing can cause the fabric pattern to emboss the face of the blanket.

Blanket washes. The blanket washes used must match the particular ink and blankets used. Repeated use of a solvent that is too easily absorbed by the rubber will cause deterioration. Washes with high kauri-butanol (KB) solvents should always be

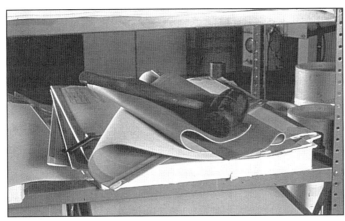

avoided. A low KB number indicates low solvency power. A high KB solvent will be absorbed by the rubber more easily (contributing to blanket wear) and will evaporate more slowly. Retained solvent on the blanket increases the tackiness and frictional coefficient of the blanket. If possible, washes for heatset blankets should have a KB number of 30 or less. The KB of the ink solvent must also be considered. Heatset ink vehicles also contain a hydrocarbon solvent, the most common one being No. 535 Magie oil with a KB of 27.5. When ordering blankets, some shops send samples of the inks or coatings to be used. This ensures that the blankets received from the manufacturer are compatible.

There is a trade-off between rate of evaporation of a solvent and environmental concerns. A low rate of evaporation forces delays in reinking after a blanket wash. However, low evaporation rates are usually correlated with lower volatile organic compound (VOC) content. Many of the modern blanket washes have low evaporation rates to reduce VOC emissions. This can slow production. An additional concern is that many of these solvents are slick if spilled. Therefore press operators must be extra careful around solvent spills.

Some substances should not be used to wash blankets because they will swell the rubber, changing the effective squeeze at the printing nip. Among these are chlorinated and coal tar solvents, ketones, and ester. Aromatic solvents like toluene, xylol, turpentine, and pine oil will also cause swelling. Benzene, carbon disulfide, and carbon tetrachloride should not be used because of their high toxicity and ability to dissolve rubber. All chlorinated solvents are toxic.

When storing a blanket out of the container, do not remove the protective tape from the bar ends. This will protect the printing surfaces from scratches.

Hand-wash the blanket with warm water after mounting it on the cylinder. This will assure that any powder, oil, or dust is removed from the surface before printing.

Automatic Blanket Washers

A blanket wash should follow all but the shortest of press stops. When ink is allowed to sit on the blanket of an idle press, the tack, or "stickiness," of the ink film increases. When the press is started up again, the sticky ink on the blanket may tear

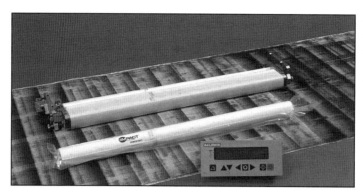

Figure 11-12. IMPACT™ automatic blanket cleaner and press cylinder cleaner. (Courtesy Baldwin Technology Co., Inc.)

the web, causing a web break. The blankets also collect paper dust and paper coatings over time, which forms a deposit on the blanket surface, called piling. Automatic blanket washers perform washes quickly and efficiently. One type of washer consists of a row of nozzles, mounted over the blanket, which spray solvent onto the rubber surface. Solvent and loosened material are carried off by the web upon start-up. An inherent danger of these blanket washers is that solvent vapors tend to follow the web into the dryer. Flame dryers with line burners located near the entrance to the dryer burn off the solvent vapors. Solvent vapors can, however, accumulate in high-velocity hot-air dryers and possibly explode. These blanket washers do not effectively remove paper coating buildup; however, they can retard buildup if used during each paster cycle. (Note: The solvent used in automatic blanket washers must be compatible with the blankets and with the blanket washer.)

Newer models of blanket washers use the solvent spray nozzles as well as a scrubbing device, which contacts the blanket and removes solvent and debris before start-up. Air bladder devices employ a pneumatic pump to fill a rubber bladder that is positioned across the blanket. The expanding bladder pushes a cloth towel against the blanket. The cloth towel sits in a dispenser and comes in a roll, which is replaced when fully soiled. Blanket washing systems like this successfully remove piling on the blanket and provide the added advantage of reducing solvent deposits.

Figure 11-13. Automatic blanket washer installed on a web offset press. (Courtesy Baldwin Technology Co., Inc.)

12 Packing and Printing Pressures

Achieving Proper Printing Pressures

Neither ink or water will transfer without proper pressure between the elements making the transfer. The plate and blanket have to run with pressure between them to transfer ink; running contact is not enough. There must also be adequate pressure between the blanket and the paper. Pressure is critical in lithography. Pressure between the plate and blanket and between the blanket and paper is specified in **squeeze,** which is measured in thousandths of an inch (0.001) or hundredths of a millimeter (0.01). This measurement represents the amount that the plate and blanket are collectively packed over bearer height. Typical squeeze values on web presses range from 0.003 to 0.005 in. (0.075 to 0.125 mm) for the plate-to-blanket pressure. Blanket-to-blanket squeeze is typically 0.008 in. (0.20 mm), with acceptable tolerances as low as 0.002 in. (0.05 mm).

The procedure for setting cylinder pressures is called **packing.** The noun "packing" also refers to the paper or plastic sheets that are put under the blanket and plate.

As discussed in chapter 9, the bodies of the plate and blanket cylinders are lower than the surface of the bearers. The exact difference in height—called the **cylinder undercut**—varies from manufacturer to manufacturer. Knowing the exact amount of undercut on the plate and blanket cylinders is essential to setting proper pressures in the printing unit.

Packing sheets are put under the blanket and plate to increase the diameters of the cylinder bodies. Packing makes it possible to use a fairly wide range of plate and blanket thicknesses on one press. Packing sheets themselves are available in a variety of thicknesses.

In order to transfer ink from the plate to the blanket, either the plate or blanket, or both, must be packed above bearer height. Pressure between plate and blanket is increased by adding packing sheets and decreased by removing packing sheets.

Plate-to-blanket squeeze. A space will exist between unpacked plate and blanket cylinder bodies. The distance of this gap equals the undercut of the plate cylinder plus the undercut of the blanket cylinder. Even with a plate wrapped on the plate cylinder and a blanket on the blanket cylinder, a space will still exist, resulting in no contact. To create squeeze at the nip, the press operator must insert packing material under the plate and blanket to effectively increase the diameter of each cylinder body. The total thickness of plate, blanket, and packing sheets on both cylinders determines the amount of squeeze.

Packing the cylinders so that the plate and the blanket are exactly even with the surface of their respective cylinder bearers theoretically creates no squeeze. In this case, the cylinders are only lightly touching, and there is still no pressure between them. Adding a 0.001-in. (0.025-mm) thick sheet of packing material under the plate will create a 0.001-in. squeeze. Adding the same thickness sheet to the blanket cylinder results in squeeze pressure of 0.002 in. (0.05 mm). The effective surface of each cylinder body will now run at 0.001 in. over its respective cylinder bearers for a total of 0.002 in.

Consider an example. Assume that the plate is to be packed 0.001 in. (0.025 mm) over its bearers and the blanket 0.004 in. (0.10 mm) over its bearers, resulting in a 0.005 in. (0.125 mm) squeeze. Assume also that the plate cylinder undercut is 0.015 in. (0.375 mm) and that the blanket cylinder undercut is 0.071 in. (1.775 mm). The plate and its packing will have to equal 0.016 in. (0.40 mm) in thickness if the plate is to run 0.001 in. (0.025 mm) over bearers. Similarly, the thickness of the blanket and its packing will have to equal 0.075 in. (1.875 mm) to exceed bearer height by 0.004 in. (0.10 mm). (See the following sidebar for another example of the "arithmetic of packing.")

The Arithmetic of Packing

This method for calculating packing assumes that (1) the bearers are set according to the method recommended by the press manufacturer, (2) the height of the plate and blanket relative to bearer height is derived from a test (described later in this chapter) in which the press is underpacked and minimal amounts of packing are added until a full-strength solid is printed, and (3) the results from this test are corrected for any significant changes in the caliper or type of stock being run.

Squeeze desired: blanket-to-blanket*	0.011		Squeeze desired: plate-to-blanket*	0.005
Paper thickness	−0.003		Required squeeze per blanket	−0.004
(Divide by 2)	0.008		Required squeeze per plate	0.001
Required squeeze per blanket†	0.004		Plate cylinder undercut	+0.015
Blanket cylinder undercut	+0.071		Total plate interference required	0.016
Total blanket interference required	0.075		Plate thickness‡	−0.012
Blanket thickness‡	−0.067		Required plate packing‡	0.004
Required blanket packing‡	0.008			

*All dimensions are given in inches.
†These figures should be checked with a packing gauge after the press is fully packed.
‡These figures should be checked with a deadweight bench micrometer before the respective materials are mounted on press.

Since pressure cannot be determined in pounds per square inch, the exact pressure created by squeeze is difficult to measure scientifically. Also, pressure does not correlate directly with squeeze. There are a variety of factors influencing the true pressure at the nip, including the hardness of the blanket or the thickness of the paper. However, it has been time-tested that the variability in pressure due to factors other than squeeze is usually not enough to adversely affect print quality. Therefore, squeeze is the most effective means to set image transfer pressure.

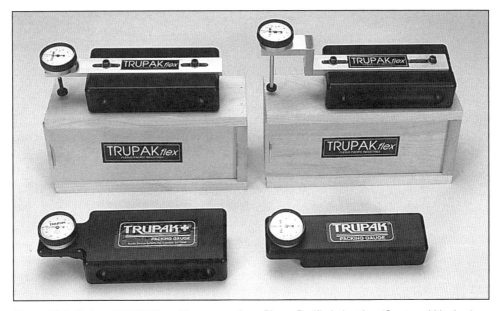

Figure 12-1. *Various TRUPAK™ packing gauges from Plexus Pacific Industries. (Courtesy Litho Inc.)*

After the press operator has installed the blankets to manufacturer's specifications and after plates are properly mounted, packed cylinders should be verified with a **packing gauge.** This device is used to ensure that proper packing has been installed. The gauge gives the press operator an accurate reading of how far over or below bearers the cylinders are packed. The general procedure for packing gauge use is as follows. Be certain to follow manufacturer's recommendations when using the gauge.

1. Pack the plate and blanket according to specifications.
2. Obtain a piece of paper slightly larger in size than the packing gauge body.
3. Set the packing gauge against a piece of paper that will protect the plate surface.
4. Set the gauge against the plate (not on the paper protective layer) and zero the gauge.
5. Move the gauge onto the bearer. The reading will be a positive (amount over bearers) or negative (amount under bearers).

This verifies that the packing has been done correctly. This same general procedure should then be carried out for the blanket cylinder.

It is good practice for the press operator to form the habit of measuring all materials used in packing the cylinders, no matter how many times a particular brand

Figure 12-2. *Press operator using a deadweight bench micrometer from E. J. Cady & Co. to measure blanket thickness.*

of plate or blanket is used. A bench micrometer (a deadweight gauge, such as a Cady gauge) is a device that will accurately measure the thickness of the materials to within a thousandth of an inch (or hundredth of a millimeter). The press operator should form the habit of measuring new shipments of plates. Though the plates are labeled with their gauge, some variability may exist. This is particularly true with blankets. Each blanket should be measured at nine different points with the bench micrometer. These readings should then be averaged to determine the working gauge of the blanket.

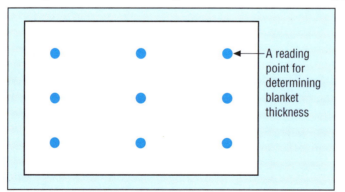

A reading point for determining blanket thickness

Figure 12-3. *The nine reading points for blanket thickness.*

Packing the blanket accurately is more difficult than accurately packing the plate because the blanket is pliable and may shift or compress. A blanket's mounting tension and running compaction decrease the blanket's original thickness. This is especially true of a new or well-reconditioned blanket. Therefore blankets should be re-torqued and checked with a packing gauge a short time after installation.

Blanket-to-blanket squeeze. In the earlier example, the plates were packed 0.001 in. (0.025 mm) and blankets 0.004 in. (0.10 mm) over the bearers. These values created a squeeze between plate and blanket of 0.005 in. (0.125 mm). With each blanket packed 0.004 in. over bearers, the running contact between blanket cylinders in the unit will create a squeeze of 0.008 in. (0.20 mm). The thickness of the paper that will run between the two blankets will add even more squeeze at the printing nip.

Most web offset papers caliper between 0.002 in. (0.05 mm) and 0.004 in. (0.10 mm). Assume that the paper is 0.003 in. (0.075 mm) thick. To achieve a total squeeze at the printing nip of 0.011 in. (0.15 mm), each blanket needs to be packed 0.004 in. over its bearers. It is not necessary to repack cylinders every time the caliper of the stock changes, especially for a 0.001-in. (0.025-mm) or 0.002-in. change in paper thickness.

Minute variations in the surface contour of the paper can make the achievement of a uniform and continuous solid difficult due to resulting pressure variations. A high degree of surface variation—rough texture—requires more pressure than a smooth-textured paper. For example, coated papers require less pressure than do pebble-finish or embossed stocks.

It is important to remember that a change in squeeze to the paper requires a change in the packing of not just the blanket, but all the cylinders in both couples. On a blanket-to-blanket printing unit, a change in the packing of one couple must be matched with changes in the cylinder packing of the other couple. If the two blanket cylinders are packed differently, the diameters of the adjacent cylinders will be different, making each one's surface speed slightly different. Thus the ability of the press operator to maintain accurate registration will be adversely affected.

Pressure variables. A given amount of squeeze at the plate-to-blanket nip creates more pressure than the same amount of squeeze applied between the blanket-to-blanket nip. This is because with the plate-to-blanket nip, a rigid surface (the plate) and a resilient surface (the blanket) are in contact. Conversely with the blanket-to-blanket nip, two resilient surfaces are squeezed together. One can assume that, for a given amount of squeeze, pressure at the printing nip is about half that at the plate-to-blanket nip. The ultimate goal is always to print good solids with minimum pressure. Minimizing pressure will assure that dot gain is kept to a minimum, thus reproducing the best halftones possible.

Testing for proper squeeze. The bearers on any press will deform over time; the amount depends on the construction of the press cylinders. This deformation will change the effective undercut of the press cylinder. Knowing this change in the radii of the bearers is important, because it increases the amount of squeeze generated in the printing nip. The following prescribed procedure for determining squeeze automatically remedies this problem. Follow this basic procedure to determine the most effective squeeze value for a given ink, blanket, and stock combination:

1. Prepare a plate that includes large solids and screen tints. Select a screen frequency that is commonly used on the stock.
2. Pack and mount the plate according to standard procedures.
3. Print until ink and water balance is achieved.
4. Remove packing from under the plate until the solids no longer print.

5. Reinsert packing, 0.001 in. (0.025 mm) at a time, until solids print full strength. The press is now properly packed with minimum pressure.

6. Measure the packed cylinder height with a packing gauge. This figure automatically accounts for any bearer deformation occurring on the press.

Plate Packing Issues

Most web presses require a plate to have a packing sheet underneath the plate for proper printing pressures. Any material having enough dimensional stability and uniformity of thickness to raise a plate (or blanket) to proper height and keep it there can be used for packing.

Kraft paper packing. Probably the most common material currently used is specially manufactured kraft paper. Kraft paper is a highly calendered (smoothed), water-resistant paper with negligible compression. It is made in a variety of thicknesses so that the press operator, by choosing the right combination of sheets, can create nearly any packed height that is required.

Packing paper is usually adhered to the plate to make it easier to handle when mounting. There are several ways to adhere packing. Oil may be applied to the back of the plate, which will seal the plate and packing together. During running, the oil has the added benefit of helping to prevent the plate cylinder body from rusting and also repelling water that otherwise creeps under the plate and softens the packing. Grease may be used instead of oil; however, it must be spread thinly and evenly to prevent lumps that cause uneven pressure between the plate and blanket.

Paste or glue are more likely to lump than are oil or grease, so these adhesives are usually along a strip at the lead edge. This allows the trailing edge of the packing to shift and conform as it is rolled unto the cylinder.

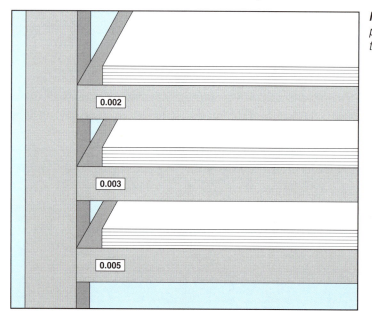

Figure 12-4. Storage of packing sheets according to caliper.

Polyester plate packing. Polyester (especially frosted polyester) is much tougher than kraft paper and is coming into wide use as a packing material under plates. Polyester has a high resistance to lithographic chemicals and good dimensional stability. It is more expensive than kraft paper but, with reasonable care, can be reused. Some polyester sheets come with a crack-and-peel adhesive backing, for adhering it to the plate cylinder.

Some shops mount the plastic packing material with spray adhesive to the plate cylinder. This method is particularly helpful if numerous plate changes are made per shift. When packing is pre-mounted in this way, the press operator must carefully adhere the polyester to prevent wrinkles or air bubbles. Once the plate cylinder has been covered with the polyester, the ends of the polyester should be trimmed to eliminate any overhang.

Importance of trimming packing exactly. Packing is usually cut to be slightly narrower than the plate on each side, so that it doesn't draw fountain solution in under the plate where it can damage the packing. Packing should extend exactly to the cylinder gap at the trailing edge. If the packing protrudes into the gap and hits the reel rod of the lockup, it may back out and wrinkle under the plate. If the packing is short, a portion of the image may not transfer due to inadequate pressure.

Blanket Packing Issues

The blanket packing should be cut so it extends all the way around the cylinder, from leading edge to trailing edge. This ensures complete plate-to-blanket and blanket-to-blanket contact all around the cylinder. Packing for blankets can be cut to various widths. Many shops cut blanket packing just to web width, otherwise ink and gum build up on blanket edges beside the running web. Since the blanket edges are slightly lower, the paper is less likely to run into one of these tacky buildup areas, which would probably break the web. Packing cut to web width or less reduces this possibility.

Cutting packing to web width also prevents inks from mixing when different colors are run in the top and bottom halves of the same printing unit. If the blankets are packed wider than the web, an ink film will build up on the blanket edges, which will then transfer to the corresponding blanket. This ink will then spread into the ink train, contaminating the color.

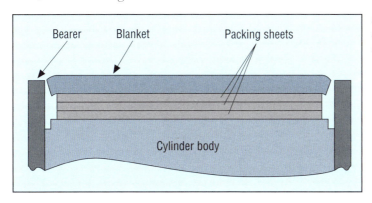

Bearer   **Blanket**   **Packing sheets**

Cylinder body

Figure 12-5. Packing cut slightly narrower than the blanket and to web width.

No-pack blankets. Some blankets, particularly those manufactured for newspaper presses, are manufactured to require no packing. Consistency is critical when tensioning/torquing no-pack blankets. When overtightened, the effective gauge (thickness) of the blanket is reduced, creating printing pressure problems.

Section V
The Inking System

13 Inking System Principles

Depositing a metered, exacting film of ink on the printing plate is essential to quality printing. The inking system of a lithographic press is engineered to accomplish this task. The system consists of a series of rollers, called an ink train, which carries the ink from the ink fountain to the plate. An entire inking system is comprised of both the ink train and the ink fountain, and in some cases specialized auxiliary equipment. In-line presses contain one inking system per unit, functioning to supply ink to the one plate in the unit. Blanket-to-blanket web offset presses contain two inking systems per unit: one supplies ink to the top plate, and another to the bottom.

Inking System Functions

The inking system of a web offset press serves five important functions: (1) ink milling, (2) varying ink input across the system, (3) storing ink for consistent ink film thickness, (4) dampening control, and (5) keeping the plate clean.

Ink milling. The many rollers in the ink train mill the ink into a thin, controlled ink film. This assures that a consistent ink film is deposited on the paper, with no variation in density. It delivers an even, controlled ink film to the plate. The thick film entering the system on each ductor stroke splits among many rollers, so that the

---

Safety Precautions

- Make sure all guards and shields are in place before operating equipment.
- Never allow cleaning of the moving rollers by hand.
- Avoid all roll pinch points.
- Never release a safe button that someone else has set.
- To avoid the possibility of dropping tools into the press or other hazardous locations, do not carry them in pockets.
- Observe and practice all safety rules, regulations, and advice given in the press manual.
- Wear clothing that will not become entangled in any part of the press equipment.
- Only clean the ink fountains while the press is stationary to avoid injury and press damage.
- Do not work on moving rollers with rags, tools, etc., because of the high risk of accident and damage.

intermittent ink feed from the ductor changes into the continuous feed required by the plate.

Varying ink input across the system. The inking system functions to deposit the ink film as it is needed on the plate. For example, the image areas of the plate may be very heavy on the left side and very light on the right. In this case, more ink would be needed in the left side of the inking system to replenish the higher level of ink consumption on the left side of the plate. To accomplish this, the system must allow the control of ink flowing into the system, both in total volume and variably, from side-to-side.

Storing ink for consistent ink film thickness. The inking system acts as a reservoir to maintain consistent ink film thickness from impression to impression. Any variability in ink film thickness affects print quality. For example, one revolution of the plate cylinder will draw a certain amount of ink from the ink system. If the ink were not stored sufficiently in the many rollers of the system, the next rotation of the plate cylinder would pick up less ink and the resulting image would be lighter than the preceding image. This condition is referred to as *ink starvation,* which is rarely a problem in ink systems that have many rollers, storing large quantities of ink.

Dampening control. The inking system helps to control dampening on the plate by picking up some water as the press runs. Some of the water mixes with the ink, forming an emulsion. Too much emulsification prevents image areas from accepting ink; however, controlled emulsification is essential to the lithographic process. Some water is also picked up by the blanket and transferred to the sheet. The rest of the water evaporates.

Keeping the plate clean. The inking system helps to clean the plate by picking up foreign matter that may collect on the plate. A roller specifically designed for this purpose is often used in one of the form roller positions.

The Ink Fountain

An ink fountain section of the web offset inking system may fall into one of two categories; conventional inking and ductorless inking. Though these variations exist, conventional ductors are by far the most common among web offset presses.

Conventional inking systems. A conventional inking system (figure 13-1) consists of (a) a *fountain* that holds a supply of ink, (b) a *ductor* or *transfer roller* that carries the ink from the fountain to the roller train, (c) a *roller train* that works and distributes the ink, and (d) *form rollers* that deposit the ink onto the plate.

In conventional fountains, the ductor roller functions to take the ink from the fountain roller and deposit it on the first roller in the inking train. The *fountain roller* is the very first roller in the system. (This roller is sometimes called the "foun-

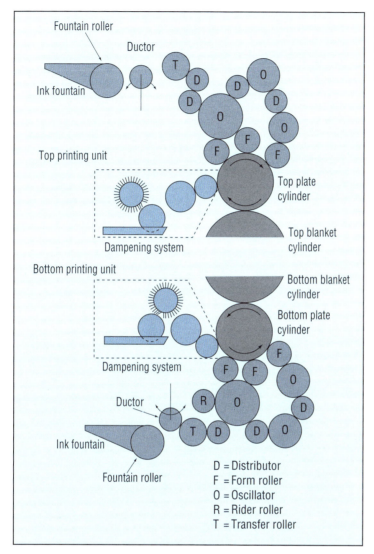

Figure 13-1. *The upper and lower inking systems on a typical commercial web offset press. (Courtesy MAN Roland, Inc.)*

Fountain roller

Ducter

Ink fountain

Top printing unit

Dampening system

Top plate cylinder

Top blanket cylinder

Bottom printing unit

Bottom blanket cylinder

Bottom plate cylinder

Dampening system

Ducter

Ink fountain

Fountain roller

D = Distributor
F = Form roller
O = Oscillator
R = Rider roller
T = Transfer roller

tain ball.") The fountain roller is typically composed of steel and rests against the fountain blade, which allows the operator to meter ink into the system. The ductor roller dwells against the fountain roller for a time, picking up ink and then swinging over to the first driven roller in the ink train, depositing the ink before swinging back into contact with the fountain roller. This cycle is repeated, with fresh ink feeding intermittently (not continuously) into the inking train as the press runs. The ductor is the only roller in the inking system that functions in this way.

The ink fountain is a trough that is formed from a wide, flexible steel strip, called a ***fountain blade,*** that is angled against the fountain roller to form a V shape. With most fountain rollers, the amount of roller rotation per cycle of the ductor can be adjusted. Some fountain rollers are driven by their own variable-speed motors; others cannot be adjusted at all but rotate at a constant speed.

To avoid roller wear, the fountain blade should not touch the fountain roller; instead, there should be a narrow gap between the two, the width of which deter-

Preparing the Inking System

The inking system of a lithographic press consists of a series of rollers that carries the ink from the fountain to the plate. Blanket-to-blanket web offset presses contain two inking systems per unit: one supplies ink to the top plate, and another to the bottom. Following is a set of procedures to properly prepare the inking system for web offset print production:

1. Check/adjust/maintain the ink fountain roller and the ink fountain roller feeding control.
   - Inspect the ink fountain roller for any grooves or nicks on the roller or in the corners.
   - Inspect the ink fountain roller (especially the ends) for overall cleanliness.
   - Check the adjustable pawl and/or ratchet system for proper operation of the ink fountain roller feeding control.
2. Install/adjust/maintain the ink fountain blade and ink fountain keys.
   - Check for warp, nicks, or other damage on the ink fountain blade.
   - Also, check for overall cleanliness.
   - Install the ink fountain blade. Inspect ink fountain keys for cleanliness.
   - Lubricate keys for proper movement and adjustment capabilities.
   - Lock the ink fountain in position relative to the fountain roller. (This procedure is sometimes referred to as closing the ink fountain.)
   - Use a feeler gauge to set the fountain blade to the fountain roller.
3. Install/adjust/maintain the ink ductor roller.
   - Inspect the ductor roller for breaks or dents in the roller.
   - Make sure it is straight, glaze-free, and ink-receptive.
   - Check the ink ductor roller for caliper and durometer (hardness).
   - Adjust the ink ductor roller so that it contacts the fountain roller and the oscillating roller along its entire length.
4. Set/adjust the ink form rollers for correct pressure.
   - Read the press manual instructions to determine the proper system and sequence of adjustment to the roller train. (Some adjustments may be automatic.) Some systems require an adjustment between the rollers and plate first; others require the first adjustment to be made between the rollers and oscillators.
   - Adjust the inking form rollers to the plate so that their weight provides the pressure, and the roller bearings are supported in the cylinder gap.
5. Perform an ink stripe test to evaluate form roller adjustments.
   - Put a uniform amount of ink on the rollers.
   - Drop the ink rollers against the dry, gummed plate.
   - Lift rollers and inch the plate cylinder to view stripe.
   - Evaluate stripe for overall condition of form rollers and requirements for setting pressure.
6. Set/adjust the ink train rollers for correct pressure.
   - Read the press manual instructions. (Some adjustments are automatic.)
   - Check for any end play in the roller train.
   - Check oscillating rollers for proper movement. (They are usually not adjustable.)
7. Perform the strip test to determine proper ink roller pressures. (The three-strip method is best.)
   - Rollers should be clean and free of ink.
   - Obtain bond paper (usually 20 lb.) and cut into strips.
   - Insert strips of paper between rollers (at each end of the rollers).
   - Adjust until the strips at both ends are gripped firmly.
   - If necessary, perform the same procedure to adjust the rollers to the plate.

mines how much ink will be carried by the fountain roller as it rotates. A row of thumbscrews or motor-driven screws sits underneath the fountain blade on most presses. These screws, called fountain keys, are evenly spaced along the entire width of the blade. The keys are adjusted to move the blade closer to or farther away from the fountain roller, controlling the thickness of the ink film across the roller. The keys are controlled remotely from the press console on systems with servo-motors (figure 13-2). In this case, an electronic signal is sent to the servo-motor from the remote control console, which then opens or closes the ink key(s).

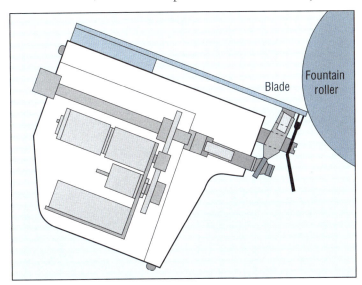

Figure 13-2. *Single ink for a remotely controlled ink fountain with a continuous blade.*

Blade   Fountain roller

    In an alternative design, the fountain blade's relative position to the fountain roller may be manually adjusted by an eccentric roller or cam that is controlled by a lever (figure 13-3). This mechanical design has been used on some newspaper web offset presses as well as presses used for digest work. One of its advantages is the ease and rapidity of adjustments and the ability to visualize the ink level by looking at the position of the cam lever.

    The amount of ink put on the ductor and fed into the inking train depends on two things: (a) the thickness of the ink film on the fountain roller, which is determined by the gap between the fountain blade and fountain roller, and (b) the rate of fountain roller rotation while in contact with the ductor. As a general rule, press operators want to meet the ink demands by setting the fountain to yield a minimum ink film thickness on the fountain roller, controlling the quantity of ink with fountain roller rotation. This decreases the need for the ink train to mill a thick layer of ink into a thin layer, increasing the likelihood of a thin, metered layer of ink on the plate.

Ductorless inking systems. An alternative to the conventional inking system is the anilox inking system, often referred to as "ductorless" inking (figure 13-4). This ink-metering system transfers ink in a continuous flow from the fountain using an engraved fountain roller. A ***doctor blade*** squeegees the surface of the rotating fountain roller, leaving ink in the cells. The ink is then transferred to a short series of rollers or, in some cases, directly to an ink form roller. As such, there is no fountain blade or ink

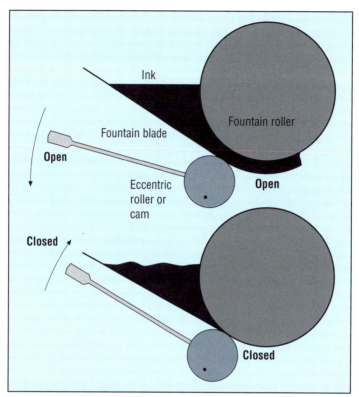

Figure 13-3. Lever- or cam-operated ink fountain.

keys for locally adjusting the amount of ink transfer. One limitation of these ductorless systems is that they may be incapable of the precise side-to-side control that is easily obtained by conventional ductor-based inking systems. As such, these systems are best for work that has very consistent ink demands from side-to-side.

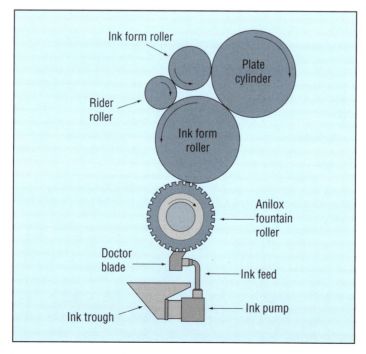

Figure 13-4. An anilox-offset inking system.

Fountain rollers for ductorless inking have either chrome-plated surfaces engraved by electronically controlled cutting mechanisms or, in other designs, the fountain rollers have ceramic surfaces, which are engraved by lasers. Ceramic rollers are composed of various combinations of aluminum oxide, chrome oxide, and titanium oxide.

The primary application of anilox systems is in newspaper printing where fairly fluid inks are used. These inks are ideal because the designs of many of these inking systems employ very tiny fountain roller cells requiring the use of inks that flow more easily.

Since conventional intermittently-fed inking systems supply surges of ink to the roller train (usually by way of a ductor roller), the ink film must be split numerous times between many rollers before it reaches the plate. This ensures smooth coverage on the plate. Anilox systems, however, continuously feed a smooth, metered ink film directly from the fountain roller; therefore, the ink train is much shorter. Furthermore, continuously-fed inking systems eliminate ink starvation and ghosting.

The Ink Train

Though the design of ink trains for lithographic presses vary to some degree from one press manufacturer to the next, the basic roller names and functions are the same. This section will present the typical rollers found in the inking train and their purposes.

Driven rollers. All systems contain a number of hard rollers, driven directly off the press drive. These hard rollers may be called *oscillators, vibrators,* or *drums.* Some of these rollers oscillate in a side-to-side motion, which distributes the ink laterally

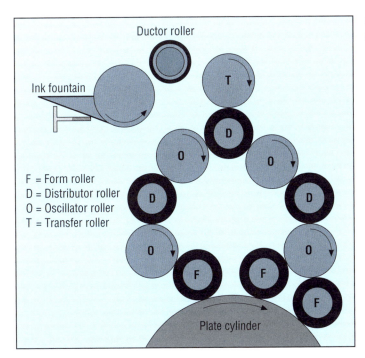

Figure 13-5. *Typical inking system.*

across the press. The side-to-side oscillation is adjustable. When ink coverage on the printing plate is consistent from side to side, more dramatic oscillation can be adjusted, which serves to more effectively even out the ink film. It is best to have minimum oscillation when printing work varies in side to side ink coverage. This assures that the variations in ink key settings will carry all the way down to the form rollers, which contact the plate. The larger-diameter oscillators perform the additional job of effectively increasing the storage capacity of the inking system by increasing the length of the ink train.

Oscillators are usually steel-based, which may then be copperplated or coated with synthetic materials like ebonite and nylon. These rollers are fixed in the press frame and cannot be adjusted in relation to the soft rollers that they contact. Because of these characteristics, these rollers are often referred to as "fixed" rollers and as "driven" rollers.

Water-Cooled Ink Oscillators

When the press is running, inking system temperatures reach as high as 165°F (74°C) or more, sufficient to set some heatset inks on the rollers. For this reason, water-cooled ink oscillators are commonly used on web offset presses. Chilled water circulating through the ink oscillators maintains roller temperatures at a level that will not set the ink. The system should be equipped to heat the inking system to the desired temperature on start-up and then to cool it to maintain that temperature during running.

Some water-cooled oscillators are single-ended, with water inflow and outflow at the same end of the roller. Others are double-ended, with the water inflow at one end and outflow at the other. Double-ended oscillators cool more uniformly across their width.

Nondriven rollers. Other rollers in the inking system are nondriven because they are not tied directly into the press drive. Rather these rollers move by contact with the driven rollers. Nondriven rollers include *distributors, riders,* and *form rollers.* Distributors and form rollers are always soft rubber rollers. They have mountings in the press frame by which they can be adjusted against adjacent rollers. Distributors and form rollers are most typically composed of synthetic rubber such as PVC (polyvinyl chloride), Buna-N (a copolymer of butadiene and acrylonitrile), or polyurethane. These substances are applied to a steel shaft.

The distributor roller sits between two oscillator rollers and serves to smooth, condition, and transfer ink, connecting the other rollers in the system. The form rollers directly contact the plate, depositing ink on the plate surface. Unlike distributors, rider rollers run against only one other roller. A rider roller lengthens the inking system and further works the ink before it reaches the plate. Riders also effectively collect unwanted foreign particles from the ink. Soft riders are usually set against hard rollers, while steel riders (often copperplated) are set against soft rollers. Usually, a rider roller is one-half the diameter of the roller that it contacts. Consequently, the rider runs twice as fast, thus stripping and holding many foreign particles such as dried ink specks or paper lint.

Some web offset presses have four form rollers while others have only two or three. Some presses have *oscillating form rollers,* which are substituted for the first and, sometimes, fourth (last) form rollers. This form roller oscillates (moves laterally, or side to side) at a rate sometimes different than the adjacent oscillator to smooth the ink film, eliminating ghosting. ***Ghosting*** is a printing problem that occurs when ink is drawn from the form rollers of the press by the plate, leaving an ink film thickness that is thinner or has a pattern created by the plate image. This unwanted pattern can then be transferred to an area of the image on the next rotation of the form roller. Oscillation of one of the form rollers helps to lessen this problem (figure 13-6).

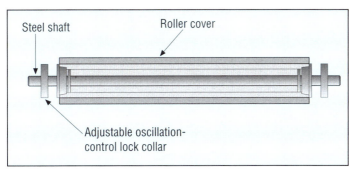

Figure 13-6. *An oscillating form roller.*

Design Principles of Ink Trains

A few important design principles are built into the inking system. Each of these helps to optimize the efficiency of the inking system for depositing an even film of ink across the entire plate image.

Length of ink train. The number of rollers and the diameter of the rollers affect the distance that the ink travels from the fountain to the plate. The length of the train serves two important purposes. First, the ink begins at the ductor roller as an uneven, thick layer of ink. As it moves through the system it becomes progressively thinner and more even. Second, the ink stored in the long ink train resists changes in ink film thickness, which could be brought on by two different sources. The ductor roller, which intermittently replenishes ink to the train, produces surges in ink thickness to rollers early in the system. The long ink train acts as a buffer to combat this surge of ink. Also, ink is constantly being drawn from the ink system by the plate, but in varying amounts. The length of the ink train assures that the ink will be replenished to the form rollers as an even ink film.

Variations in roller diameters. The rollers in the inking system vary in diameter, with rider rollers having the smallest diameter and oscillators the largest. Because the diameters vary, the RPMs (revolutions per minute) of the rollers vary as well. This design principle helps eliminate ***repeat flaws.*** This can be understood by considering a series of rollers that have the same diameter, any flaw in the ink film would transfer from one roller to the next, moving throughout the system to the plate. The diameter variations, coupled with roller oscillation, eliminate any repeat flaws.

Alternation of hard and soft rollers. Rollers in the ink train generally alternate between hard nylon or copperplated steel rollers and soft, synthetic rubber rollers.

This allows precise pressures to be set between rollers, with the soft rollers able to pivot on journals against the hard, stationary rollers. Accurately setting the pressures between rollers is critical, as it affects the even transfer of ink throughout the system.

Inking Systems for Waterless Printing

Waterless printing (printing without dampening solution using special plates) requires a printing press that is equipped with a temperature control system. Two types of press temperature control systems are used: an ink oscillator cooling system and a plate cylinder cooling system. With the ink oscillator cooling system, a standard inking system is used with the exception that chilled or heated water solution flows through hollow vibrator rollers on the press. These temperature control systems allow the press operator to maintain ink temperature within a narrow range of only a couple of degrees Fahrenheit.

It is not unusual for each of the inks on the press to perform best at a slightly different temperature. For example, a black ink might operate best at, say, 72–74°F (22.2–23.3°C), while a cyan might operate best at, say, 68–70°F (20–21.1°C). Therefore, the temperature of each inking system on the press is independently controlled by using a zone control unit that blends hot and cold water to the proper temperature for the ink being used. In addition, infrared sensors monitor the temperature of each printing unit, providing immediate feedback to maintain the proper temperature level.

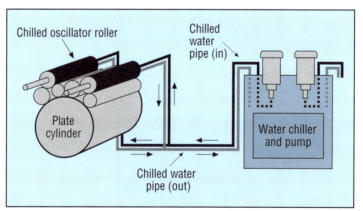

Figure 13-7. The ink roller cooling system used for waterless lithography. (Courtesy Toray Marketing & Sales (America), Inc.)

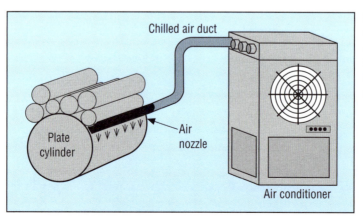

Figure 13-8. The plate cylinder cooling system used for waterless lithography. (Courtesy Toray Marketing & Sales (America), Inc.)

Setting Roller Pressures

Proper ink roller pressure settings assure good ink transfer and allow the rollers to ride and drive properly. Heavy or uneven pressure settings can cause ink distribution problems, increased roller wear, and, through excess heat (created by friction), setting of the ink on the rollers. About one-half of the power used to drive a sixteen-page blanket-to-blanket press goes into driving the inking train.

The proper setting of rollers requires time and care. Press operators should check and reset roller pressures on a regular basis to maintain quality printing conditions. All rollers in the inking system should be set parallel to one another. The oscillators should always be parallel to one another, because they are fixed in the press frame. Consistent roller wear on only one side of the press or the inability to get equal pressure settings are indicators that the oscillators are out of parallel. Worn oscillator bushings or bearings can be one cause. Accidentally running an object like a sponge or rag through the system is another. For a major malfunction like this, the press operator should call in the manufacturer.

A new roller has a greater diameter than a worn old one, so every time a roller is changed in the inking system all rollers should be checked for proper pressures. The difference in diameter can be enough to change the roller's pressure setting significantly.

There are two common methods of setting roller pressures. One involves feeling the resistance, or drag, created by the pressure between rollers. Another involves visually inspecting the rupture in the ink film created by the separation of the two inked rollers. This rupture will form a visible stripe that can be used as feedback on the roller setting.

The resistance method. To roughly check the pressure setting between rollers, use the following procedure:

1. Prepare three sets of three strips of paper, 0.004–0.005 in. (0.10–0.13 mm) thick and about 12 in. (300 mm) long. Two of the strips should be about 2 in. (50 mm) wide and the third about 1 in. (25 mm) wide.

2. Insert "paper sandwiches" between the rollers being set at the center and at both ends of the pair of rollers. The narrow strip should be sandwiched in between the two wider ones.

Figure 13-9. *The use of a paper sandwich for setting rollers.*

3. Pull the narrower middle strip and feel the drag on the strip. This will indicate approximately how much pressure there is between the two rollers. The pull in all three positions should be the same.

A roller-setting gauge provides accurate, measurable results. The metal tongue of the gauge is inserted between two paper strips. When the tongue is pulled out, the gauge gives a reading that indicates the maximum amount of drag on the tongue.

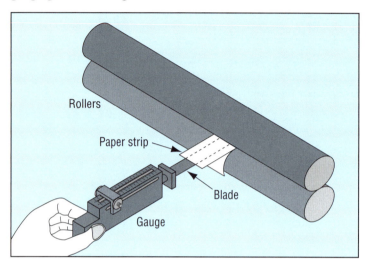

Figure 13-10. *The use of a roller-setting gauge.*

Rollers

Paper strip

Blade

Gauge

Visual stripe method. Another common means of checking roller pressures involves inking the press with a light ink, like opaque white or yellow, inching the press to separate rollers, and looking at the resulting rupture stripe. Follow these directions when setting pressures this way.

1. Ink the rollers with a light colored ink.

2. Inch the press slightly to create an ink film rupture between the two rollers in question.

3. Measure the rupture with a gauge.

4. Adjust the roller pressure more or less to achieve an even rupture mark from side-to-side at the width recommended by the press manufacturer.

5. Repeat the procedure described in steps 1 through 4.

Roller Setting Concerns

It is recommended that the roller pressures be set starting at the ductor roller and ending with the form rollers. The order of roller settings is critical because a change in one roller can affect the next roller in line.

Ductor roller settings. Setting the ductor roller involves three adjustments:

1. Setting the ductor to the fountain roller.

2. Setting the ductor to the transfer roller.

3. Rechecking the ductor-to-fountain roller setting.

Accurately setting the ductor to the transfer roller is particularly critical. If the ductor roller is set too hard against the transfer roller, the adhesion supplied by the ink slows the transfer roller, which is running at press speed. GATF has conducted tests on instrumented presses with the ductor roller set too hard against the transfer roller. The load created by the friction between the ductor and transfer roller was enough to slow the unit slightly, which then affected web tension before and after the printing unit. This could result in registration problems, among other things. Such a load also contributes to gear wear.

Even when the pressure is set properly, the impact of the ductor striking the transfer roller causes a phenomenon called ***ductor shock.*** This describes the vibration sent through the inking system. Printing problems associated with this ductor shock include doubling and slurring. This problem can be minimized with the proper roller pressure setting.

The second adjustment—setting the ductor to the transfer roller—may change the setting between the ductor and the fountain roller; therefore, the press operator should recheck the ductor-to-fountain roller setting.

Setting distributors and rider rollers. The settings of the distributors and rider rollers are usually less critical than those of the ductor. This can best be shown by an example. If the distributor roller marked "X" in figure 13-11 were removed from the inking system, enough ink could be forced down the left side of the roller train to meet the demands of the plate. However, this transfers more ink over these rollers than they were designed to carry, which reduces the length of the ink train, thus affecting problems like ink starvation.

For this reason, inking systems should not be run with rollers missing; however, it is important to note that a press running with a distributor roller with no pressure

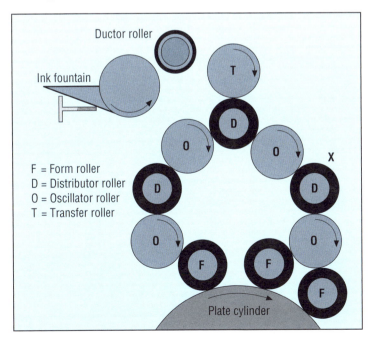

Figure 13-11. *Typical ink system to show the importance of properly setting distributor rollers.*

or light pressure to the oscillator may exhibit the same effect. For example, assume that the distributor marked "X" is not in good contact with the oscillator above it. The ink will move erratically down through the train from roller to roller or may stop altogether at the poorly set distributor. In addition, proper inking system washup becomes extremely difficult.

On the other hand, setting the rollers too heavily wears rollers and roller bearings excessively. Also, excessive heat will build up in the roller train, causing ink problems. In setting the distributors, the press operator strives to equalize pressure across the roller and achieve the stripe recommended by the press manufacturer. This condition automatically assures that the rollers are parallel to the oscillators.

Some distributors have spring-loaded mountings or simple sockets. On such rollers, well-maintained mountings and adequate lubrication are critical to proper functioning.

Form roller settings. The form rollers contact the plate and the oscillators. These rollers are not driven but derive their rotation from the adjacent oscillators. It is important to note that the oscillators drive the form rollers, not the plate. Ink form rollers should be set heavy enough against the oscillators so that they are properly driven, and as light as possible against the plate, while still being able to transfer a full charge of ink.

All form rollers have adjustable mounting sockets to simplify their settings against the plate and oscillator. The design of the sockets permits the press operator to adjust the forms to the plate without disturbing the setting between form roller and oscillator.

Form roller settings should be checked frequently. A variation of the visual ink stripe method described above should be used to check the form roller to plate pressure. Follow these general instructions to achieve the desired ink form roller pressure.

1. Ink the press, covering all rollers with a normal ink film.
2. Stop the press.
3. Drop the form rollers onto the properly packed plate.
4. Pull the form rollers back off of the plate. This will leave an ink stripe image for each form roller contacting the plate.
5. Measure the width of each stripe.
6. Adjust the pressure as required to achieve the desired stripe. Repeat steps 1 through 5.

Figure 13-12. Stripes on the plate from the form roller. Notice the unevenness of the stripes, indicating that both form rollers must be adjusted.

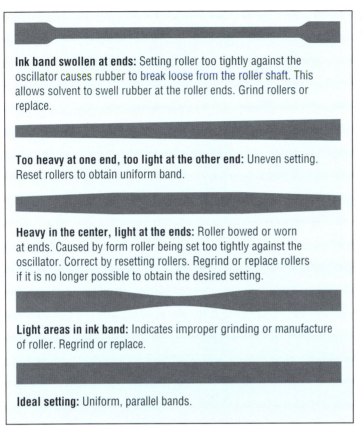

Figure 13-13. *Roller stripes and their meanings.*

Ink band swollen at ends: Setting roller too tightly against the oscillator causes rubber to break loose from the roller shaft. This allows solvent to swell rubber at the roller ends. Grind rollers or replace.

Too heavy at one end, too light at the other end: Uneven setting. Reset rollers to obtain uniform band.

Heavy in the center, light at the ends: Roller bowed or worn at ends. Caused by form roller being set too tightly against the oscillator. Correct by resetting rollers. Regrind or replace rollers if it is no longer possible to obtain the desired setting.

Light areas in ink band: Indicates improper grinding or manufacture of roller. Regrind or replace.

Ideal setting: Uniform, parallel bands.

Some press operators use the following rule to determine the required width of the ink stripe: $\frac{1}{16}$ in. (1.6 mm) for every 1 in. (25 mm) of form roller diameter, with the stripe of the last form over the plate always set narrowest, regardless of its diameter. For example, a roller with a diameter of 3 in. would produce a proper stripe of $\frac{3}{16}$ in. ($\frac{1}{16} \times 3$). Other press operators set the first form roller over the plate heaviest. Each form roller after the first is then set to lay down a narrower stripe than the one before it. While these rules of thumb may have merit, it is best to follow manufacturer's recommendations on ink stripes.

Roller Pressure Load and Roller Hardness

The ink stripe created by the pressure from a soft roller to a hard roller is not a good predictor of actual load on the roller. This is due to changes or variations in roller hardness. To take an extreme example, consider the amount of pressure necessary to create a $\frac{1}{8}$-in. (3-mm) ink stripe between two steel rollers; an enormous load would be required (figure 13-14). Similarly, a harder rubber roller requires more pressure to create the same stripe that a softer rubber roller requires. When a roller becomes too hard and the pressure load increases to maintain the same stripe, several problems can occur. First, excess friction could be created from the additional load, increasing heat and adversely affecting the ink in the system. Also, ink will not transfer as well from harder rollers.

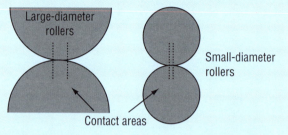

A roller's durometer is a measure of its hardness or softness. Durometer strongly affects roller settings when the ink-stripe roller-setting method is used. As an extreme example, the amount of pressure required to get a ⅛-in. (3-mm) stripe between two solid steel rollers is many times greater than that required to get the same stripe between two rubber rollers of the same diameter. The harder the rollers, the more pressure required to get a stripe of a particular width.

Large-diameter rollers

Small-diameter rollers

Contact areas

Another factor affecting roller settings with the ink stripe method is the diameter of the rollers. For a given pressure, large rollers will produce a wider stripe (contact area) than will small rollers. These are two major reasons why there is no single standard for durometer and for the size of roller stripes for all presses. Roller composition and size vary from press to press as does durometer and, therefore, the proper size of the stripe when setting ink rollers. Consequently, the press manufacturer's recommendations for roller settings should be followed.

Figure 13-14. Durometer and roller settings.

Reasons for roller hardening. Soft rubber rollers become harder over time for a variety of reasons. Paper coating that accumulates and works its way into the inking system will cause rollers to harden over time. Oxidation of the roller surface will cause hardening, as will some solvents used for ink roller cleaning.

Measuring roller hardness. Roller hardness is measured with an instrument called a durometer. This instrument has a flat bottom with a center pin sticking out a short distance. The body of the durometer extends to a dial indicator. When the durometer is pressed against a surface, the pin is retracted in relation to the relative hardness of the surface, resulting in a movement of the dial. Thus, the harder the surface, the higher the reading. Durometers can be used to measure the hardness of very rigid surfaces, like soft metals. There are versions of durometers with different hardness scales. The proper scale used for the measurement of rubber rollers is the shore A scale, which corresponds well with the tolerance for rubber roller hardness. Though durometers are made to measure flat surfaces, a recommended procedure that should achieve repeatable results using the durometer on round rollers is as follows:

1. Set the durometer foot against the roller.
2. Rotate the durometer to a ten o'clock position, back to a two o'clock position, and then to the twelve o'clock position, all within 3 seconds.
3. Take the reading from the dial immediately. The durometer reading will begin to drop the longer the instrument is held in place.

Figure 13-15. *Press operator attaching 1-kg weight to type A durometer.*

Form rollers should read between 25 and 35 on the shore A scale. Distributor rollers and rider rollers can be a bit harder, ranging from 35 to 45. When roller hardness exceeds these readings by 10 points, rollers should be replaced.

Operation Concerns

In the day-to-day operation of an inking system, the press operator performs three different operations: setting the ink fountain at the start of a job, filling and adjusting the ink fountain during running, and washing up.

Setting ink levels. After putting ink in the ink fountain, the press operator sets the ink fountain for the job to be run. In other words, the ink film thickness on the fountain roller is set according to the estimated amounts of ink required by the form. In

---

Operating the Inking System

The operation of the ink fountain is essential to the operation of the press. Following is a set of procedures for operating the inking system.

1. Read the work order to learn proper color and variety of ink to be used.

2. Select the ink for the job.

3. Determine the characteristics needed for the ink to be compatible with the paper stock. These characteristics include body, length, tack, and drying capabilities.

4. Put ink in the fountain.

5. Adjust ink flow from the ink fountain to the ink ductor roller. (Do not allow ink ductor to touch ink train.) Place ink ductor roller against ink fountain roller. Turn the ink fountain roller continuously and, as the roller turns, adjust the ink fountain to lay an even ink film thickness across the entire length of the rollers.

6. Ink the rollers. Start the press. Complete the ink fountain adjustments. Measure ink densities on the sheet and determine when the proper ink film thickness is placed on the roller train. (Establish ink film thickness for standard densities using an ink film thickness gauge. A range of 0.2–0.4 mil should be obtained by reading the last hard roller in the ink train.)

7. Determine the proper setting for inking the plate. Obtain a rough setting relative to the plate image. Obtain a fine adjustment relative to the plate image.

8. Set the swing and speed of the ink fountain roller.

9. Control and monitor the ink system during the pressrun. Monitor ink color with a densitometer and color control QC devices. Match the ink color to the OK sheet.

areas of heavy coverage, the fountain keys are opened more. For lighter coverage, the blade should be set closer to the ink fountain roller. Alternatively, ink transfer can be regulated by using ink fountain dividers that allow the ink to lie in only part of the fountain. The section with no ink demand should be lubricated. Special compounds known as open-pocket compounds are available for this purpose.

The setting of the fountain keys and blade determines the variations of ink from side to side due to the form. The amount of ink fountain roller rotation determines the volume of ink fed into the inking train. The amount of rotation can be adjusted by the press operator. Once the press is printing, the press operator visually evaluates the original settings by inspecting the first makeready signatures coming off the press and adjusting the keys as required. It is far easier to control ink flow starting with too little ink on the rollers; therefore, initial ink settings should be light.

Treatment of the fountain blade. An ink fountain blade treated properly can last the lifetime of the press. When opening the blade across the fountain, the press operator should always start with the fountain keys at each end and work alternately toward the center (figure 13-16). In setting the blade closer to the fountain roller, the press operator should always start at the center and work alternately between each end. These procedures prevent a buckle from forming in the blade (figure 13-17), which can damage the fountain.

The ink fountain blade should never be set so tight that it actually scrapes the fountain roller surface completely clean of ink. Such a setting scrapes the roller and the blade, eventually leading to expensive repairs.

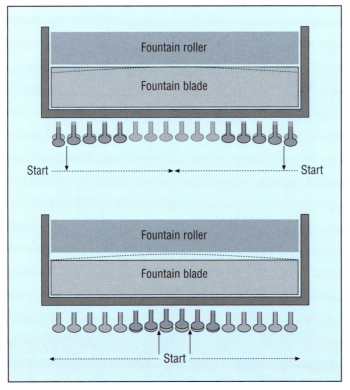

Figure 13-16. Setting the fountain blade of the inking system. To open fountain keys (top illustration), start at the ends and work toward the center. To close fountain keys (bottom illustration), start at the center and work towards the ends.

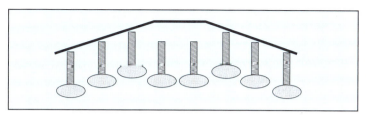

Figure 13-17. A fountain blade that has buckled because the keys were opened in the incorrect order.

Washup Concerns

Solvents for washup. During ink system washup, a solvent that meets EPA and OSHA regulations is used to cut the ink, thin it out, and simplify its removal from the rollers. Washup solvents must be compatible with the ink. In addition, the solvents should be approved by the blanket and roller manufacturers. This is necessary to prevent premature aging of blankets and rollers.

In washing up the inking system, solvents dissolve the ink, removing it from the rollers. Today, solvents are usually classified as one-step or multistep, depending on the number of distinct cleaning stages their application requires. Multistep solvents involve the use of two or three separate solutions. When using them, the press operator should never substitute one solution for another and should always use them in their proper sequence. One-step solvents are typically used for daily washups; multistep cleaners are more often used for thorough cleaning on a weekly basis.

The roller manufacturer should always be consulted for instructions as to the choice and frequency of use of any solvents. Regardless of which solvent is used, the press operator must ensure that the press is run until no solvent remains on the rollers after washup is completed.

Inking System Washup Procedure

Wash up the inking system using a multipart cleaning solution and following the procedures recommended by the press and roller manufacturers:

1. Stop the press.
2. Remove the ink from the ink fountain.
3. Open the ink fountain, then clean the fountain roller and blade.
4. Clean the ink knife.
5. Install a washup blade on the press, or obtain blotter sheets.
6. Set the washup speed of the press.
7. Apply #1 cleaning solution to dissolve the ink.
8. Adjust the washup blade against the distributor roller to remove the ink from the rollers. (If a blotter sheet is used, place the inking rollers in the "on" position against the plate cylinder.)
9. Apply #2 cleaning solution to remove the #1 solvent and any ink pigment particles remaining on the rollers. If using blotter sheets, place a new blotter sheet on the press and apply #2 cleaning solution. Lower ink rollers against the plate.
10. Apply #3 cleaning solution to remove all traces of the #2 solvent and to condition the ink rollers. If using blotter sheets, place a new blotter sheet on the press and apply #3 cleaning solution. Lower ink rollers against the plate.
11. Move washup blade away from the ink rollers and remove the final blotter sheet.
12. Stop the press.
13. Remove and clean the washup blade and tray.
14. Inspect rollers for any remaining solvents and for overall cleanliness.

Washup operations. It is important that the rollers be completely free of ink and solvent after washup is completed. If allowed to set, any solvent/ink combination left on the rollers can complicate the subsequent startup and damage the soft inking rollers.

The use of a washup device (washup machine) and solvents reduces washup time and helps greatly in keeping rollers in good condition. A washup device usually consists of a Teflon or molded rubber blade with a drip pan attached below it. Thumbscrews at each end of the blade allow for pressure adjustment between the blade and the oscillator. Minimal pressure between the blade and roller allows the blade to skim solvent and ink off the roller. As ink is removed from the roller, more ink from other rollers replaces it, and the whole inking train is cleaned in a short time. Both sides of the blade and the drip pan of the washup device should be thoroughly cleaned as soon as possible after use. Some press operators will place rags in the bottom of the drip pan to catch the ink.

It is common practice to leave press plates mounted while the inking system is being cleaned. If the plates are to be used again it is necessary to apply gum to prevent the washup solvents from sensitizing them. Multistep washup or roller conditioning rinses usually dissolve gum so extra care must be taken to prevent solvent dripping onto the plates. Otherwise, the plates must be removed before starting the washup or rinse operation.

The ink fountain blade, ink fountain roller, and ink ductor roller must also be thoroughly cleaned. The fountain blade should be removed or swung away for washup so that its underside can be cleaned. This also improves access to the fountain roller, especially the fountain roller ends.

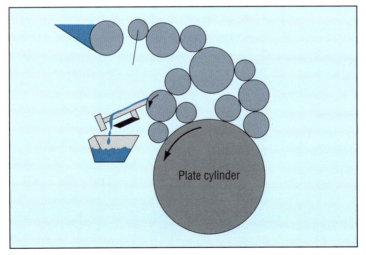

Figure 13-18. A washup device. The device's blade skims ink from the oscillator. The ink and solvent collect in a drip pan mounted beneath the blade.

Plate cylinder

Roller Maintenance and Storage

A clean, properly adjusted inking system in good mechanical condition operates efficiently at maximum speed. An improperly maintained inking system that is either dirty or improperly set can produce a variety of ink distribution problems, which result in unsatisfactory print quality and potentially increased roller and plate wear.

Care and Maintenance of the Inking System

A clean, properly adjusted inking system in good mechanical condition operates efficiently at maximum speed. Maintenance of the inking system is the responsibility of the press operator.

1. Check with your roller manufacturer for instructions as to the choice and frequency of use of any cleaning solvents.

2. Perform daily washups with approved, one-step solvents.

3. Perform weekly washups with approved multistep solvent, to prevent roller glazing.

4. Chemically treat copper oscillators periodically with a copperplating solution to combat stripping.

5. Scrub stripped oscillator rollers with pumice and water, followed by an etching solution to keep oscillators ink-receptive and water-repellent.

6. Clean fountain frame regularly to prevent accumulation of ink, dirt, and paper lint.

7. Clean ink buildup from tie bars (connecting the two sides of the press frame) regularly.

8. Remove fountain keys and soak in solvent.

9. Oil each key before replacing in the fountain. Replace oiled fountain keys, preferably into same holes from which they were taken.

10. After thorough cleaning, store rollers vertically, in their original covers, in a closed cabinet.

Rollers are subject to surface buildup of dried ink, gum, and, sometimes, paper coating material. This condition is known as *glazing.* When glazing occurs, the roller loses its ability to properly carry and transfer ink. Periodic washup with multistep solvents is usually enough to prevent glazing on the press. In extreme cases, the rollers have to be removed from the press and scrubbed by hand.

Because of the lateral movement of the oscillators, ink rollers invariably develop ink buildup on the ends, especially after the washup. If these ink cuffs are not removed during every washup, the ink dries and then has to be chipped away. If they are not removed, dried ink cuffs may chip during running. This debris can be a major cause of hickeys (figure 13-19).

Oscillator rollers are subject to a condition called *stripping,* in which the roller becomes water-receptive because gum is adhering to its surface. When this occurs, the roller will resist being ink-receptive. The usual cause of roller stripping is too much gum in the dampening solution. If the dampening solution pH is too low, roller

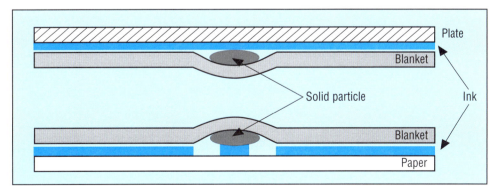

Figure 13-19. *A solid particle (such as dried ink), enlarged, on the blanket, producing a hickey on the printed sheet.*

Figure 13-20. A roller that is stripping.

stripping increases greatly because acid improves gum's ability to stick to the roller surface. The cure for a stripped roller is to vigorously rub the roller with pumice and water (warning: this will damage soft, rubber composition rollers). Most oscillators are made of steel tubing covered with copper, ebonite, nylon, or some other ink-receptive material that is resistant to roller stripping caused by dampening solution chemicals. Although this reduces roller stripping, the press operator should not run excess gum or acid in the dampening solution.

Clean roller surfaces do not indicate that the entire inking system is adequately clean. There are many unseen places in the system where ink, dirt, paper lint, and other foreign matter can collect. Eventually, these accumulations fall onto the rollers if the press operator doesn't remove them. The fountain frame (the heavy metal block in which the ink fountain blade and keys are mounted) is one of these places; ink builds up on the underside and should be removed regularly. Ink also collects on the tie bars near the roller train that connect the two sides of the press frame. These bars should also be cleaned regularly.

The ink fountain keys can become difficult to adjust because of dried ink. They should be removed and soaked in solvent while the metal block is cleaned. Ideally, the cleaned keys should then be returned to the same holes from which they were taken.

Roller storage. Roller storage is also an important part of maintenance. Rollers purchased from the manufacturer will be covered with a paper sleeve. Always leave the sleeve on the roller until the roller is ready for use. This will help prevent accidental damage and protect the roller surface from oxidation. Always store rollers vertically. Storing them horizontally causes them to bow. Also never set the roller down on a flat surface for a long period of time. This will create flat spots on the roller body. Used rubber rollers should be cleaned before storage, and care should be taken to prevent exposure to sunlight, because sunlight prematurely ages the rubber. Rollers should not be stored near a source of heat or ozone. Ozone (a form of oxygen and a powerful oxidizing agent) is usually found around a source of electrical discharge such as a motor armature. Ozone can cause fine cracks to develop on the rubber surface.

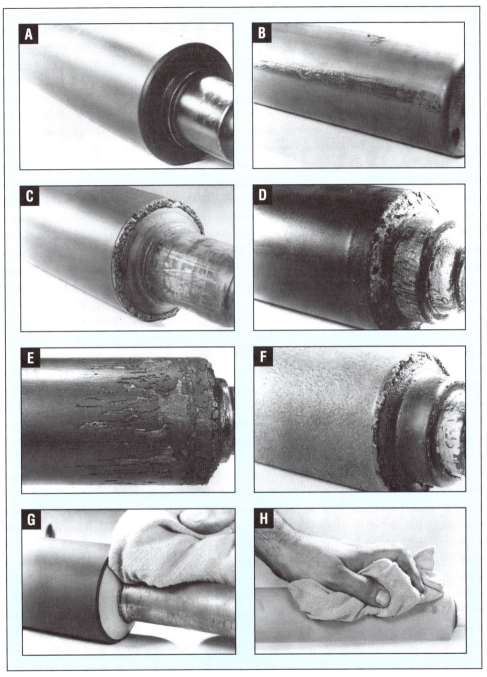

Figure 13-21. The importance of roller care. (A) The surface of a properly maintained roller should look and feel velvety. (B) Residual ink or solvent left on the roller overnight can cause the roller to pick apart when the press is started. (C) Ink buildup on the roller ends can get rock-hard, in time cracking and splitting the roller ends and generating hickeys. (D) Buildup of dried ink and varnish can create minute cracks and wrinkles in the roller surface. (E) End-picking can be caused by ink buildup and by running narrow webs without lubricating the roller ends. (F) A roller in as poor condition as this one will have to be re-covered. (G) The minimal time needed to keep roller ends clean is more than paid back in increased roller life, press productivity, and improved print quality. (H) Periodic hand cleaning of rollers is an essential part of roller care.

Ink System Auxiliary Equipment

Ink agitators. Certain inks back away from the rotating fountain roller. If this goes unnoticed, the ink system will stop receiving ink and ink film densities will drop below acceptable quality levels. The agitator is a cone that extends into the ink mass close to the fountain blade and roller. The cone revolves as the agitator moves back and forth across the width of the fountain. Ink agitators keep ink from backing away by keeping ink in the fountain in constant motion. This lowers ink viscosity (allowing it to flow more easily).

Figure 13-22. *Conical ink agitator.*

Automatic ink level controller. An ink level controller is a device that checks the level of the ink in the ink fountain. It signals an ink pump to pump ink when a certain level is reached. Ink level controllers eliminate manual ink replenishment and may reduce the waste caused by color variations and print density inconsistencies.

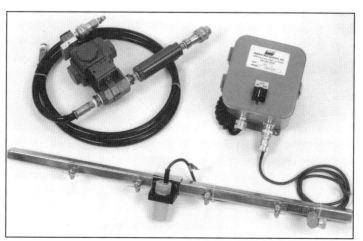

Figure 13-23. *AWS Inkqualizer™ automatic ink level controller. (Courtesy Applied Web Systems, Inc.)*

Ink pumping systems. In plants that print in low volume, inks may be purchased in cans and manually fed to the press with an ink knife. High-volume web printers usually have large containers (e.g., drums) or reusable tanks from which the standard process colors of ink can be pumped to the press.

Typically, ink pumping systems for dispensing ink from drums employ air-operated ram-type pumps with piping going to the gear side of the press and then extension hoses to the different printing units. Flow meters can be placed in the ink lines so that readings can be taken at the start of a job and again at the end of the job to determine how much ink was used. Piping can also be designed so that the ink sequence can be changed as dictated by job characteristics.

Figure 13-24. *A PileDriver pump used for ink distribution on web presses. (Lincoln Industrial Corporation)*

Figure 13-25. *Pumping system for color inks at the Baltimore Sun newspaper. (Courtesy Sun Chemical/ GPI Division)*

Figure 13-26. An installation of a four-color heatset ink pumping system. (Courtesy Flint Ink Corporation)

Remote control console. Most press manufacturers offer free-standing remote control consoles with their presses. A remote control console is a computerized device that enables the press operator to control most of the important press functions without leaving an inspection table. Typical control functions include inking, dampening, and image register. Most remote control consoles also allow the press operator to adjust the position of the plate cylinders.

The console usually includes a remote set of ink fountain keys, usually in a tumble switch or push-button array design, and each numbered to the position of the key

Figure 13-27. The PECOM central control console, which controls and monitors the entire printing process. The right-hand monitor is optional for the PECOM Video System. (Courtesy MAN Roland Druckmaschinen AG)

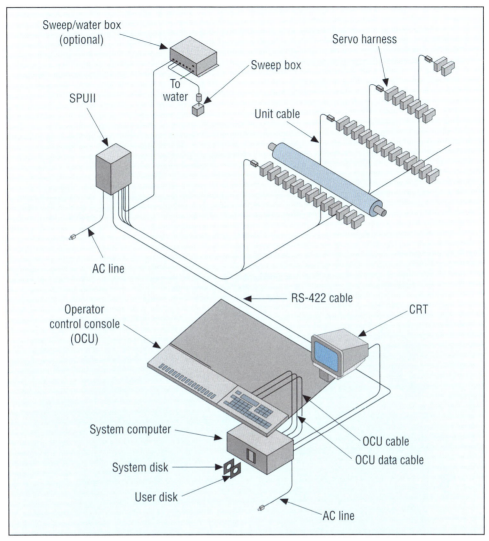

Figure 13-28. *Microcolor II system for adjusting ink key settings from a remote control console. (Courtesy Graphics Microsystems, Inc.)*

along the fountain blade. The press operator can increase or decrease the level of ink at each fountain key by pushing the corresponding button. Some systems display the fountain blade profile through an array of light-emitting diodes (LEDs). Usually, in case of electrical failure, the fountain keys on press can be manually adjusted as well.

The fountain settings for a given job can be recorded on magnetic disk, allowing the press operator to preset the fountain automatically the next time the job is run. A recent trend is to use data generated in prepress to provide ink key presets, which greatly reduces the amount of time required to achieve color.

Plate scanner. A plate scanner is a device that scans the plate before plate mounting, measuring the ink levels required in each ink key zone to preset the inking control console. This decreases the amount of time necessary to bring the ink to the correct level. The information is recorded, often on magnetic disk, so that it can be used to preset the ink fountain directly after scanning, or at a later date. Many plate scanners also produce a printout that graphically displays the ink density of each individually controlled ink zone.

Closed-loop systems. The goal of the closed-loop system is to set and maintain ink densities within strict tolerances quickly and efficiently. These systems employ an on-line densitometry-based CCD video camera or a combo video/spectrophotometry system, each capable of scanning and measuring a color bar on the moving web, continually feeding data to the ink control console for ink film thickness compensation.

Those using these systems today are finding them not only useful for maintaining consistent color quality, but as an efficient means of real-time data collection for statistical process control and customer quality assurance. Many systems also monitor dot gain, print contrast, and trapping.

Figure 13-29. Color-Quick, an on-press color measurement system that provides closed-loop color press control (Courtesy Graphics Microsystems, Inc.)

Figure 13-30. The operator control station of QTI's Color Control System features a touch-sensitive screen. (Courtesy QTI)

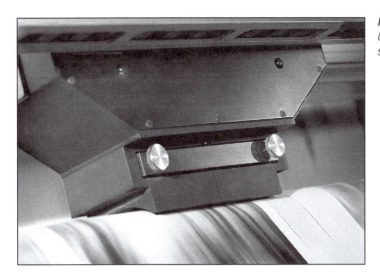

Figure 13-31. QTI's Color Control System scanner. (Courtesy QTI)

Coaters

Various types of coatings are added to the surface of the web to provide luster and to enhance wear and functionality of the printed product. Coatings can be formulated with a variety of characteristics, including high gloss, rub resistance, and oil and moisture resistance. Adding coating units in-line on web offset presses also results in better control over the process and faster turnaround of finished goods. In-line coaters apply a variety of coatings, including water-based, solvent-based, catalytic, or ultraviolet (UV). Each type of coating can be applied on all types of paper.

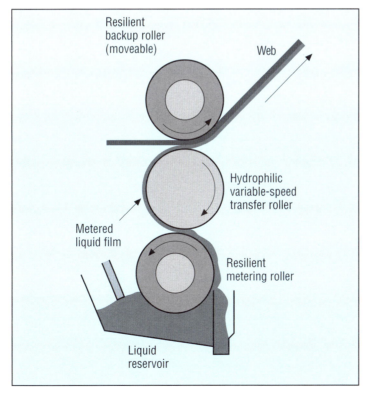

Figure 13-32. A three-roll compact coater. (Courtesy Dahlgren USA, Inc.)

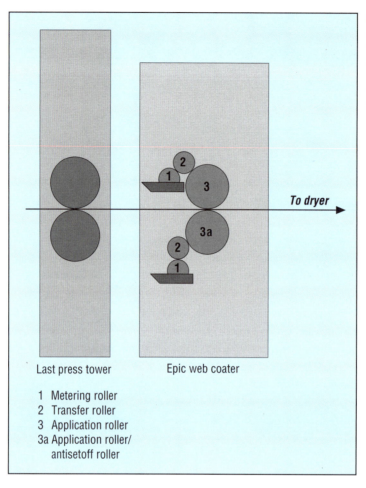

Figure 13-33. In-line web coater that can apply water-based, solvent-based, catalytic, or UV coating over wet ink. (Courtesy Epic Products International Corp.)

Last press tower   Epic web coater

1 Metering roller
2 Transfer roller
3 Application roller
3a Application roller/
    antisetoff roller

To dryer

14 Ink

Unlike other printing processes, lithography employs the use of both water and ink, with the image and nonimage areas differentiated chemically. Lithographic ink must be formulated to work with water, which is the principal ingredient in fountain solution. The ink must also stand up to high shearing forces on press and high temperatures, and must properly transfer to paper at high speeds.

Although there are many classifications of printing inks, web offset printing principally employs six: news, nonheatset, heatset, waterless, ultraviolet (UV), and electron-beam (EB). Each type of web offset ink is made from a pigment or a blend of pigments dispersed in a viscous liquid called a vehicle. The pigment provides the ink with its color, and the vehicle acts as a carrier to transport the pigment to the paper. However that is where the similarities end. Each different type of ink has its own unique additives designed specifically for the intended press, paper, and end-use products.

Ink Manufacture

Ink manufacturing is complex and requires extensive chemical and printing process knowledge as well as specialized ink manufacturing equipment use. While there are a number of manufacturing methods, most share these basic phases of production: formulating, premixing, milling, filtration, and testing.

Formulating

The formulation of inks involves the selection and proportioning of ingredients. These decisions are made in accordance with the intended use of the ink. A printer running news ink in a heatset press would not get very far. However, there are some general categories of ingredients that are common to most lithographic inks, including pigments, vehicles, resins, driers, and waxes.

Pigments. Pigments are finely divided solid materials that give inks color. There are two categories of pigment used in ink manufacture. Carbon black, used in the manufacture of black inks, is categorized as a dry pigment. Pigments used for color inks fall into the category of flushed pigments. Dry pigment is manufactured as a powder, which must be fully dispersed in the vehicle as the first step in manufacture.

Flushed pigments are formulated in aqueous (water) based reactions. To ready the pigment for printing ink manufacture, the water is "flushed" from the pigment under vacuum at high temperatures, and then the pigment is fully dispersed in a vehicle base.

Pigments can be either opaque or transparent. Transparent pigments tend to be very fine and are required for process-color printing. Opaque pigments are larger in size and are typical for spot colors and metallic inks.

Vehicles. The solid pigment particles are suspended in a potentially complex chemical mixture known as the vehicle. The nature of the vehicle determines most of the working properties of the ink and some of its optical qualities. Typical vehicles are derived from vegetable oils (like linseed oil or soy oil) or petroleum. The vehicle contains the binding agent that bonds the pigment particles to the paper; it also determines the conditions required to dry the ink.

Resins. Ink vehicles contain one or more synthetic resins and modified natural rosin that thicken the body of the ink (making it more paste-like) and help bond the pigment particles to the paper. Heatset resins are melted by the heat from the dryer and then set and hardened by the chill rolls.

Solvent. Heatset offset inks contain solvents as well as oils and resins. These inks dry when the solvent evaporates from the rest of the ink and the hot, resinous binder is chilled. Conventional heatset vehicles contain a solvent that maintains the ink in a semiliquid state. The solvent evaporates rapidly when heated, making high press speeds possible.

Driers. Driers are added to inks to help them form a hard dry surface film. There are many types of driers. Nonheatset inks require driers that react with oxygen, speeding the oxidation process in a process called *oxidation polymerization.* The driers used here are typically metal salts of manganese and cobalt. UV and EB inks require dryers, called acrylate monomers that serve as a catalyst for reactions with a UV light source or electron beam radiation. Even heatset inks may contain small amounts of metallic driers, like those found in nonheatset inks. This will assure a completely dry film for offline finishing and shipment.

Waxes. Wax is added to ink as a slip compound that improves scuff resistance of the printed ink film and reduces friction as the web slides across rollers and around cylinders. Animal waxes, like beeswax, and mineral waxes, like paraffin, are commonly used as slip compounds.

Premixing

Premixing is particularly important when inks are made from dry pigments. The dry pigment is formed of large clumps of particles called aggregates or agglomerates. These aggregates must be broken down and reduced in size as the pigment is wet with the vehicle. The initial dispersion involves premixing the dry pigment with the vehicle using a mechanism consisting of rotating mixing blades. This process thoroughly wets the dry pigment. The second stage of premixing involves adding more oil and resin to form a "mill base."

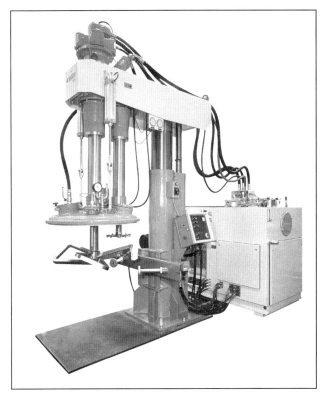

Figure 14-1. Heavy-duty twin-shaft disperser. (Courtesy Buhler Inc.)

Unlike dry pigments, flush pigments are predispersed in the vehicle and therefore require less rigorous premixing. Inks made from flush pigments may go directly to milling after mixing the primary ingredients.

Milling

The mill base still contains large agglomerates that must be further broken down. Also, the addition of other ingredients like waxes and driers can be done at this stage. In many cases, this is done in a shot mill, which is made up of a chamber, rotating disks, and shot pellets. The ink is forced into the chamber and the rotating disks move the metal pellets through the ink, breaking the pigment down.

Another device used for milling is the three-roll mill, which shears the ink as it passes between the rolls, which rotate at different speeds. This shearing breaks up the pigment agglomerates into microscopic particles so that each becomes completely surrounded and wet by the varnish. The mill allows finely dispersed pigment to flow through the mill, but will retain coarser pigment particles. The ground ink is taken off the high-speed roll by a doctor blade. A thorough job of grinding may require as many as three passes through the mill, depending on whether the pigment is soft or hard and how easily it is wet by the varnish.

Milling is costly, and inks can be made less expensive by reducing the amount of milling. This, however, causes several problems including reduced color strength and coarser pigment. Large pigment particles in the ink may cause premature plate wear. Since the pigment is the most expensive part of the ink, and uniform dispersion is accomplished by extensive milling, high-quality inks cost more.

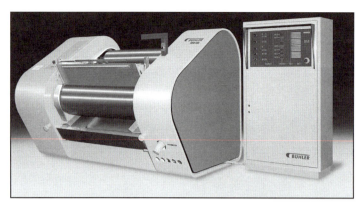

Figure 14-2. "Viva" programmable three-roll mill. (Courtesy Buhler Inc.)

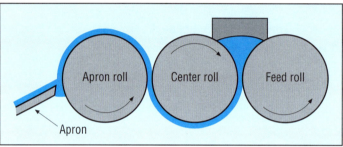

Figure 14-3. The grinding action of a three-roll ink mill.

Filtration

After milling, ink may be put through a series of filtration steps to remove any oversized particles. The filtration system consists of bag filters that have decreasing pore sizes, from 150 microns down to about 10 microns. In some cases (particularly when manufacturing inks from dry pigments), an electromagnetic filter is used as a part of the filtration system to eliminate metal fragments.

Testing

The finished ink can be tested for a wide variety of properties. Those particularly important to the web offset printer include color, viscosity, tack, fineness of grind (pigment particle size), and water pickup (emulsification rate). Once the product is assured of meeting requirements, the ink is packaged for shipping.

Web Offset Inks

As already explained, several different inks are used in web offset printing. Heatset inks are used in much general commercial work. News inks are used in web offset printing of newspapers, newspaper inserts, business forms, directories, and direct mail. Nonheatset inks are used in business forms printing, while ultraviolet (UV) and electron-beam (EB) inks are typically used with in-line presses producing advertisements and direct mail.

Heatset inks. Heatset ink vehicles consist of resin dissolved in a solvent, and drying takes place principally by evaporation. In printing with heatset, the dryer must raise the temperature of the ink enough to evaporate the solvent, leaving the resin to bind the pigment to the paper. After the dryer, the web passes over chill rolls to cool the ink film. The cooling hardens the ink and is necessary to prevent setoff or marking in the delivery.

Heatset inks usually contain thermoplastic resins with no oxidizing properties, in which case no metallic driers are used. Many types of these inks lack rub and scuff resistance but are still suitable for many types of products. Heatset inks may incorporate drying resins to produce tough, rub- and scuff-resistant films.

The solvents used in heatset inks are usually narrow-cut petroleum fractions. Their boiling points range from about 450°F to 600°F (230°C to 315°C), and the solvent selected depends on printing conditions. Solvent with a high boiling point resists evaporation and premature setting under the heat of the press rollers. However, more heat is required in the dryer to set the ink. Solvent with a low boiling point requires less heat to burn off in the dryer, but may set on rollers that are not water-cooled.

The solvent selected should be varied depending on the characteristics of the dryer, speed of the press, ink coverage on the job, and the ink receptivity of the stock. The printer should discuss these factors with the ink manufacturer when selecting the ink.

Quickset inks. The vehicles in quickset inks are dispersions of high-molecular-weight resins in a drying oil and a solvent. When printed, part of the solvent and oil quickly penetrate the paper, leaving the resin concentrated in the ink film. The loss of only a fraction of the solvent and oil greatly increases the viscosity of the ink, so that the printed film becomes relatively setoff-free in a very short time.

Offset news inks. Web offset is the leading method for printing newspapers. Inks for this purpose are composed of pigment (usually carbon black), mineral oil, resin, and sometimes drying oil. News inks may contain a solvent. These inks never really dry, and the pigment is easily rubbed from the paper. Adding a modified drying oil to these low-rub news inks improves their rub resistance by forming a film that better retains the pigment.

Waterless inks. Waterless offset inks contain many of the same ingredients that conventional lithographic inks contain. The difference between them is that waterless inks have vehicles that allow them to have higher initial viscosities than those of the conventional inks. Waterless offset inks also may have slightly lower tacks than those of conventional offset inks. The viscosity and tack differences between the two inks have to do with the differences between the two plates. Waterless offset plates use a silicone material for nonimage areas of the plate. This silicone material has a low surface energy that resists ink if the ink's viscosity is high enough for it to be more attracted to itself than to the silicone material.

The viscosity of liquids changes rapidly with temperature changes. In conventional lithography, the presence of water in the dampening solution cools the ink, helping to maintain a workable viscosity. The viscosity of waterless offset inks is maintained on press by a press temperature control system, either a plate cylinder cooling system or ink oscillator (vibrator) cooling system.

Waterless offset inks are formulated at varied viscosities in order to be adaptable to the geographic temperature environment of any press. Zone controls for each printing unit assure that each ink stays at a customized temperature.

Ultraviolet (UV) inks. The UV ink drying system involves specially formulated inks and ultraviolet lamp stations on press. The inks contain photoinitiators so that when UV inks are exposed to ultraviolet radiation, polymerization (hardening) occurs almost instantaneously. These inks contain no solvent, so VOC emissions from the ink are eliminated. However, ozone is a by-product of the UV drying process and must be vented from the pressroom.

Many UV ink users are purchasing equipment with interstation drying, so that they can accomplish dry-trap printing. This increases the trap percentage, expanding the color gamut reproduced. Interstation drying also lowers dot gain. Dot gain is the result of the wet ink film being squeezed between the paper and the blanket. When wet ink films pass through multiple units, the pressure from each unit's nip may cause more distortion. With interstation drying, the film is dry after printing and the pressure from succeeding units will have no effect on the image.

Electron-beam (EB) inks. Electron-beam inks are similar to UV inks, except that they do not require photoinitiators. The source for the polymerization is a stream of electrons that bombard the ink film. EB inks have all of the advantages of UV inks; they cure rapidly, dry hard, and present no emission control problems. Some claim that because these inks lack photoinitiators, they are more stable and more easily handled. EB inks do tend to be relatively expensive.

Optical Properties

Color. Color perception is determined by psychological, physiological, and environmental factors. Light is reflected, transmitted, or radiated from an object to the eye. Light stimulates the nerve cells that identify color. Color identification varies between individuals. The eye is basically sensitive to red, green, and blue light. All other perceivable colors are combinations of these three frequencies of light. In printing, the reflections of these three colors (red, green, and blue) from the surface of the paper are respectively controlled by application of the transparent inks known as cyan, magenta, and yellow.

Objects appear colored when they either reflect or transmit light. Pigments appear colored in white light because they absorb certain wavelengths and reflect or transmit others. As such, color is a result of the spectral composition of the light. The three process color inks are cyan, yellow, and magenta. Ideally each absorbs light from one-third of the spectrum and transmits light from the remaining two-thirds (figure 14-4). Cyan has its characteristic color because it transmits blue and green light while absorbing (or filtering out) red light. Yellow transmits red and green light

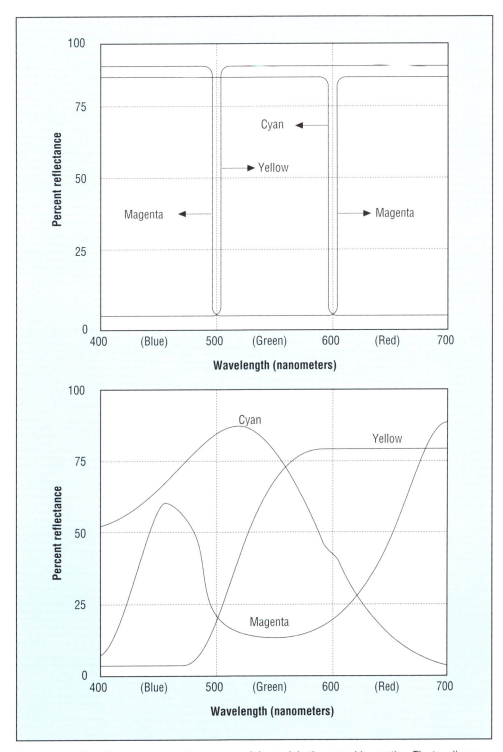

Figure 14-4. *The difference between how process inks work in theory and in practice. The top diagram shows how a printed surface with an ideal set of inks would reflect light over the visible spectrum. The lower diagram represents the light reflectance pattern of such a surface printed with an "average" set of process inks. The discrepancy between the two is made up through color correction techniques carried out in prepress.*

and absorbs the blue. Magenta absorbs green light and transmits the red and blue. In actuality, these theoretically "true" colors cannot be manufactured.

If examined spectrally, yellow, magenta, and cyan appear slightly gray and off cast. Wide disparities in the "trueness" of process colors may be found when comparing different manufactured batches. So, for example, two magenta inks may be very different in color. This is due to different pigment types and concentrations in the ink. Process colors can be measured for comparison with a densitometer, determining their hue error and grayness. Any color of ink can be measured with a spectrophotometer, which shows the spectral distribution of the color. A colorimeter can also be used to plot the color in a color space. In these ways, color inks can be compared and tested for repeatability.

Because process inks are transparent, process colors can be overprinted without changing the way each individual ink film absorbs or transmits light. For example, magenta overprinted by cyan produces blue, because magenta filters out green light and cyan filters out the red.

Ordinary non-process inks are typically opaque and work in a slightly different manner. An opaque yellow ink, for example, reflects (rather than transmits) red and green light and absorbs blue light. When opaque inks are overprinted, the top color hides the bottom color, making them unsuitable for process color printing. The color of a printed ink film is a combination of both masstone and undertone. **Masstone** is the color of a thick film of the ink. It is the color of light reflected by the pigment. **Undertone** is the color of a thin film of ink. It is the color of light reflected by the paper and transmitted through the ink film.

Color strength. The color strength (tinctorial strength) is a measure of grade, concentration, and dispersion of pigment in the ink. Color strength must be sufficient to produce a fully rich color in a printed ink film of normal thickness. The relative color strength of an ink is measured by mixing it with opaque white (called a bleach test) in the proportion of one part to fifty parts. A drawdown test is performed to compare inks side-by-side.

Inks with high color strength may be printed as thin films, thus increasing ink mileage. Furthermore, a thin ink film is less likely to emulsify, since less fountain solution is required to maintain ink/water balance. When ink strengths fall below specifications, the ink film thickness must increase to maintain the same density, resulting in higher dot gain and lower print contrast.

The web printing industry has accepted the Specifications for Web Offset Publications (SWOP) for ink color strength. GATF is the verification center to test inks to ensure that they conform to SWOP specifications.

Opacity. Opacity is the covering power of an ink—the ability of its printed film to hide what is underneath. Some pigments have high opacity while others, particularly process ink pigments, are transparent. Opacity can be measured by making drawdowns over black or a contrasting color.

Even minor opacity in a process ink can considerably affect color reproduction. An opaque yellow run on the third unit can cover cyan and magenta. For this reason, if an opaque process color ink must be run, it should be run in the first unit.

Gloss. Gloss is the ability of an ink film to reflect light specularly. High gloss is obtained when the resins in the ink film dry in a smooth, unbroken surface. Coarsely ground pigment particles protrude from the dried film surface and reduce gloss, scattering light rays. An ink that has a high pigment concentration may have reduced gloss because pigments cannot be thoroughly covered by the vehicle. Highly porous paper can absorb and drain off enough resin to allow pigment particles to protrude above the surface. The highest gloss is obtained on the more tightly coated stocks.

Working Properties

Offset inks are classified as paste inks, as opposed to the fluid inks used in flexography and rotogravure. Paste inks are viscoelastic because they behave both like fluids and like elastic solids. Viscosity, length, thixotropy, and tack are four interrelated ink properties that strongly affect how the ink transfers and distributes on the rollers.

Viscosity. Viscosity is defined as the degree to which an ink resists flow. High-viscosity inks will not flow easily, while low-viscosity inks will flow readily.

Sir Isaac Newton, who theorized that liquids flow in direct proportion to force, first described viscosity in the seventeenth century. For example, the force required to stir water is constant at one speed. If the force applied to the stirrer doubled, the speed of the water would double as well. Water and fluid inks behave in this way and are called "Newtonian liquids." Many materials, including lithographic inks, do not behave like Newtonian liquids. These materials are defined as non-Newtonian liquids and have thixotropic qualities. With these liquids, an increase in the force applied to the stirrer is not proportional to the speed of the liquid. The viscosity of ink increases when it is not being worked and decreases when a consistent shearing force is applied. For example, it would take a lot of force to stir a can of lithographic ink initially, but as the ink is worked, the force required to continue stirring would decrease.

Temperature also affects viscosity of lithographic inks. As the temperature of the ink increases, the viscosity decreases. This is why waterless inks must be temperature-controlled. If their temperature falls out of the proper range, their viscosities increase or decrease to the point where ink will not transfer properly.

Thixotropy. A thixotropic ink becomes more fluid as a result of working and less fluid when standing. Offset ink that sits for some time may act like a solid, exhibiting no flow characteristics. However, after being worked on a slab, the same ink may flow quite freely. If then left to stand for a while, the ink gradually sets up and eventually regains its original solid consistency. During printing, the thixotropic ink is in a more fluid state when applied to the paper than it was in the ink fountain. This is a result of being worked by the press rollers.

Length. The length of an ink is a measure of its ability to be stretched into a thread. A long ink will stretch without breaking for a longer distance than a short ink. Length is measurable by several methods. The tap-out test, which measures elastic flow, involves placing an ink knife on a puddle of the ink to be tested and then raising the knife. The length of the string formed indicates the relative length of the ink. Measuring ink flow down an inclined slab indicates length under the force of gravity. Long ink flows further than short ink. This fluidity is directly related to viscosity and the thixotropic state of the ink. This flow test indicates how the ink will behave in the ink fountain. Short inks flow poorly and tend to back away from the fountain roller, requiring more steady agitation. Long inks form a flat-surfaced puddle in the fountain, indicating low viscosity and low thixotropy.

Lithographic inks are usually formulated to be long. The inks will become shorter on press due to emulsification with fountain solution. Elastic flow, as measured by the tap-out test, remains relatively constant on the press. Inks that start out too short are sensitive to water pickup and tend to become waterlogged. Short inks also tend to cake on the rollers or will pile on the blanket. Long inks tend to mist and lose their ability to print sharply. When running at high speeds, rollers in the ink fountain are turning very quickly. Misting occurs when small strings of the ink break from the ink body and become airborne.

Figure 14-5. Comparison of long ink (left) and short ink.

Tack. Tack is the pulling force required to split an ink film between two surfaces. Lithography is the only major printing process that requires tacky inks for proper image transfer. Tack influences many printing performance factors including dot gain, paper picking, trapping, and image sharpness. Several on-press factors can either increase or decrease ink tack. Three key factors that affect tack and may be controlled by the press operator are ink viscosity, press speed, and ink film thickness. Understanding the relationship among these factors will provide the press operator with insight into tack-related problems.

Inkmakers formulate low-viscosity inks to accommodate high-speed presses. High-viscosity ink run on a high-speed press will result in high tack at the printing nip, possibly causing picking. In this case, the picking may disappear by reducing press speed.

Stefan's Equation

In the late 1800s, J. Stefan, a scientist, studied the forces related to tack and developed an equation to predict and explain tack. Stefan's equation can be used by the press operator to understand the forces influencing tack.

$$F = \frac{VSA}{t^3}$$

where F is the tack of the ink, V is viscosity, S is speed of separation, A is area of the surfaces being separated, and t is thickness of the ink film

To understand the basis of the formula, consider that the value of tack (F) will decrease if the thickness (t) of the ink film is increased. If the viscosity (V) of the ink increases, the tack (F) will increase as well. Increasing the speed of the separation will also increase ink tack. Knowing these relationships, it can be predicted that increasing the speed of the press, and thus the speed of ink separation, will increase the effective tack of the ink proportionately. Also, decreasing the ink's viscosity (as with a reducer) will result in a less tacky ink.

Tack and color strength are inversely proportional to ink film thickness; that is, a thick ink film has low tack but high color strength, and vice versa. The press crew adjusts ink film thickness to produce the correct color or optical density. Variation in color strength from ink to ink requires the press operator to adjust ink film thickness. Tack should successively decrease from unit to unit to effect proper ink trapping. Excessively increasing ink film thickness to achieve a specific optical density lowers tack; an ink film of higher tack on the blanket of the succeeding unit may lift the previously printed ink film from the paper. Excessively decreasing an ink film to reduce its color strength to a specified optical density increases tack; the ink may pick the paper or lift the previously printed ink film from the paper. For these reasons, color strength should be consistent so that the ink may be printed at the appropriate thickness every time.

Thixotropic setting can increase tack when the ink is left undisturbed in the fountain or on the rollers, plate, and blanket during a shutdown. When the press starts up, the effective viscosity, and therefore tack of the ink, will be much higher than when the press is shut down. This can result in web breaks.

Finally, solvent evaporation can also increase ink tack. Heat generated by friction in the printing units, pressroom temperature, and air flow can evaporate the solvents in heatset inks before they reach the dryer. The ink sets, and the tack increases. A temperature increase without a loss of solvent greatly lowers tack.

The tack numbers displayed on the ink can by the manufacturer are based on Inkometer readings taken under specified laboratory conditions. Many press factors will change the tack, including ink film thickness and emulsification. A new ink having the same tack number as the old ink performs roughly the same if the press factors are not changed. Tack number should not be relied on as a sole indicator of ink performance in cases involving changes in press, speed, plates, and paper.

Water pickup. Emulsification is the mixing of the ink vehicle and water, some of which must occur for good printing conditions. Ideal lithographic inks are formulated to pick up a percentage of their weight in water (usually about 30–40%), after which they should stop picking up any more. Emulsification is necessary to allow the ink on the form rollers to absorb some of the water from the plate, keeping the blanket and paper dryer. The emulsification also lowers tack values at the nip, reducing picking problems. Excessive water pickup significantly lowers tack. If this occurs, low image sharpness, excessive dot gain, and poor trapping may result. The amount of water picked up by the ink depends on the nature of the ink vehicle and the condition of the ink/water balance in the printing couple.

Drying

Heatset inks. The boiling point of the solvents in heatset ink is critical. A low-boiling solvent excessively evaporates on press. A high-boiling solvent requires excessive dryer heat, generating a whole series of dryer-related paper problems: low moisture content, cracking in the folder, blistering, and loss of strength.

The resins in the ink must have the proper melting point. If dryer temperature is too high, gloss will be reduced because the liquefied resins will seep into even the most tightly sealed stocks. In extreme cases, too much resin is absorbed by the paper and dry, loose pigment particles remain on the web surface (a condition called chalking).

Some heatset inks contain some metallic driers to aid in the final drying stage, which occurs off the press. If the ratio of drying compound to resins is too high, the resins will not properly combine and solidify into an ideal ink film.

UV inks. The dwell time for UV inks must be sufficient for polymerization of the ink film to take place. Running the press too fast will decrease the dwell under the UV lamps, resulting in unset ink that will mark in the finishing stages. For this reason, presses that run UV inks generally run no faster than 1200 feet per minute. Ink film thickness plays a key role here as well. The thicker the ink film, the longer the required dwell time under the UV light source.

Nonheatset conventional inks. Many factors will affect the drying of conventional inks. Fountain solution that is formulated to be too acidic will neutralize the metallic driers in the ink, causing much longer drying times. Temperature increases will speed drying. An infrared dryer can be used to heat the ink film to speed oxidation polymerization. Ink films that are too thick will take much longer to oxidize.

Section VI
The Dampening System

15 Dampening

The printing process of lithography requires that a thin layer of dampening solution, composed primarily of water, be applied to the plate surface. The lithographic printing plate has properties that make it ink-receptive in image areas and water-receptive in nonimage areas. The wet dampening film applied by the dampening system serves to repel ink from the nonimage areas of the plate. The dampening systems that carry dampening solution to the plate vary more widely in design than do inking systems. This chapter describes the characteristics of dampening solutions (many printers refer to dampening solution as "fountain solution") and presents the major options in dampening systems along with common operating features.

Basic Functions of Dampening Solution

The primary function of dampening solution is to desensitize the nonimage areas of the printing plate to ink. A surface that is "desensitized to ink" is a surface that ink will not adhere to. An area sensitized to ink will accept ink readily. The dampening solution also helps to keep the plate, rollers, and blanket cool and washes away much of the dirt and debris on the plate or blanket, which might otherwise build up as a kind of surface crust known as *piling.* The dampening solution is also absorbed into the ink on the form rollers, reducing the tack (or stickiness) of the ink film. This is important because it reduces the chance of surface fibers or previously printed ink

---

Safety Precautions

- Make sure all guards and shields are in place before operating equipment.
- Never release a safe button that someone else has set.
- To avoid the possibility of dropping tools into the press or other hazardous locations, do not carry them in pockets.
- Observe and practice all safety rules, regulations, and advice given in the press manual.
- Wear protective gear for eyes, ears, head, hands, and feet where necessary to protect against injury.
- Wear rubber gloves, aprons, and face shields when mixing fountain solutions or working with other chemicals.
- Eyewash stations must be nearby due to the acid content of fountain solutions.
- Wear clothing that will not become entangled in any part of the press equipment.
- Only clean the dampening rollers and fountains while the press is stationary to avoid injury and press damage.
- Do not work on moving rollers with rags, tools, etc., because of the high risk of accident and damage.
- Wash hands thoroughly with soap and water after working with fountain solutions and chemicals.

films from being pulled from the web, a problem known as ***picking.*** However, the dampening solution must not excessively emulsify the ink or prevent proper ink drying. The dampening system may also help to reduce static problems that can occur in the delivery end of the press.

Dampening Solution Ingredients

Dampening solutions are usually sold as concentrated solutions that are diluted with water to the proper concentration. The proper mixture of chemicals making up the solution is critical for quality printing. Though there may be many chemicals that make up a given manufacturer's dampening solution concentrate, the general ingredients common to most are described below.

Water. The primary ingredient in dampening solution is water, which makes up about 95–99% of the solution in weight. Water serves to repel the oily ink from the nonimage areas of the plate surface and to help cool roller and cylinder surfaces of the press by evaporation.

Gum. Gum arabic (a natural substance) or synthetic gum is a critical ingredient in dampening solution. The gum dissolves in water and coats the plate surface, replenishing the desensitizing film on the plate. Without gum, the plate might print clean for a short time, but soon the nonimage areas would begin to pick up ink, a problem called ***scumming.*** In fact, gum can work its way into the inking system over time, coating the inking rollers with a thin coating of dried gum. When this occurs rollers become stripped, meaning that they are desensitized and no longer accept ink well.

Acid. Citric and phosphoric acid are common types of acid added to dampening solution. Gum will become a desensitizing film only when in the presence of acid. A gum and water mixture would not work to desensitize the plate surface if the solution were not acidic. When pH (the measure of acidity) is reduced to optimum levels of about 3.5 to 4.5, the gum molecules are converted into their acid counterparts, allowing the formation of a desensitizing layer on the plate surface. The proper concentration of acid in the solution is critical.

Buffers. The pH of dampening solution may change over the course of a press run. Alkaline papers may act to raise pH levels, while plate chemicals applied during the run may reduce it. Buffers act to keep the acidity of the dampening solution from exceeding a certain level. For example, a dampening solution that is properly mixed might have a pH (acidity level) of 4.5. As long as a buffer is present, more acid may be added to the solution but the pH will still remain at 4.5.

Wetting agents. Wetting agents are added to dampening solution to allow the water to spread as a thinner film on the surface of the plate. The principle of surface tension can best be understood by considering an example. If a drop of water were placed on the hood of a freshly waxed car, the drop would form a bead. This means that the surface of the wax coating is not allowing the edges of the drop to spread out and thus, the surface tension is very high. Over time the hood of the car loses the wax coating and becomes oxidized. The same drop of water will now spread out on the surface of the hood very easily. Thus, the surface tension is low.

For many years, isopropyl alcohol was used in dampening solutions to reduce surface tension and increase viscosity so that the plate would stay clean with less water. Isopropyl alcohol evaporates quickly to form **volatile organic compounds** (VOCs), the release of which is now limited by U.S. Environmental Protection Agency laws. Because of potential environmental damage, alternate wetting agents have been developed, most commonly referred to as alcohol substitutes.

Antifoaming agents. Defoaming agents are added to reduce problematic foam buildup. Dampening solution acts a bit like soapy water in that the solution tends to foam when worked. Foam can adversely affect the even transfer of solution in the dampening system.

Fungicides. Fungicides help prevent the formation of fungus and bacteria in the dampening system. Fungus and bacteria can form very quickly in moist environments including dampening system pans (fountains), recirculation tanks, and water lines.

Alkaline Dampening Solutions

Alkaline dampening solutions are commonly used by web offset printers in newspaper production. News inks that are compatible with alkaline solutions must be selected. Unlike the acid solutions, alkaline solutions do not contain a desensitizing gum but rather contain sodium carbonate or sodium silicate. Most of these solutions also contain a sequestering agent, a substance that retards the calcium in the solution from precipitating. Wetting agents are also common in these solutions.

An advantage to using alkaline dampening solutions includes the elimination of problems related to gum use, including roller stripping and glaze buildup. Also, fungus tends not to form in the fountain pan. Use of alkaline dampening solutions for commercial work results in a number of problems including excessive water emulsified in the ink and the release of ink pigments, causing tinting. **Tinting** occurs when the dampening solution becomes dyed with the ink pigment. This results in a slight tone of color in the nonimage areas of the printed sheet.

pH and Conductivity

pH. The degree of acidity or alkalinity of a substance is measured on the pH scale in increments from 0 (highest acidity) to 14 (highest alkalinity), with 7 being neutral (figure 15-1). pH stands for "potential of Hydrogen" and is a measure of the concentration of hydrogen ions in a solution. The pH scale is exponential, with each successive number indicating a tenfold increase in potency. For example, a solution with a

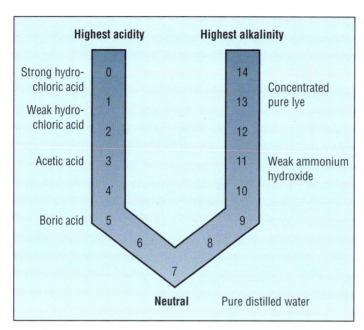

Figure 15-1. The pH scale.

Highest acidity   Highest alkalinity

Strong hydro-   0
chloric acid
                 1   13   Concentrated
Weak hydro-   pure lye
chloric acid
                 2   12
Acetic acid   3   11   Weak ammonium
                                   hydroxide
                 4   10
Boric acid   5   9
                 6   8
                     7

Neutral   Pure distilled water

pH of 4 is ten times more acidic than a solution with a pH of 5 and one hundred times more acidic than a solution with a pH of 6.

The pH of a solution can be measured with chemically coated indicator strips or with electronic meters. This test strips are sold as thin tape-like rolls or as individual strips. The strip is dipped into the solution and some predetermined amount of time is allowed to elapse. The color of the strip is next compared to a color-reference table to determine the pH. A pH meter achieves a much higher level of accuracy than paper strips. Normally, readings in 0.1 pH increments can be made. These electronic meters are convenient and accurate, though some maintenance may be required. The instruments may need to be calibrated with deionized (neutral) water.

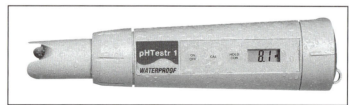

Figure 15-2. A pH meter for use in laboratory or plant. (Courtesy Cole-Parmer Instrument Company)

Conductivity. Conductivity is the ability of a substance to conduct electricity. Pure water itself is a very poor conductor of electricity. As materials dissolve in water they form ions, which conduct electricity. Conductivity correlates well with pH because an increase in ion concentration (pH being a measure of hydrogen ions) also results in an increase in conductivity. Some partially ionizable materials, like gum arabic and alcohol, reduce conductivity. Conductivity of dampening solutions is measured in ***micromhos/centimeter,*** a very sensitive scale for measuring electrical conductivity. The unit of measurement for the resistance of electrical flow is "ohms." The unit of measurement for the conduction of electricity is "mhos" (mho is ohm spelled backward). A micromho is a very small increment equal to one millionth of a mho.

Electronic conductivity meters are used for measurements (figures 15-3 and 15-4). These instruments have a probe, which is immersed in the solution. Accurate and dependable readings can be made.

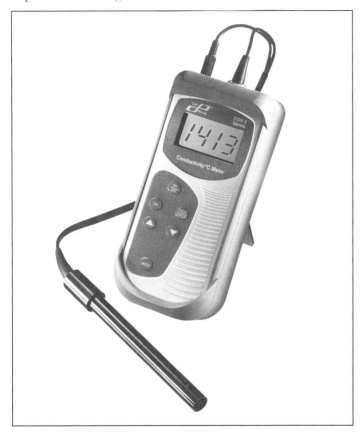

Figure 15-3. Conductivity meter. (Courtesy Cole-Parmer Instrument Company)

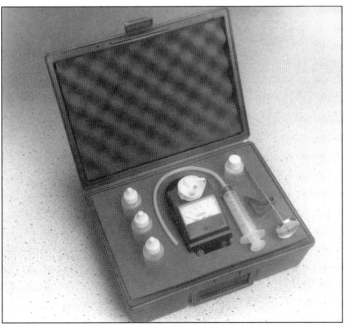

Figure 15-4. A combination conductivity/pH meter, Model M6/pH, with Model PLK Litho-Kit™. (Courtesy Myron L Company)

Effects of acidity. Acceptable dampening solution pH may range from 3.5 to 4.5, depending upon the fountain concentrate ingredients and the condition of the water used. The gum in dampening solutions will desensitize plates at much lower pH values; however, the higher acid content may cause the gum to desensitize not only plate images but also ink rollers, causing roller stripping. The gum may also attack the ink, neutralizing ink driers and causing poor ink drying.

When the acidity of the dampening solution is too high, say over 5.5, the gum no longer acts as a desensitizing agent. The nonimage areas of the plate will become sensitized (ink receptive), and scumming will result.

Unfortunately, pH readings alone do not accurately indicate whether the dampening solution has been mixed in the correct concentration. Because of the buffers added to dampening solutions, the pH reading will plateau at a certain level and remain relatively unchanged when more acid is introduced. So even though the increased acid is not affecting pH, all of the other ingredients in the fountain concentrate (e.g., chemical salts and antifoam agents) are also increasing in concentration.

Conductivity is unaffected by the buffer and proportionally increases as the amount of fountain concentrate is increased; therefore, solutions may be monitored more accurately using conductivity readings. Conductivity readings also indicate the concentration of a neutral solution, which pH readings do not.

Benchmarking conductivity. If the conductivity of a dampening solution at different concentrations is known, it is easy to measure the strength of a working solution by measuring its conductivity. The following procedure can be used to develop a graph (figure 15-5) that plots conductivity and pH against concentration.

1. Measure the conductivity and pH of the water normally used to make the dampening solution. Place water in a clean 1-gal. (3.8-l) bottle.

2. Add 1 oz. (29.6 ml) of fountain solution concentrate. Remeasure both conductivity and pH. Record these values. (To provide more steps, 0.5-oz. increments may also be used.)

3. Add another ounce (2 oz. total) of concentrate, and remeasure both conductivity and pH. Repeat this process until the amount of fountain solution concentrate added exceeds the manufacturer's recommendations.

4. Plot these values on a graph that has concentration (oz./gal. or ml/l) on the horizontal axis and conductivity and pH on the vertical axis.

The most important factor in preparing dampening solution is to make sure that it is the proper concentration. As already discussed, most acidic dampening solutions are buffered so that, as the amount of concentrate increases, the pH drops initially but then levels off, while the solution's conductivity increases in a straight line. Thus conductivity is better than pH for determining the amount of concentrate in the dampening solution, but the dampening solution must still be in the proper pH range recommended by the fountain concentrate manufacturer.

A similar graph can be developed if alcohol or an alcohol substitute is also used in the dampening solution. Pure alcohol has little effect on pH but does lower conductivity. Most substitutes in the proper amounts have little effect on pH or conductivity.

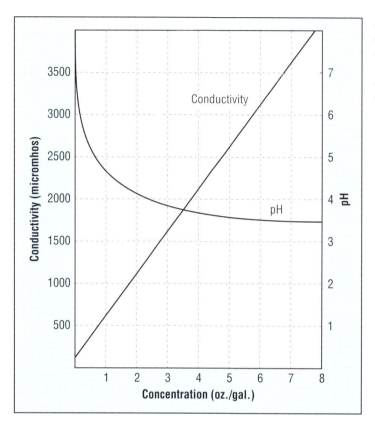

Figure 15-5. *Graph of concentration vs. pH and conductivity for a hypothetical combination of dampening solution concentrate and water.*

Any unusual conductivity readings justify rechecking the conductivity of the water and the fountain solution concentrate. It is normal for the conductivity to increase during the pressrun because materials from the ink and paper contaminate the fountain solution. Therefore, conductivity measurements should be made before the dampening solution is used on press and also during the pressrun. A sharp increase in conductivity will warrant flushing the system and replacing the dampening solution.

Water Quality

The consistency and quality of the water used in dampening solution is very important. Poor water quality can be the root cause of many printing problems. Pure water approaches a conductivity of 0 micromhos, containing no ions to conduct electricity. Typical tap water might have a conductivity of 200 micromhos or more. As the amount of dissolved matter increases in water (and thus ions), the conductivity increases directly in a straight line. Conductivity then is commonly used as a measure of water purity.

If the water quality varies from batch-to-batch in conductivity and pH, the printer may not be able to control the dampening solution. As a rule of thumb, if the conductivity of the incoming water varies less than ±50 micromhos, consistent dampening solution can be mixed. Day-to-day fluctuations of 200 micromhos indicate that some type of water treatment equipment may be needed to keep incoming water constant. Water quality can vary considerably among geographic regions. Many impurities may be introduced into water at treatment plants including small amounts of chloride, chromium, copper, fluoride, iron, and sulfates.

The hardness of water relates to the amount of calcium and magnesium in the water. Relative hardness is indicated in parts per million (PPM) of calcium carbonate. Generally, soft water has a conductivity of 0–225 micromhos, and hard water has a conductivity greater than 450 micromhos. But the relationship between water hardness and conductivity varies somewhat, depending upon the specific minerals and compounds in the water. Water hardness may have no relationship to conductivity in many cases. Very hard water used in dampening solution can lead to the formation of calcium salts, which are produced by combining calcium phosphates (found in hard water) with acids found in dampening solutions. Calcium soaps can also be formed by the interaction of calcium and the resins found in lithographic inks. Calcium soaps are oil-loving and may lead to deposits of ink in unwanted areas, like water form rollers or cloth-covered dampening rollers. Calcium salts tend to have the opposite effect, desensitizing (stripping) ink rollers.

Water purification. Printing plants that have trouble with city water should consider their own water treatment system. There are four primary types of water treatment systems including water softening systems, distillation systems, deionization systems, and reverse osmosis systems.

- *Water softening.* Water softening involves filtration, which removes the calcium and magnesium from the water. While water softening will help prevent problems associated with the formation of calcium soaps, it may not change the conductivity of the water.

- *Distillation.* This process involves boiling tap water in a water still. The steam that rises from the boiling water is almost free of mineral matter present in tap water. The steam is next fed through condenser coils where it is converted back to liquid water.

- *Deionization.* This process removes dissolved ions from water, which are formed from mineral salts in the water. The chemical process is complex and uses two ion exchange resins to remove minerals from water. Importantly, this water purification process is much less expensive than distillation. Deionized water has a neutral pH and a conductivity close to zero. This will stabilize conductivity readings if wide fluctuations in the conductivity of untreated water exist.

- *Reverse osmosis.* Reverse osmosis systems remove 90–95% of impurities from water, stabilizing conductivity while also softening water. In this process, the water is filtered through a membrane to remove most of the positive and negative ions, dissolved solids, suspended matter, and bacteria.

Water purification takes some of the guesswork out of troubleshooting dampening solution problems, because the water quality remains constant. If tap water is used in the preparation of dampening solution, it is good practice to measure the pH and conductivity of the water to track fluctuations. This information can also be acquired from the local POTW (publicly owned treatment works).

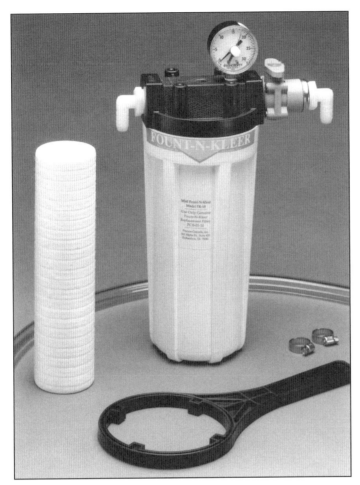

Figure 15-6. Fount-N-Kleer, a fountain solution filtration system. (Courtesy Procam Controls, Inc.)

Dampening System Categories

The dampening systems used for web offset lithography can be divided into two major categories: noncontacting and contacting. With a noncontacting dampening system, there is no physical contact between the supply of the dampening solution and the rollers that apply the dampening solution to the lithographic printing plate. The two noncontacting dampening systems used in web offset lithography are the spray dampening system and the brush dampening system. These will be presented later in this chapter.

A contacting dampening system has some type of physical link between the supply of dampening solution and the rollers that apply the dampening solution to the printing plate. Contacting systems can be further broken into two types: conventional and continuous. With a conventional system, dampening solution flows through the system intermittently. That is, a ductor roller moves back and forth between the fountain pan roller and an oscillator roller in cycles. This system can also be called an intermittent-contact, intermittent-flow, or ductor dampening system. Conventional systems are becoming rare in the printing industry and are almost never found on new press models.

The second broad grouping of contacting dampening systems is the continuous-flow dampening system. As its name implies, this system supplies a constant flow of dampening solution (and has no ductor roller). These systems are the most common on newer press models.

Conventional Dampening Systems

In general design, conventional dampening systems (figure 15-7) resemble inking systems. The basic parts of the system include the following:

- *Fountain pan.* The fountain pan holds a supply of dampening solution. The pan may incorporate an intake and return hose for recirculating dampening solution from a central storage tank.
- *Pan roller.* A brass, chrome-plated, or ceramic fountain pan roller rotates in the pan. It is usually driven by its own motor, separate from the main press drive, or by a variable-speed drive off the press.
- *Ductor roller.* A ductor roller transfers dampening solution from the fountain pan roller to the surface of a metal oscillator. This synthetic rubber roller is covered with a cloth or paper sleeve to increase its water-holding capacity.
- *Oscillating roller.* The oscillating roller is made of chrome-plated steel or a ceramic material. The roller moves back and forth to evenly distribute the dampening film.
- *Dampening form roller.* The dampening form roller directly contacts the plate and is made of a plasticized rubber material. The roller may be cloth-covered or noncovered. *Bareback* is a common term for noncovered dampening form rollers.

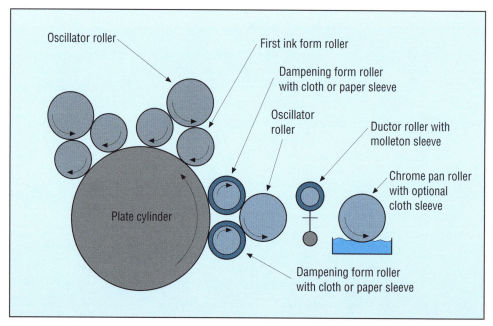

Figure 15-7. The conventional dampening system.

Preparing the Conventional Dampening System

1. Position the water pan or fountain and check it for cleanliness, proper fit, and levelness.
2. Check fountain roller for cleanliness, and clean it following the manufacturer's recommendations.
3. Place a cloth, molleton, or paper cover over the ductor roller (and over the form roller, if required) to increase its solution-holding capacity. (Note: This task is typically done once a week, when the covers become heavily inked or the nap is worn away.) (1) Remove the roller from the press. (2) Pull cover tightly over roller, making sure surface is smooth and even. (3) Presoak covered roller in water, squeeze excess from cover, and blot by rolling the roller in a sheet of paper. (4) Place covered roller in press.
4. Prepare/mix fountain solution for the press. (1) Determine the proper pH (acidic/neutral/alkaline) and conductivity needed for the solution. (Check manufacturer's recommendations.) The proper pH and conductivity must be maintained throughout the pressrun. (2) Determine if the solution's pH is compatible with the ink, plate, and paper used. (3) Add fountain solution concentrate chemicals, including alcohol or an alcohol substitute, to water in the proper proportions to obtain a working fountain solution.
5. Test fountain solution mixture for the proper pH and conductivity levels. (1) Test the pH level of the fountain solution with a reliable electric or battery-operated pH meter. (2) Test fountain solution conductivity with a reliable conductivity meter.
6. Place the fountain solution in the press.

As in the inking system, the dampening ductor swings back and forth between the fountain pan roller and the oscillator. Some slippage has to be built into the dampening system. The fountain roller cannot run at press speed without spraying solution all over the press area, but other dampening rollers do run at press speed. The ductor speeds up and slows down during its cycle to compensate for this speed difference.

The dampening oscillator may be composed of aluminum, chrome-plated steel, or ceramic-coated. The roller is fixed (cannot be pivoted for pressure settings) and is driven from the press drive.

The dampening form rollers directly wet the plate. Most presses have one dampening form roller, although some have two. Dampening form rollers are non-driven, rubber rollers, usually covered with cloth or paper sleeves.

Operation principles. The conventional dampening system is an intermittent-feed system, meaning that the dampening solution does not continuously flow from the fountain pan to the plate. The amount of solution contained in the dampening system is lowest just before the ductor contacts the oscillator. Then, water surges through the system as the ductor contacts the oscillator. These surges are more difficult to control than ink surges, because the flow of dampening solution to the plate is short.

Absorbent cloth and paper dampening covers help control the surges by increasing the storage capacity of the rollers. The roller cover acts like a sponge, absorbing excess water and giving it up when the supply becomes low. This produces a more consistent dampening flow to the plate. The problem with large-supply dampening systems is their slow response to changes. To increase the amount of water, the speed of the pan roller is increased. To cut back on the level of dampening, it is necessary to turn the dampening supply off and let the running press draw the excess

---

Operating a Conventional Dampening System

1. Check alignment of the oscillator periodically. It must be parallel to the fountain roller and the plate cylinder. If it is not, have a press mechanic make the adjustment.

2. Set the pressure between the form roller and the oscillator at the plate cylinder gap. (1) Use smooth strips of packing paper to set dry rollers or use plastic of the same thickness (0.002 in. or 0.05 mm) to set wet or somewhat damp rollers. (2) Place several sandwiches of three strips between the form roller and the oscillator.

3. Use sandwiches of strips to set the form roller to the plate. If the roller is set properly, it will not "bump" at the plate cylinder gap.

4. Set the ductor using the following procedure: (1) Inch the press until the ductor is at maximum pressure against the oscillator. (2) Insert paper sandwiches at three points between the two rollers. (3) Test for drag. (4) Inch the press until the ductor touches the fountain pan roller. (5) Again, insert sandwiches at three points between the rollers. Equal drag indicates correct setting and alignment.

5. Control the dampening solution fed to the plate in one of three ways, according to the press used, as follows: (1) Adjust the fountain pan roller for faster or slower rotation. (2) Adjust the length of time the ductor dwells against the fountain roller. (3) Set special tabs, squeegees, or rollers called water stops against the surface of the fountain roller. The pressure exerted by the water stops controls the flow of dampening solution across the rollers and is equivalent to adjusting the blade in the ink fountain. Water stops are commonly used to reduce the amount of solution reaching heavily inked areas of the printing plate.

6. Monitor pH and conductivity levels throughout the pressrun. Check pH and conductivity at the first sign of ink/water balance problems.

7. Perform proper maintenance regularly by following the manufacturer's recommendations and any established pressroom procedures.

---

amount of dampening solution from the system. In both cases, skill is required by the press operator to change the setting without producing waste. In order to increase this storage capacity even further, press operators may put a double cover on the roller, with the undercover made of cotton or flannel.

Roller covers. The most widely used material for covering dampening form rollers is cloth. Molleton is a woven fabric that is slipped over the roller and sewn tight at the roller ends. Molleton has a relatively long nap and is good for storing water; however, it releases lint when new and easily becomes ink-covered. In addition, dirt or ink picked up can mat down the molleton, changing the effective diameter of the roller. This change in roller diameter requires new pressure settings.

Figure 15-8. Two dampener roller covers: parchment paper strip (left) and paper dampening sleeve (right).

Because of the inadequacies of cloth covers, paper covers were developed. The paper used is a special vegetable parchment, first sold as strips for winding around the roller and currently available as tubular sleeves. When these roller covers get dirty, they are easier to clean or replace than cloth, but a major drawback to using paper covers is that they have limited water-storage capacity.

Roller pressures. Setting the pressure between the various dampening rollers is critical. With too little pressure, the rollers drive inefficiently. Too much pressure wrings water from the rollers, and insufficient dampening solution reaches the plate. To set pressures, plastic strips are inserted between rollers at three points across the length of the rollers. The strips are pulled by hand to measure drag. For more accurate results, the metal tongue of a roller-setting gauge may be inserted in place of the middle plastic strip.

The primary reference point used in setting the rollers is the oscillator, which is fixed and parallel to the fountain roller and to the plate cylinder. In most cases, dampening form-to-plate pressure should be less than that between the ductor and the oscillator. Too tight a setting can squeegee water from the plate and also increase plate wear. It is best to set just enough pressure so that adequate water transfers smoothly to the plate.

To set the ductor, the printing unit is inched until the ductor is at maximum pressure against the oscillator. A plastic strip is inserted at three points between the two rollers and tested for drag. The printing unit is then inched until the ductor touches the fountain pan roller. Again, the strips are used in three positions along the ductor to ensure correct setting and alignment.

Another means of adjusting water feed is also possible (besides adjusting the speed of the pan roller). This involves the setting of special tabs, squeegees, or rollers called water stops against the surface of the pan roller (figure 15-9). This modulates the flow of water across the press much like closing the ink keys in the ink fountain. Water stops are especially useful when running narrow webs where dampening solution tends to build up on the roller ends if water stops are not used to hold it back.

Flap ductor systems. Another variation of the intermittent-feed system is the flap ductor (figure 15-10). Its fountain roller has a removable covering of canvas flaps, and the oscillator is mounted close enough to the flap roller to touch the ends of the flaps. In most cases, one molleton-covered form roller wets the plate. The flapper applies solution to the oscillator. Increasing the flap roller rotation transfers more solution to the oscillator. A major advantage is that the system eliminates ductor shock. Ductor shock, as described in chapter 13, is the vibration sent through the unit rollers when the ductor strikes the oscillator. This shock can cause printing problems. The major disadvantage is that the system makes it difficult to modulate water feed across the press.

Maintenance of conventional dampening systems. Keeping roller covers clean is the biggest maintenance problem on the conventional dampening system. They become ink-covered during the pressrun and then fail to carry water adequately.

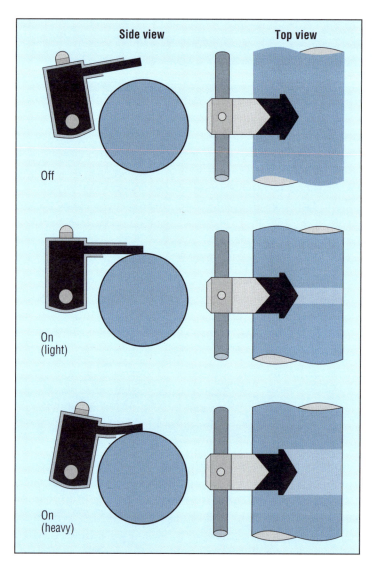

Figure 15-9. One design of a water stop for regulating water flow.

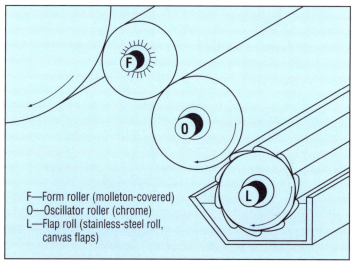

F—Form roller (molleton-covered)
O—Oscillator roller (chrome)
L—Flap roll (stainless-steel roll, canvas flaps)

Figure 15-10. The Levey flap ductor dampening system.

After removing a covered roller from the press, the press operator must not reverse the ends when reinserting the roller. If the press operator does, the roller rotates in the reverse direction and the cover tends to twist, resulting in the covering linting and becoming baggy. To prevent this, some press operators paint a reference mark on one end of all covered rollers.

Cloth- or molleton-covered rollers need to be checked on a regular basis for roundness and if found faulty should be reground or replaced. Low areas on the roller surface will not transfer dampening solution evenly. The smooth metal surface of the oscillator can also become ink-covered, breaking up the film of water on the oscillator. Such a roller should be thoroughly rinsed with ink solvent. The solvent should then be washed away with water, and the roller gum-etched, which will make the roller more sensitized to water. Gum etch, a powerful desensitizing solution, is typically a mixture of one part 85% phosphoric acid to thirty-two parts 14° Bé gum arabic, called 1:32 gum etch. The gum etch should be rubbed to a smooth film and allowed to dry. The roller should then be wet down to see if the water beads. If it beads, the cleaning operation should be repeated. If it doesn't, the roller is ready for operation.

Continuous-Flow Dampening Designs

Conventional systems are becoming rare in the printing industry and tend to be found only on older presses. Continuous-flow systems eliminate some of the dampening control problems associated with conventional systems because the dampening solution is no longer supplied intermittently; these systems do not include a ductor roller. A very important advantage of these systems is their rapid response to changes in fountain settings due to the absence of covered form rollers. This superior design is incorporated into almost all new press models.

The two basic categories of continuous-flow dampening systems are integrated and segregated. With integrated ("inker-feed") systems, the dampening solution is fed through a portion of the inking system; hence, the term "integrated." Segregated ("plate-feed") systems feed dampening solution directly to the plate. Some continuous-flow dampening systems enable the operator to switch between segregated and integrated mode by moving a bridge roller.

Continuous-flow dampening system elements. Though variations exist, the basic parts of a continuous-flow dampening system are as follows:

- *Fountain pan.* The fountain pan holds a supply of dampening solution. The pan also may contain an intake and return hose connected to a dampening solution recirculation tank.
- *Water fountain roller.* The fountain, or pan, roller sits in the fountain pan and picks up dampening solution as the roller rotates. This may be a resilient roller or a hard roller, depending upon the system.
- *Metering roller.* The metering roller sits against the transfer roller. This roller can be adjusted for pressure against the transfer roller. In some designs, one roller functions as both the pan roller and the transfer roller.

Preparing a Continuous-Flow Dampening System

1. Mix the fountain solution according to the manufacturer's instructions for the system used on the press.

2. Check alcohol or alcohol substitute, pH, and conductivity.

3. Fill the circulating tanks with solution, but do not fill them completely.

4. Fill the water pans, and turn on the circulation pump to allow the fountain solution to be pumped to the water pans.

5. Start the drive motor.
   - Make sure the metering roller gear is engaged.
   - Adjust the individual unit moisture control to the normal operating speed.

6. Make preliminary metering roller adjustments.
   - Adjust the metering roller so that it is almost touching the chrome roller on each end. The metering roller must be parallel to the chrome roller to adjust overall feed correctly.
   - Tighten the metering adjustment screw on the gear side until the heavy film of water disappears.
   - Tighten the metering adjustment screw on the operator's side until the heavy film of water disappears.
   - Tighten both sides an additional one-half to three-quarters of a turn.
   - Adjust the lateral water balance between metering roller and chrome roller by moving the screws at each end of the metering roller. (This affects each end of the plate.)
   - Adjust the skewing pressure for the center of the plate. First, loosen the locking mechanism of the skewing device, then increase the amount of water and move the metering roller toward the parallel position.

7. Check/adjust the speed of the fountain pan roller.
   - Set speed after the metering roller is properly adjusted for balance across the plate.
   - Adjust the moisture control (speed) to compensate for pressrun variables, such as temperature, humidity, alcohol evaporation, and ink drying.

- *Transfer roller.* The chrome transfer roller is pressed against the metering roller to form a nip. This point of contact meters a thin film of dampening solution to the form roller.

- *Form roller.* The form roller is a plasticized rubber roller, which takes the dampening film from the transfer roller and deposits it onto the plate surface.

Metering nip. All continuous-flow dampening systems have a metering nip formed by pressure between a hard transfer roller and a soft metering roller. This nip distributes the dampening solution into a thin, even film. The thickness of the metered dampening film at the nip exit is dependent on (1) the hardness of the resilient roller, (2) the pressure exerted between rollers (determined by roller settings), and (3) the viscosity of the dampening solution. Viscosity is defined as the resistance to flow, so a high-viscosity fluid flows slowly and is "thicker." An increase in the viscosity of the dampening solution results in a thicker metered film, while a decrease in viscosity results in a thinner film. The viscosity of the dampening solution is affected by temperature. Thus many dampening systems are refrigerated to control the viscosity of the dampening solution.

Operating a Continuous-Flow Dampening System

1. If an inker-feed system is used, set the inking form roller heavier against the plate so that it leaves a stripe recommended by the press and roller manufacturers.

2. Control the amount of dampening solution delivered to the plate by adjusting the speed of the fountain pan roller and the soft roller that transfers the dampening solution to the first form roller.

3. If a combination system is used, make the adjustments according to the requirements of the system. For example: one system used rollers that cannot be skewed to control dampening distribution. Instead, an air-flow system is used to achieve even distribution across and around the plate cylinder.

4. Adjust the pressure between the chrome transfer, or pan, roller and the resilient roller that together form the metering nip.
   - This will change the thickness of the metered dampening film at the nip.
   - This thickness is also dependent on the viscosity of the dampening solution. Increasing the viscosity of the dampening solution will result in a thicker metered film and vice versa.

5. To clean the system, disengage the dampening unit and secure the system.
   - Remove the standpipe from the water pan and turn off the circulating pump discharge valve. This allows the circulating pump to draw fountain solution back to the tank.
   - Turn off the circulating pump motor.
   - Remove the remaining solution from the water pan by wiping the pan toward the drain. Continue until pan is dry.
   - Clean the metering roller, as follows:
     a. Soak the rag with appropriate cleaner.
     b. Place roller in stop position and scrub from end to end.
     c. Move roller one-eighth of a turn and scrub the roller again. Repeat until the entire roller is scrubbed.
   - Rotate the roller slowly and pour fresh alcohol on the roller from one end to the other.
     a. Alcohol evaporates any water remaining on the transfer roller.
     b. Alcohol prevents drops of gum from forming on the outside of the transfer roller when the system is off.
   - As the transfer roller begins to dry, shut off the moisture control.
   - Turn the metering roller adjustment screw until the surface of the metering roller no longer touches the surface of the transfer roller. Do not separate the gears.
   - The form roller is washed up with the press rollers and requires no special cleaning.

Reverse slip nip. With most continuous-flow dampening systems, one of the rollers at the metering nip rotates clockwise and the other rotates counterclockwise. As a result, the surfaces of the two rollers are traveling in the same direction at the point of contact.

However, there are several dampening systems in which both rollers rotate in the same direction (both clockwise or both counterclockwise). Consequently, at the point of contact, the two rollers are rotating in opposite directions, producing a *reverse slip nip.* The objective of this system is to prevent the dampening solution being fed to the plate from returning to the nip. Theoretically, all of the metered dampening solution is carried to the printing plate, and all of the return solution is carried to the fountain pan.

Systems incorporating a reverse slip nip can operate with a relatively low concentration of wetting agents. In addition, response to changes in dampening feed rate is quickened because all of the solution reaching the plate is used and not returned to the metering nip. Eliminating the returning dampening solution from plate to nip results in a linear relationship between the speed of the metering rollers and the dampening level on the plate.

Integrated systems. One of the principal features of an integrated dampening system is the use of the first inking form roller to deliver both ink and dampening solution (figure 15-11). A soft rubber metering roller coupled with a hard chrome or ceramic roller delivers solution to the form roller. In most designs, the hard roller rubs against the first ink form roller, which then transfers dampening solution directly to the plate. The pan roller speed can be increased or decreased from an independent motor. An increase in roller speed will increase the flow of dampening solution to the plate. Secondly, the nip between the rollers, called the ***metering nip,*** can be adjusted. When the two rollers are moved closer together, a decrease in dampening film will result.

Integrated systems often require setting the inking form roller with more pressure to the plate, chiefly because the rollers in these systems are considerably softer and larger than those in a conventional system. Follow the manufacturer's specifications for stripe widths.

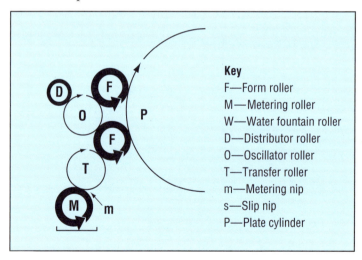

Figure 15-11. Dahlgren dampening system, an example of an integrated continuous-flow dampening system.

Key

F—Form roller
M—Metering roller
W—Water fountain roller
D—Distributor roller
O—Oscillator roller
T—Transfer roller
m—Metering nip
s—Slip nip
P—Plate cylinder

Segregated systems. There are several segregated continuous-flow dampening system designs (figure 15-12). Unlike an integrated dampening system, which uses the first ink form roller for dampening, segregated systems all have dampening form rollers completely separate from the inking system. As with the integrated systems, each has a metering nip formed between a soft metering roller and a hard chrome or ceramic roller. Because these two rollers are driven independently of the press, there is also a slip nip. Slip nips occur when the surface speed of the rollers are different, causing a slip. This slip action functions to squeegee the dampening film from the roller surface.

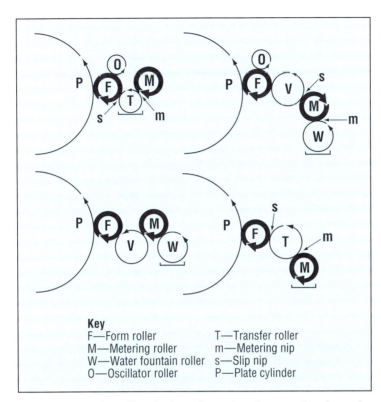

Figure 15-12. *Four examples of segregated continuous-flow dampening systems.*

Key
F—Form roller
M—Metering roller
W—Water fountain roller
O—Oscillator roller
T—Transfer roller
m—Metering nip
s—Slip nip
P—Plate cylinder

Usually, the roller farthest from the plate can be skewed to modulate the water feed across the press. When the roller is skewed, more dampening solution is forced inward toward the center.

Combination continuous-flow systems. A combination continuous-flow dampening system incorporates features of both integrated and segregated systems. In a combination system, an oscillating or vibrating bridge roller contacts both the dampening form roller and the first ink form roller.

The Epic Delta system consists of an oscillating bridge roller and a form roller that is driven at a slower surface speed than the plate. The differential speed results in a slip action on the plate, giving the system a hickey-elimination feature. Hickeys are unwanted specks that appear in solid areas as a tiny speck surrounded by a white halo. Dried ink and other debris on the plate surface cause them. The bridge roller can be used either as a rider (when disengaged) or as a connection between the dampening form roller and the inking system.

Non-Contacting Dampening Systems

Brush dampening system. Brush dampening systems are categorized as noncontact systems because the system is unconnected at some point along the flow of dampening solution. One major brush dampening system design employs a brush roller mounted above the fountain pan roller (figure 15-13). The bristles of the brush roller ride in contact with a variable-speed fountain pan roller. The brush roller rotates at a constant

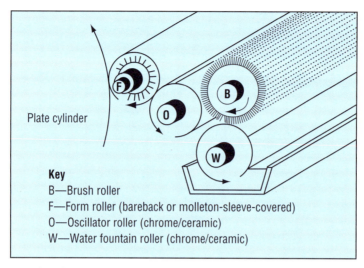

Figure 15-13. A brush dampening system in which the brush roller rotates at a constant speed and is set at a constant pressure against the fountain pan roller. This pressure is great enough to flex the bristles, which flick solution at the oscillator.

speed and is set at a constant pressure against the fountain pan roller. This pressure is great enough to flex the bristles. The bristles flick solution at the oscillator, which is not in contact with the brush. The amount of solution fed is varied by changing the speed of the fountain pan roller, and water flow can be modulated across the press by using water stops. This system uses a single form roller that is often run bareback (without a cover).

A second brush system design incorporates a row of flicker fingers pressing against a brush roller mounted in the pan (figure 15-14). The fingers flex the bristles to "flick" the water to the oscillators, and they also perform a metering function; the amount of flex determines the amount of water flicked. The amount of feed can be varied across the plate in this way, because each finger is independently adjustable. The rotation of the brush roller can also be varied to transfer more or less water to the plate. Because of the relatively uneven dispersion of water by the brush, these systems require a large amount of storage in molleton covers to ensure uniform dampening.

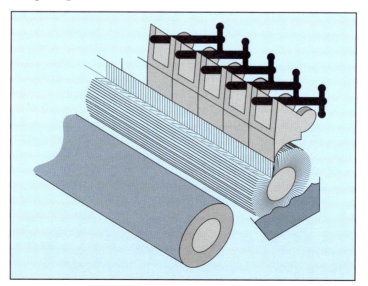

Figure 15-14. A brush dampener using flicker blades that press against a brush roller mounted in the fountain pan.

The settings between oscillator roller and form roller and between form roller and plate on all brush systems are determined by the plastic strip method, which is discussed earlier in this chapter.

In operation, the brush roller should be set light initially, then advanced to its proper settings. Heavy settings can cause undue wear, which can break down the brush roller. When this happens, the press operator begins to find bristles on the plate, and those still on the roller lose their flexibility.

Some press operators change the conventional dampening system by running bareback dampening forms. Brush systems are often run this way, because the entire system responds more rapidly to changes in fountain settings. Isopropyl alcohol or an alcohol substitute is sometimes added to the dampening solution to help the bareback form rollers run more efficiently. This is necessary because the bareback roller will apply less dampening solution, and a wetting agent will help a thinner film of dampening solution keep the plate clean.

Spray-bar dampening systems. Several dampening systems spray fine mists of solution directly onto the rollers of their respective inking systems (figures 15-15 and 15-16). The spray comes from a row of nozzles mounted on a bar across the press. Each nozzle can be independently metered. Unlike other dampening system designs, there is no recirculation of dampening solution. This is a distinct advantage in that fresh dampening solution constantly dampens the plate. The reduced exposure to press conditions virtually eliminates any changes in the dampening solution. This eliminates one of the big drawbacks encountered with continuous-flow and conventional dampening systems that recirculate the solution.

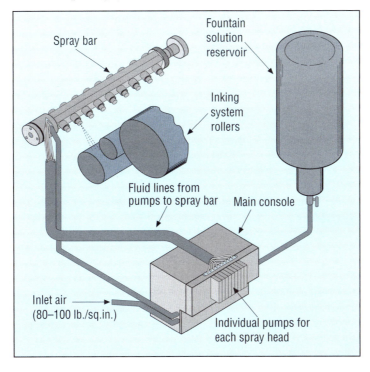

Figure 15-15. Smith dampening system. The pumps in the main console convert the fluid stream into short pulses that come out of the spray bar as a fine mist. The spray bar itself is mounted directly over the inking system rollers.

The biggest problem with running these systems is the possible contamination of the individual pumps and spray nozzles by minerals that are present in most water supplies. Treating water so that it will not clog either the pumps or the spray nozzles helps to solve this problem.

The following illustration shows two different configurations for the Smith dampening system. In both cases, the spray bar is located so that it sprays down into an in-running nip. Depending upon the physical location of the rollers and the inking system, this may or may not require the addition of extra rollers as indicated in the figure 15-16.

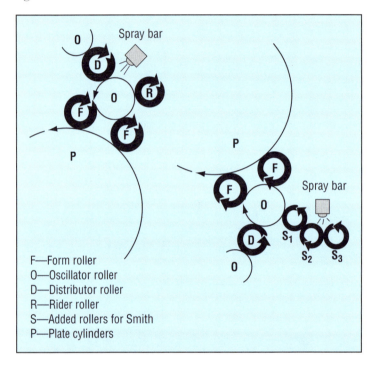

Figure 15-16. *Two configurations of the Smith dampening system.*

Dampening Solution Recirculation Systems

Dampening solution recirculation systems pump solution into each fountain pan from one or more central tanks. These systems allow all fountains to be serviced from a single point, and they also filter the solution to remove solids that may impair print quality. Some systems have a refrigeration unit to keep the solution cool, in order to minimize changes. This is an especially valuable feature in dampening systems running alcohol, because alcohol evaporates at a lower rate when chilled. This reduces alcohol use and saves money.

Automatic fountain solution mixing devices are often incorporated into recirculation systems. The units mix in small quantities, constantly adding fresh solution and maintaining greater uniformity.

Solution control devices are also especially useful if alcohol is being used. Controls are available that can maintain alcohol content to within ±1% by monitoring the specific gravity of the solution and adding alcohol as required.

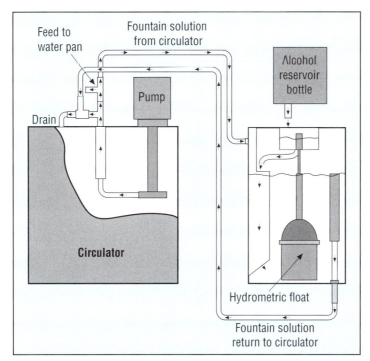

Figure 15-17. *Fountain solution recirculation system (left) with a unit to control alcohol content. The hydrometric float measures the specific gravity of the solution and adds alcohol (or alcohol substitute) as needed. Solution constantly circulates through both systems. (Courtesy Baldwin Technology Co., Inc.)*

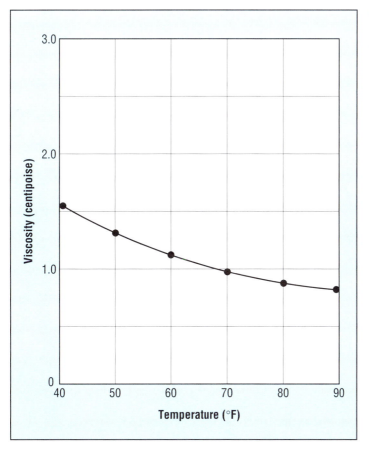

Figure 15-18. *The effect of temperature on the viscosity of water.*

The concept of cooling fountain solution is by no means a new one. The advent of continuous-flow dampening systems in the late 1960s significantly increased the demand for refrigerated lithographic dampening systems.

Continuous-flow dampening systems require temperature control much more than conventional systems, because the continuous system includes a metering nip in which performance is proportional to the viscosity of the dampening fluid. Viscosity is highly sensitive to temperature. For this reason, variations in the temperature of the dampening solution produce variations in viscosity, which result in variations in metered film thickness.

The use of refrigeration offers a second benefit in that the temperature of the dampening system is held constant. This can prevent problems that could affect the quality of a job. If enough cooling capacity is available, the fountain solution can be run quite cold, as low as 40°F (4°C). With systems running alcohol, this can mean a reduction in alcohol consumption of as much as 50%. However, running cool benefits any dampening system, because potentially adverse chemical changes occur more slowly at lower temperatures.

Section VII
Dryer and Chill Rolls

16 Dryer and Chill Rolls

The need for a heated dryer and chill rolls depends on the ink used and how it dries. Heatset inks currently dominate the commercial web offset field, thus this chapter is devoted to this topic. Web offset forms printing and newspaper work, on the other hand, chiefly rely on quickset and absorptive inks that do not require drying and chilling capacity built into the press. Web presses that use UV (ultraviolet) and EB (electron beam) inks do not require heated dryers.

Alternatives to Heatset Offset

Quickset inks (including web offset newspaper inks) dry chiefly by absorption. With these inks, pigments and resins are dispersed in a thin hydrocarbon solvent. After impression, the solvent penetrates the paper, leaving pigment and resins on the paper surface. This drying process, especially typical of newspaper inks, leaves the print with a dull finish, which limits the application of quickset ink in commercial printing. Other quickset inks oxidize after the solvent-absorption phase. Upon exposure to oxygen, the resins cross-link to form a dry ink film. Although the gloss of these inks better suits

---

Safety Precautions

- Keep all fire exits clearly marked, clear of obstructions, and easily accessible.
- Know the locations of all fire extinguishers, and how to correctly use them.
- Check fire door exits for proper working order.
- Be thoroughly familiar with the procedure for coping with a web on fire, and dryer and exhaust fires and explosions.
- Know how to sever the web if it catches on fire.
- Purge the dryer before ignition.
- Check to make sure dryer doors and vents are closed when dryer in ON.
- Check oven and exhaust frequently for the presence of undesirable ink fumes and solvent accretions.
- Make sure the dryer is OFF before inspecting or working around it.
- Remove torn or severed web or loose particles from web before starting up the dryer.
- Never adjust dryer controls while the dryer is in operation.
- Be careful of burns when taking web temperatures in the dryer.
- Do not drop the pyrometer on the moving web.
- Do not smoke in the pressroom.
- Place the press on SAFE when working on the dryer.

commercial printing, their use is restricted by the need for highly absorbent papers. This makes them incompatible with many coated stocks commonly run in commercial web offset. The use of UV inks is growing in the web offset field. These inks can be used to print coated or uncoated stocks and dry to a high gloss when exposed to ultraviolet radiation. The UV radiation causes the vehicle in the ink to polymerize, resulting in a dry film. Another alternative is the electron-beam ink. This ink dries almost instantaneously when exposed to a stream of electrons. Each of these ink drying methods are discussed in more detail in chapter 14.

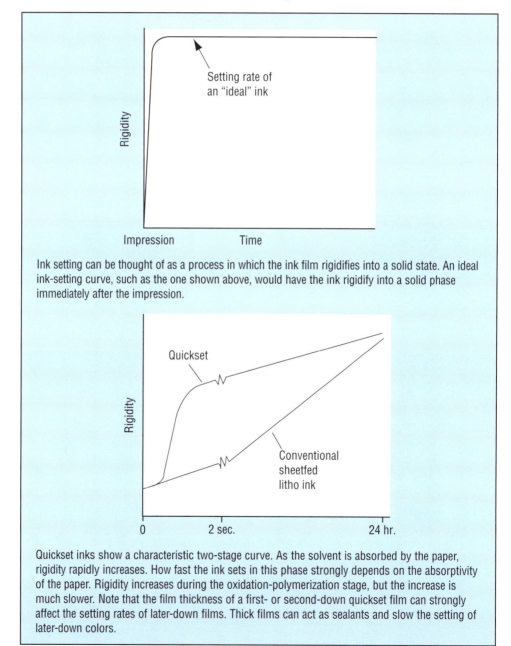

Ink setting can be thought of as a process in which the ink film rigidifies into a solid state. An ideal ink-setting curve, such as the one shown above, would have the ink rigidify into a solid phase immediately after the impression.

Quickset inks show a characteristic two-stage curve. As the solvent is absorbed by the paper, rigidity rapidly increases. How fast the ink sets in this phase strongly depends on the absorptivity of the paper. Rigidity increases during the oxidation-polymerization stage, but the increase is much slower. Note that the film thickness of a first- or second-down quickset film can strongly affect the setting rates of later-down films. Thick films can act as sealants and slow the setting of later-down colors.

Figure 16-1. Setting of quickset ink.

Heatset Principles

Heatset inks dry with a high gloss and, in principle, are compatible with all stocks. Drying equipment is required on the press to evaporate volatile solvents from the ink. Chill rolls follow the dryer to cool the heat-softened binding resins. Solvents in the ink reduce the ink's viscosity (making it runnier), and after its removal, solid pigment particles are left embedded in semi-soft resins. Final solidification or setting occurs with the cooling of the binding resins. Absolute drying occurs off the press and consists of an oxidation process several hours to several days long.

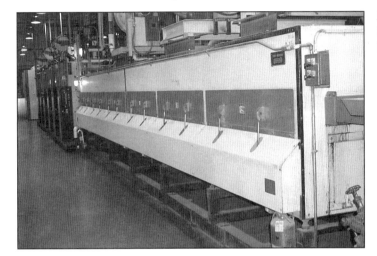

Figure 16-2. A hot-air dryer for a heatset web offset press.

Figure 16-3. The chill roll section of a heatset web offset press.

The ink solvent evaporates quickly, as air temperatures in the dryer may reach 500°F (260°C). The average time the web spends in the dryer is only about 0.7 sec., but it exits with a surface temperature of up to 300°F (149°C). Less than a second later the chill rolls cool the web to about 75°F (24°C). The ink must be well set after leaving the chill rolls. Ink leaving the chill rolls only partially set produces wasted signatures due to marking in the folder, which leads to a press shutdown and folder cleanup (figure 16-4).

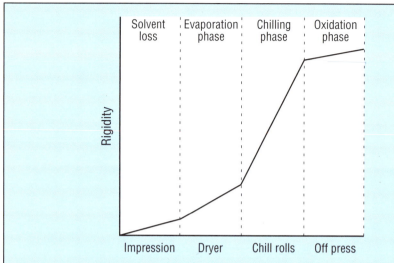

After impression, some of the solvent in heatset inks is absorbed by the paper, resulting in a decrease ink the ink's fluidity. This effect is more pronounced with some stocks than others. Thixotropic setting—the formation of weak molecular bonds in the ink—tends to occur. This setting process aids ink stability through successive nips in multicolor printing. (Highly thixotropic inks are harder to handle on press, often requiring steady agitation to maintain fluidity in the fountain.) The evaporation and chilling phase contribute the most to the increase in ink-film rigidity. Today's solvents have boiling points from 470°F (243°C) to 630°F (350°C). The temperatures required to remove the necessary amount of solvent are also high enough to soften the resins and other materials used to bind the pigments to the paper, making chilling necessary. After cooling, the resins and embedded pigment particles form a tough, solid film. The oxidation phase proceeds slowly over a matter of hours and is accomplished by the combining of atmospheric oxygen with the ink film.

Figure 16-4. Setting of heatset ink.

All blanket-to-blanket presses employ floating dryers, so called because the web makes no contact, but "floats" through the dryer. This is necessary because the web has wet ink on both sides. Floating dryers are categorized according to the method used to heat the web. There are three basic kinds of floating dryer:

- **Direct-impingement dryer.** The direct-impingement dryer is the oldest type of dryer. This dryer uses an open flame, positioned an appropriate distance from the web, to evaporate the ink solvents. There are very few direct-impingement dryers operating in the United States today.

- **Hot-air dryer.** High-velocity hot-air dryers, the type used on most modern web presses, blow hot air at the web through nozzles (figures 16-5 and 16-6). Air blowing devices called air knives direct a high-velocity stream of air at an angle to the web. Exhaust ducts are spaced between the nozzles to vent the solvent-laden air and help to maintain air circulation in the dryer. The mixture of hot air and ink solvent vapor is then recirculated through a combustion chamber, which burns off much of the solvent.

- **Combination dryers.** This dryer uses both the open-flame and medium-velocity hot-air techniques. The first half of the dryer usually contains the flame nozzles, and the second half the hot-air section.

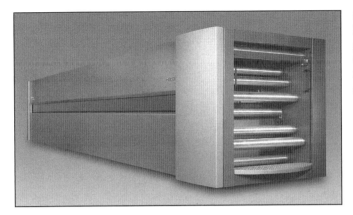

Figure 16-5. The Ecocool, a web offset dryer with integrated afterburner pollution control, cooling section, and chill rolls. (Courtesy Heidelberger Druckmaschinen AG)

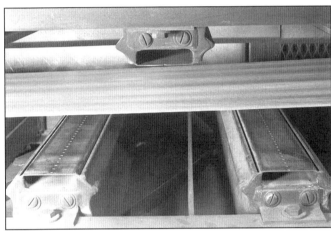

Figure 16-6. Nozzles of a high-velocity hot-air dryer.

Comparing dryer designs. The high-velocity hot-air dryer has one very important advantage over the flame dryer. As the web moves through the dryer, a thin layer of air moves along with it. This boundary layer is part of the moving web's aerodynamics. When solvents are evaporated, they accumulate in the boundary layer until it is saturated. Once the air is saturated, a higher level of heat is needed to continue solvent evaporation. This lowers dryer efficiency. To counteract this potential problem, high-velocity hot-air dryers introduce turbulence into the boundary layer, which prevents saturation (figure 16-7).

If the boundary layer reaches the chill roll saturated with solvent, the solvents may condense and collect on the chill roll surface. Solvent buildup at the chill rolls can resoften the ink and lead to marking on the chill rolls and smearing in the folder. Devices variously called scavengers, scrubbers, or air knives, disturb this layer and exhaust solvents before they leave the dryer.

Electrostatic-assisted drying. New research is being done in the area of electrostatic-assisted drying. It has been found that electrostatic energy helps to break up the solvent-laden air at the moving web's surface. This reduces the air knife velocity required to remove the solvent-saturated air and increases dryer efficiency. New dryers incorporating this technology are being developed, but as of the writing of this book, they have yet to enter the market.

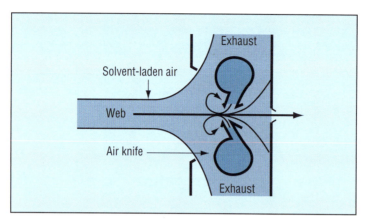

Figure 16-7. The removal of solvent-laden air from the web.

Controlling the web in the dryer. Initial designs of high-velocity hot-air dryers had a problem with web control that resulted in longitudinal wrinkles caused by the long span of web in the dryer. These wrinkles are called corrugations. This problem has been solved by alternately spacing nozzles to put a controlled ripple in the web. A web deformed in this way does not flutter or corrugate, even at high speeds.

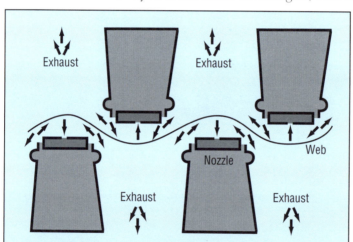

Figure 16-8. Putting a controlled ripple in the web. (Courtesy MEGTEC Systems)

Early high-velocity hot-air dryers often caused the web to flutter in the dryer, due to the force of the air jets beating against the web. Later, it was found that by allowing a controlled ripple to form in the web, the web would show even greater stability in the dryer than if left undisturbed. The ripple is created through careful spacing of opposing blowers and exhausts in the dryer. This principle has since been incorporated into most present-day hot-air dryers.

Pollution control. Heatset web offset presses generally have made use of pollution control equipment, such as condensers or afterburners, to meet environmental regulations. The evaporated solvents are channeled to devices that destroy a high percentage of the volatile organic compounds (VOCs). Excessive amounts of VOCs in the atmosphere lead to the formation of smog, among other environmental problems.

The primary technology for the destruction of contaminants such as carbon monoxide (CO) and VOCs is oxidation. Thermal oxidizers incorporate afterburners that heat the exhaust gas from the dryer to 1,400–1,500°F (760–815°C), at which temperature the hydrocarbon solvents are burned to harmless carbon dioxide and water (figures 16-9 and 16-10). Residence times (duration of the exposure of the fumes to heat) are up to two seconds. In some systems, the heat is recovered for use with the high-velocity, hot-air dryer. One system burns the solvents directly to heat the high-velocity hot air used to dry the web.

Figure 16-9. A catalytic/ thermal oxidizer, an air pollution control device that can operate in either the catalytic or thermal oxidizer mode. (Courtesy Catalytic Products International)

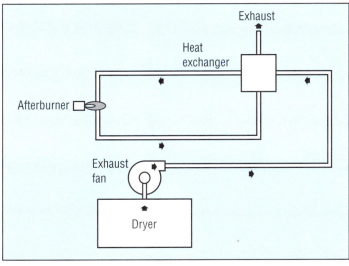

Figure 16-10. Thermal incinerator. (Courtesy MEGTEC Systems)

Catalytic oxidizers (figure 16-11) accomplish the same oxidation processes as thermal oxidizers. However, a catalyst initiates flameless combustion of the pollutants at the considerably lower temperature of 600–800°F (315–425°C). Catalytic oxidation uses less energy, has lower operating costs, and causes less thermal stress on the equipment.

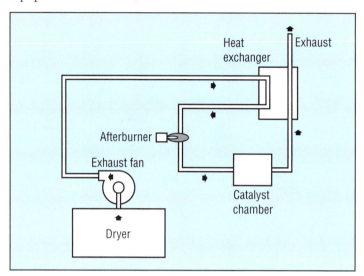

Figure 16-11. Catalytic incinerator. (Courtesy MEGTEC Systems)

Chill Rolls

The chill roll section, where heatset inks are cooled and set, is an assembly of driven steel drums positioned after the dryer. The chill rolls are cooled by refrigerated water circulating through them. The number of chill rolls needed is dependent upon the press speed. The faster the web press runs, the more chill rolls that are needed. For example, a web press operating at 3,000 ft./min. (15 m/sec) might require as many as nine chill rolls. This assures that the duration of contact time (paper in contact with chill roll surface) will be the same as that of a slower press with fewer chill rolls.

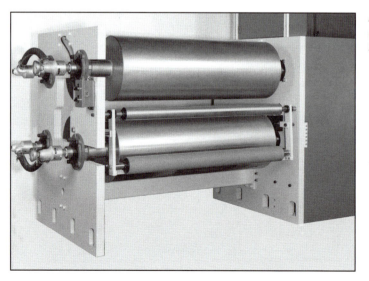

Figure 16-12. Chill stand. (Courtesy Jardis Industries, Inc.)

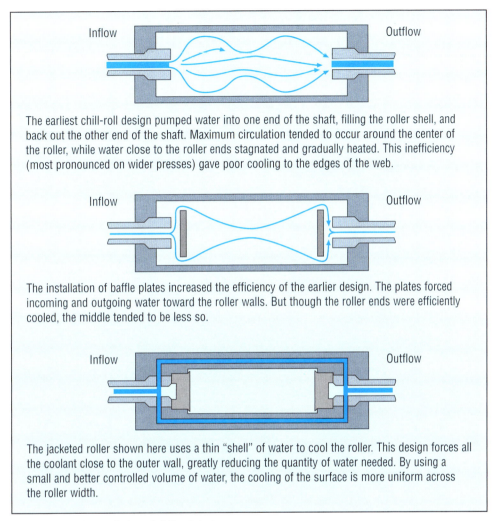

The earliest chill-roll design pumped water into one end of the shaft, filling the roller shell, and back out the other end of the shaft. Maximum circulation tended to occur around the center of the roller, while water close to the roller ends stagnated and gradually heated. This inefficiency (most pronounced on wider presses) gave poor cooling to the edges of the web.

The installation of baffle plates increased the efficiency of the earlier design. The plates forced incoming and outgoing water toward the roller walls. But though the roller ends were efficiently cooled, the middle tended to be less so.

The jacketed roller shown here uses a thin "shell" of water to cool the roller. This design forces all the coolant close to the outer wall, greatly reducing the quantity of water needed. By using a small and better controlled volume of water, the cooling of the surface is more uniform across the roller width.

Figure 16-13. The evolution of chill roll design.

Most chill rolls are driven by the press drive through a variable-speed transmission (similar to infeed metering rollers), which runs slightly faster than the last web unit. With this variable-speed control, the chill rolls regulate web tension in the long span from the last printing unit through the dryer. Even with nip rollers holding the paper tightly to the chill rolls, there is often slip between the web and the rolls. Web slip over the chill rolls can exceed 1% of press speed, which can then be made up by increasing speed. Excessive slip can be a major cause of marking on the chill rolls.

Water circulation in the chill rolls should be through a closed system, with the water used having minimal impurities. This practice reduces the accumulation of mineral deposits inside the chill rolls, which can impair chilling efficiency. In addition, chill roll monitoring equipment should measure temperature and the flow rate of the water. This will provide the press operator with important feedback should there be drying and marking problems. The press operator can control the temperature of the web leaving the chill rolls in two ways: changing the temperature of the circulating water or by increasing or decreasing the rate of circulation.

Temperature monitoring. A pyrometer (figure 16-14) can be mounted on the chill stand to read temperatures of the paper exiting the chill roll section. This device accurately measures temperature from a specific distance optically.

The temperature of the chill rolls is critical. If the web is too warm leaving the chill rolls, the resin coating will not set and ink will smear and mark in the folder. On the other hand, if the temperature is more than 15°F (8°C) lower than room temperature, condensation could accumulate on the chill rolls.

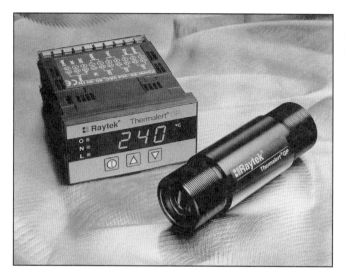

Figure 16-14. THERMALERT® GP optical pyrometer. (Courtesy Raytek Corporation)

Figure 16-15. Pinnacle Series central chiller for chill roll and ink temperature control. (Courtesy AEC, Inc.)

Chill roll tacking. The temperature difference between the heated web and the chill rolls has a tendency to force the web out of contact with the metal drums, reducing cooling efficiency. Chill roll tacking systems have been developed that use electrostatic force to maintain contact between the paper and the drum. With this system, a high voltage is applied to the web just before contact with the chill roll, creating a strong electrostatic attraction.

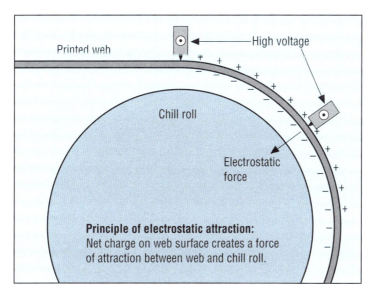

Figure 16-16. Chill roll tacking system, which uses electrostatic force to reduce the gap between the web and the chill roll. (Courtesy Hurletron Incorporated)

Dryer and Chill Roll Operation

The press operator has the means to control web temperature through the dryer and chill roll sections. Dryer and chill roll temperatures are only indicators of the actual paper temperature. The fact that there is a difference between paper temperature and dryer and chill roll temperature is very important. For example, it is possible to change the effectiveness of the dryer by changing only press speed, and thus dwell. **Dwell** is the amount of time that any given point on the web remains in the dryer. If the web is slowed down and dryer temperature remains the same, the paper temperature will increase because the dwell time is increased. Similarly, if dwell is decreased with dryer temperature constant, web temperature will decrease. These speed changes will affect proper setting of the ink. The dryer temperature should be set according to the paper, the inks, the press speed, and the length of the dryer.

Furthermore, the press operator must look at the dryer and chill rolls as a single system. For example, increasing dryer temperature to counteract smearing in the folder may only aggravate the problem. This is because a properly set dryer only removes solvent from the ink; it is the chill rolls that set the ink. If the chill rolls were not cold enough, the resins will remain warm and soft, regardless of dryer temperature. Furthermore, if the boundary layer leaves the dryer in a solvent-saturated state, the chill rolls will condense the vapors (back into liquid) on the web surface. As the solvent works back into the ink, the web begins marking and tracking.

Paper factors. It is good practice to set the dryer temperature by taking paper factors into account. A number of factors affect dryer temperature decisions:

- **Basis weight.** A change of 20 lb. in basis weight of paper can require a significant change in dryer temperature or running speed from one job to the next. This is due to the heavier paper generally requiring more heat to effectively dry the ink.

Controlling and Operating the Dryer

1. Check the job instructions for drying requirements before setting the dryer temperature.
   - Review roll paper specifications for weight and ink coverage.
   - Check the number of webs involved in the job run.
   - Check the number of colors to be printed on each web.
   - Check the web speed specified for the job.

2. Set the dryer temperature according to the paper, the inks, the press speed, and the length of the dryer.

3. Check to make sure dryer doors are sealed before turning dryer on and exhausting the chamber.
   - Manually open and close doors to assure proper operation.
   - Lock dryer doors, assuring the proper functioning of the doors and locking mechanisms.
   - Lubricate all moving parts in the dryer.

4. Check the mechanisms that automatically control dryer operations.
   - At press startup, check the following: (1) door closes, (2) burners move into position, (3) gas ignites burners, (4) correct heat maintained as press accelerates, and (5) automatic control of air-gas mixture.
   - During press stops, check the following: (1) gas automatically turns off, (2) burners swing away from web, and (3) dryer cooled to prevent scorching of web.

5. Check the control functions of the dryer.
   - Check height of the flame in relation to the web position.
   - Check flame size, color, and point conformation for most effective heat.
   - Use thermometer gauges to indicate dryer temperature.
   - Review press manual for controlling the mechanism that produces correct air and gas mixture.

6. Check the air knives for proper functioning.

7. During the production run, check these problems as they relate to the functioning of gas dryers: (1) blistering of stock and scorching of paper resulting from excessive flame, and (2) paper becoming brittle, cracking, or flaking when folded due to lowered moisture content of stock.

8. Check the proper functioning of the dryer exhaust system to eliminate danger of flash fire resulting from improperly exhausted solvents and ink fumes.

- *Paper temperature limits.* Other paper factors place an upper limit on temperature. Paper burns at 451°F (223°C), but scorches at temperatures considerably lower. More importantly, the dryer removes significant quantities of water from the web, reducing its moisture content. One result is that paper shrinks appreciably in the dryer. On most web offset presses, this creates a continuous potential for problems in tension and cutoff variation. Eventually, dryer heat can drive out enough moisture to make the paper brittle.

- *Coated stocks.* Coated stocks may require slightly higher dryer temperatures because of lower solvent absorption. On the other hand, a heavily coated stock may blister (formation of air bubbles in the paper) if heated excessively in the dryer. Paper blisters when a sudden blast of dryer heat vaporizes internal moisture faster than it can escape. It usually occurs when the paper is tightly sealed by heavy coating and printed with solids on the top and bottom of the web.

Controlling and Operating the Chill Rolls

1. Lubricate all chill rolls.

2. Turn on the chilling system refrigeration and allow it to "run in."

3. Check the temperature of water at inlet.

4. Set the chill rolls to the web.

5. Start web through chilling system in accordance with plant procedure.

6. Check the printed job for proper setting of ink.

7. Check the water temperature at both inlet and outlet locations.

8. Make adjustments in the water temperature to meet plant operating levels.

9. Have head press operator check all settings and adjustments.

Ink factors. Ink characteristics can dictate dryer temperature. Two inks that have two different solvent boiling points can appear to be the same, but dry at significantly different temperatures. Some resins release solvents less readily than others and require more heat. The appearance of the ink will not indicate this, but rather an understanding of the ink ingredients and trial and error with drying are required.

Dwell. The duration that the paper spends in the dryer must coincide with press speed and dryer length. Higher press speed reduces the time for the heat to release the solvent. With a short dryer, press speed must be slowed to adequately dry the web. Dryers often hold running speeds considerably below the rated speed of the web press.

Run minimum heat. The dryer should be set to the minimum temperature required to burn off the ink solvent. When the paper is chilled, folded, and delivered, it should not mark, smear, or setoff (setoff occurs when the wet ink from one sheet transfers to an adjacent sheet). Minimum heat lessens the chance of blistering, reduces moisture loss, and, in many cases, improves printed ink gloss. Minimum heat can be maintained by judiciously matching press speed to dryer temperature, and vice versa. As already explained, increasing press speed while maintaining dryer heat lowers web temperatures.

To determine minimum dryer temperature, set press speed at the desired rate and the dryer at a safe temperature—neither too high nor too low. Check web temperature after the last chill roll; make sure that it is at room temperature or below. If the temperature is too high, several remedies are possible: reduce dryer temperature, reduce dryer temperature and press speed, or reduce chill roll temperature. Do the latter either by lowering the temperature of the circulating water or by increasing the rate of circulation.

Never raise the dryer temperature without first making sure that the web is being adequately chilled. When press speed is slowed or dryer temperature is increased, always make sure that the web temperature after chilling is no higher than room temperature—about 75°F (24°C).

Chill Roll Plumbing Configurations

The arrangement of chill rolls and the sequence of water flowing through them affects chilling efficiency. The following research is presented to provide the press operator with insight into the factors affecting chill roll design. Figure 16-17 shows several different methods for plumbing three- and four-roll chill systems. At the top of the illustration are drawings that show the physical layout of the chill rolls. Note that the roller labels A, B, and C (and D for the four-roll system) do not represent the physical location of the rollers but rather the sequence in which the web out of the dryer runs over the various rollers. For clarity, in the four systems marked I, II, III, and IV, the rollers are shown from left to right in the order in which the web contacts them.

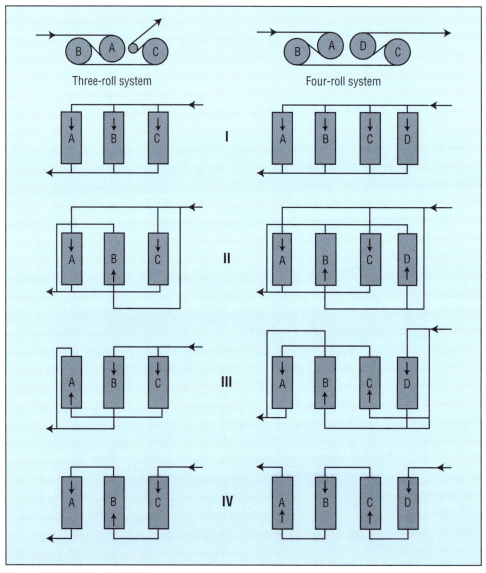

Figure 16-17. Chill roll plumbing.

Table 16-I shows the flow rate in gallons per minute or liters per minute through each chill roll for the various systems illustrated. These rates assume a total water supply of 25 gal./min. (95 l/min.). In the case of diagram IV, it is further assumed that it is possible to pump the entire water supply through all of the chill rolls hooked in series.

Diagram	Three-Roll System		Four-Roll System	
I	8.3 gpm	31.4 lpm	6.2 gpm	23.5 lpm
II	8.3	31.4	6.2	23.5
III	12.5	47	12.5	47
IV	25	95	25	95

Table 16-I. Flow rate per chill roll in gallons (liters) per minute with a 25-gpm (95-lpm) supply.

For the purpose of the calculations the following assumptions are made:

- The paper coming out of the dryer is at a temperature of 270°F (132°C).
- The paper temperature off of the last chill roll is 75°F (24°C).
- The chill rolls in the three-roll system are large enough to provide the same amount of surface contact with the web as that for the four-roll system.
- The temperature of the water being supplied to the chill system is 55°F (13°C).

Analysis of results: overall cooling efficiency. For both three- and four-roll systems (table 16-II), plumbing diagrams III and IV provide slightly better cooling because of the increased velocity of the water running through the individual chill rolls. The four-roll system illustrated by diagram III works nearly as well as diagram IV, although the flow rate of water through the chill roll is only 12.5 gal./min. (47 l/min.) for diagram III as opposed to 25 gal./min. (95 l/min.) for diagram IV.

Diagram	Three-Roll System		Four-Roll System	
I	76.5°F	24.7°C	76.6°F	24.8°C
II	76.5	24.7	76.6	24.8
III	75.6	24.2	75.3	24.1
IV	75	23.9	75.0	23.9

Table 16-II. Average web temperature after chilling.

Analysis of results: side-to-side temperature. Table 16-III shows the side-to-side web temperature variations in the web after chilling. The temperature variations for diagram I are significantly larger than those for diagrams II, III, and IV. The problem with diagram I is the plumbing method; all of the chill rolls are cold on the same side of the press and relatively warm on the opposite side of the press. With diagrams II,

Diagram	Three-Roll System		Four-Roll System	
I	8.1°F	4.5°C	8.0°F	4.4°C
II	2.7	1.5	0.1	0.1
III	2.0	1.1	0.0	0.0
IV	0.9	0.5	0.0	0.0

Table 16-III. Side-to-side temperature variation after chilling.

III, and IV, the cold water input is alternated from side to side as the web goes across the chill rolls. This could be problematic; if it is necessary to lower the web temperature to 75°F (24°C) in order to avoid marking in the folder, the 8°F (4°C) side-to-side temperature differential will mean that the cold side of the web must be chilled to 67°F (19°C) in order for the warm side to be down to 75°F (24°C). This extra cooling requirement limits press production speeds.

Analysis of results: order of chill rolls. As indicated in all four diagrams, the chill rolls are plumbed from the last to the first. Chill roll temperatures should decrease successively from the first roll that the web touches to the last. Furthermore, the first roll should reduce the web temperature by 40–50%; the remaining rolls should lower the web to room temperature. If the temperature is more than 15°F (8°C) lower than room temperature, condensation could accumulate on the chill rolls. Also, on a set of chill rolls that are 40-in. (1016-mm) wide, it should be possible to limit side-to-side temperature variation to less than 8°F (4°C).

The use of a portable infrared (IR) pyrometer helps to control and monitor the correct temperature needs. This device is pointed at the surface of the running web, showing the temperature. This allows the press operator to make changes according to this important feedback.

Remoisturizers

Dryers remove moisture from the web. Overdried paper tends to increase in size across the grain when it reabsorbs moisture, causing color register problems. The overdried paper also tends to accumulate free electrons more readily, building up sizable charges of static electricity. The best way to minimize both of these problems is to replace the moisture removed from the web during drying.

Paper growth problems. Dimensional instability can be a real problem if paper grows after binding and trimming. Wavy edges result in the finished product due to the confinement of the tightly bound sheets and the subsequent expansion of the paper. A more common problem is created by uneven growth of different signatures in a heatset product, such as a magazine, after it is trimmed. In this case, if the grain of the paper parallels the backbone of the sheet, an uneven edge forms along the front of the magazine.

Static electricity problems. The problem of static electricity is not so apparent on the press unless the product is being sheeted. It does, however, make signatures more difficult to handle in the bindery operation and increases bindery waste considerably. This is due to sheets or signatures sticking together due to static attraction. Excessive amounts of static can lead to jams in bindery machinery, resulting in expensive downtime.

Remoisturizing is accomplished by physically applying moisture to the surface of the web after it is chilled. Figure 16-18 shows a three-roll system in which a resilient metering roller turns in the water pan. A water-attracting hard roller contacts the metering roller on one side and the web on the other, transferring the liquid. The units can be run individually or in pairs—one for each side of the web.

A mixture of silicone and water is sometimes applied to the web by the remoisturizing unit. The silicone (in a water-based solution) helps to prevent marking by giving the web a more slippery surface, while the water remoisturizes the web.

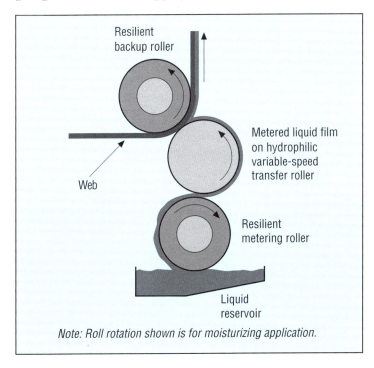

Resilient
backup roller

Metered liquid film
on hydrophilic
variable-speed
transfer roller

Web

Resilient
metering roller

Liquid
reservoir

Note: Roll rotation shown is for moisturizing application.

Figure 16-18. *Three-roll system for coating or remoisturizing the web. (Courtesy Dahlgren USA, Inc.)*

Section VIII
The Delivery System

17 Functions of the Delivery System

The delivery system must convert the printed web into a product ready for the next process: a folded signature, rolls, or flat sheets. Although the folder is the most common delivery option, some web printers, such as such as ticket, stamp, label, and forms printers, print ***roll to roll,*** in which case the finished product is re-rolled using a rewinder. Some web printers may also use a ***sheeter*** at the delivery to cut or "sheet" the web into sections. Sheeters are important in the commercial printing market for posters, broadsides, direct-mail pieces, and covers. This chapter focuses primarily on folders.

Most folded products are made up of a number of folded sections, which are bound together in some manner. Each folded section is called a ***signature.*** When producing folded products, each unfolded printed sheet is made up of a number of

---

Safety Precautions

- Recognize that the two hazardous areas at and around the drag roller are the slitters and the press frame (or adjacent rollers). The slitters are really rotary shears that can inflict serious cuts or even sever a finger.
- Use extreme caution around slitters when rewebbing the press.
- Loose clothing should never be worn when working on web presses.
- Webbing should be done at inching—not running—speed.
- The crew member responsible for feeding the ribbons into the pinch roller should insist on a tapered lead, which enables the feeder to maintain a taut and aligned ribbon whereas a full ribbon obscures vision and will tend to "ball."
- Maintain a clear path of vision, which can prevent accidents of a most serious nature.
- Crews should handle one ribbon at a time, where multiple ribbons are to be fed into the pinch roller.
- Never attempt to recover and insert a dropped lead.
- One crew member should always be positioned at the nearest button station to stop the press if an operator should get caught.
- Noise level in the folder area can be significant. Operators should wear ear protection at all times.
- The press should be put on SAFE or locked out when working on the knife or cutting cylinders.
- The safe button is mandatory when adjusting the tucker blade, folder brush, or replacing collating pins.
- Never make folder brush adjustments or lean over the folder to make adjustments while the press is running.
- Never adjust the folder while it is in motion.

arranged pages. Depending upon the size of the press sheet and the dimensions of the page, the signature may become a 2-, 4-, 8-, 16-, 24-, or 32-page section. Most web offset presses are equipped to deliver folded signatures in-line. Folders for a given press are selected depending primarily on the nature of the finished work to be produced. Page size and format, number of pages, and grain direction requirements in the finished product are primary considerations.

There are three primary types of folder designs used on web presses including (1) combination folders, (2) ribbon folders, and (3) double former folders. The operational principles of each are described in this chapter.

Folder capabilities are dictated by the repeat length (or circumference) of the printing units. With most presses, the cutoff (or printing circumference) of the press is fixed and cannot be changed. Although the web width can be varied, the signature sizes producible by a given web offset folder cannot vary much. Therefore, all jobs planned for particular web equipment must fall within a relatively narrow range of page dimensions.

Elements of the Folder

Though there are a number of variations to in-line web offset folders, most contain the same basic folding elements. Combination folders will contain all three of these types of folding mechanisms.

Former board. A former fold results when the web is pulled over the nose of a triangular metal former board. This folds the web in half—in the direction of web travel and parallel to the grain. Many folders are arranged to first make a former fold in the web, followed by additional folds. In some specialized folders, only a former fold is made, followed by a knife cutoff and delivery of the finished products.

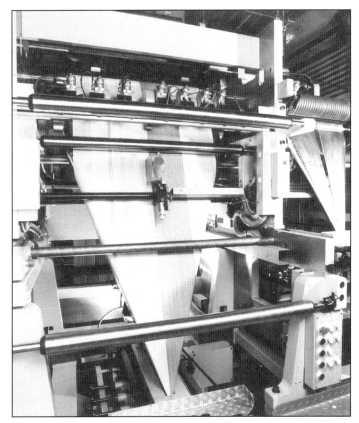

Figure 17-2. The folder superstructure for a ROTOMAN press. Notice the web being folded by the former board. (Courtesy MAN Roland Druckmaschinen AG)

Jaw folder. In a jaw fold, the paper web flows around a cylinder and is tucked into the jaws of a second cylinder by the tucker blade from the first cylinder. A knife in the same set of cylinders cuts the web into individual signatures. If a second jaw fold is made after the first, a double-parallel folded signature is the result. Jaw folds are always made across the web at right angles to the grain.

Chopper folder. A chopper fold, sometimes called a quarter fold, will be the final folding operation if it is used. After jaw folding, the signature is conveyed along a track with the folded edge forward. It passes under a blade that drops and forces the signature between the nip of two rotary folding rollers. The resulting fold is parallel to the paper grain. Chopper folding speeds are normally somewhat limited in comparison to other folds because the folding method does not maintain a positive grip on the signature. For this reason, high-production folders will divide the stream of signatures leaving the jaw fold section. Multiple chopper folders deliver signatures in two or four streams. If two streams of signatures leave the jaw folder, the chopper folders need to work only half as fast.

The Combination Folder

The combination folder is so named because it combines former, jaw, and chopper folding in one machine (figure 17-3). Its popularity is attributed to its versatile folding capabilities. Primarily designed for web widths in the popular 34–40 in. (864–1,016 mm) range, a combination folder on a press this size would deliver the following signatures (sizes are approximate):

- Tabloid signatures—eight 11×17-in. (432×279-mm) pages per web cutoff
- Double digest signatures—thirty-two 5.5×8.5-in. (140×216-mm) pages per web cutoff, or sixteen pages—two-up
- Magazine-size signatures—sixteen 8.5×11-in. pages per web cutoff

Changeover of the machine for each of these different product sizes varies with the design of the folder, but in most cases is relatively simple.

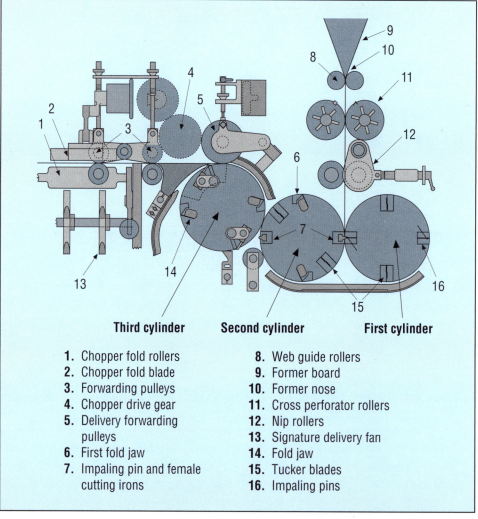

Third cylinder	Second cylinder	First cylinder

1. Chopper fold rollers
2. Chopper fold blade
3. Forwarding pulleys
4. Chopper drive gear
5. Delivery forwarding pulleys
6. First fold jaw
7. Impaling pin and female cutting irons
8. Web guide rollers
9. Former board
10. Former nose
11. Cross perforator rollers
12. Nip rollers
13. Signature delivery fan
14. Fold jaw
15. Tucker blades
16. Impaling pins

Figure 17-3. The arrangement of the major components of a typical combination folder. (Courtesy Heidelberg USA, Inc.)

Producing a tabloid signature. Assume that the press has a 22-in. (559-mm) plate cylinder circumference (and therefore a 22-in. cutoff) and the running web is 34 in. (864 mm) wide. The former fold (folding the web in half) would in effect be creating a four-page, 17×22-in. (432×559-mm) signature. This is the size of the signature at cutoff after the former board.

The cut signature passes to the first jaw fold, where the first parallel fold is made. The jaw fold halves the 22-in. (559-mm) length of the signature. The product after a single jaw fold is now an eight-page, 17×11-in. (432×279-mm) signature, called a *tabloid.*

Producing a digest signature. Alternatively, the 17×11-in. (432×279-mm) signature may receive a second fold parallel to the first. This involves another jaw fold, where the eight-page tabloid is folded into a sixteen-page, 17×5.5-in. (432×140-mm) signature. This product, called a *double digest,* would be sent to the bindery, where it would be cut halfway along the 17-in. (432-mm) side, producing two sixteen-page, 8.5×5.5-in. (216×140-mm) signatures, called a *digest fold, two-up.* Most often, the product coming out of the folder delivery is two-up and then cut into two identical 8.5×5.5-in. digest signatures.

Producing a magazine signature. The third alternative is to pass the signature through a chopper fold instead of the second parallel fold. The chopper folds at right angles to the jaw fold and parallel to the former fold. It takes the eight-page, 17×11-in. (432×279-mm) signature and folds it into an 8.5×11-in. (216×279-mm), sixteen-page signature. This signature is called the *quarter,* or *magazine, fold.*

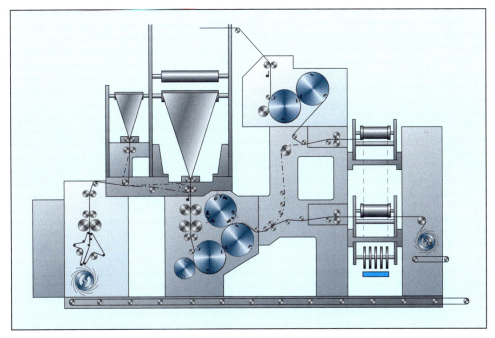

Figure 17-4. *A 2:3:3 pin-type combination folder with an additional folder and a cutter. (Courtesy MAN Roland Druckmaschinen AG)*

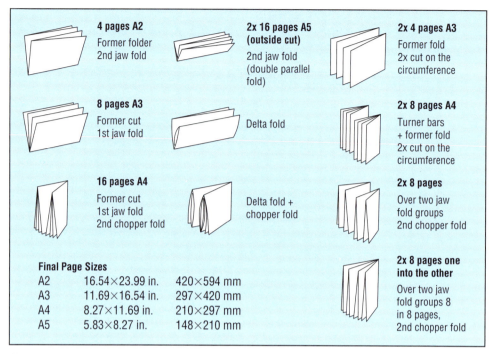

Figure 17-5. *Examples of just a few of the folding types possible using a 2:3:3 folder. (Courtesy MAN Roland Druckmaschinen AG)*

A magazine format—8.5×11-in. (216×279-mm) or 8.25×11-in. (210×279-mm) trim size—is delivered by sequential folds: a former fold, a single parallel (jaw) fold, and a chopper fold. The untrimmed page height of the signature as delivered is one-half the folder cutoff length, and its page width is one-quarter the initial width of the web. Paper grain in the finished signature runs parallel to the backbone.

Producing landscape signatures. In addition to the regular tabloid format, signatures of oblong proportion (landscape) can also be produced on the combination folder by laying out the press plates on a two-up (rather than four-up) basis. For example, two signatures with trim size of 8.5-in. (216-mm) backbone by 11-in. (279-mm) page width could be trimmed out of a signature delivered in tabloid form and later cut apart in the bindery. The difference would be a change in the paper grain direction.

Categories of combination folder. There are two categories of combination folder: single chopper and double chopper.

The single-chopper folder may be further divided into two types: single delivery and double delivery. With single-delivery types, the folded signatures are all delivered on one belt. The double-delivery type has two side-by-side delivery belts to deliver the signatures.

The double-chopper divides the stream of signatures in half before reaching the chopper fold. Because the chopper fold is the slowest of the folding mechanisms, the double-chopper design allows for faster folding. With two streams of signatures, the chopper need only work half as fast compared to a single stream.

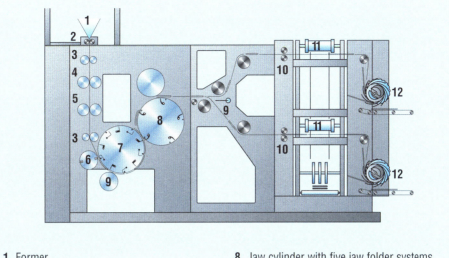

1. Former
2. Former infeed rolls
3. Pull rollers
4. Cross perforation
5. Paced lineal perforation
6. Cutting cylinders with two blades in their circumference
7. Pin-type and folder cylinder each with five choppers and pin systems

8. Jaw cylinder with five jaw folder systems for the first cross fold and another five sytems for the second cross fold and delta fold, respectively
9. Copy diverter for the splitting of production
10. Deceleration station
11. Folders for the quarter fold
12. Paddle wheel delivery for cross fold products

Figure 17-6. *A 2:5:5 double-chopper folder for long-grain production. The stream of signatures is divided in half (9) before reaching the chopper fold (11). (Courtesy MAN Roland Druckmaschinen AG)*

Collect vs. noncollect folders. Collecting folders incorporate a collect cylinder that is designed to hold two successively fed signatures before passing the two collated signatures to a jaw folder. This type of folder can be used when the circumference of the plate image is twice the folder cutoff, allowing the top half image to be collected with the bottom half image. The two signatures, generated from the same plate image, are then sent to the jaw fold together. Noncollect folders deliver a single signature from the plate image. This folder is used when the cutoff is the same as the plate cylinder circumference.

Ribbon Folder

Ribbon folders are designed to produce signatures of the size and format found in publication work. These folders are equipped with angle bars instead of former boards. The web is first slit into several ribbons of a width required by the desired product size. (Angle bars lie across the path of the web at a 45° angle. When the web is wrapped around the bar, it changes direction by 90°. See figure 17-8.) Each ribbon is turned over an angle bar so that all ribbons align with each other ahead of the jaw-folding section. Each angle bar is perforated for low-pressure air so the web never actually contacts the bar. Smearing of the printed image is eliminated due to the web riding on a cushion of air.

The ribbons of paper are collated (assembled) in the desired order and brought down to the cutoff knives and folding jaws in either one or two streams. In this manner, the press simultaneously delivers either one or two sets of signatures of the same size. If two streams are delivered, they can have either equal or different numbers of pages according to the way the ribbons are webbed through the folder.

Ribbon folders are often made with a collect/noncollect feature so that two successive signatures can be collected (put together) after cutoff, then jaw-folded together. In this way, two sixteen-page signatures are inserted together upon delivery. In noncollect mode, four eight-page signatures can result. Such folders are usually designed to produce two sets (two inserted signatures) from one conveyor when in collect mode, or

Figure 17-7. LISA shear-cut slitting wheel with housing, used to slit the web into ribbons. (Courtesy Converter Accessory Corp.)

four sets (of single signatures) from two conveyors when in noncollect mode.

As explained earlier, the collect feature requires a folder cutoff length that equals one-half the cylinder circumference of the printing units. The total page output from the folder depends on the page input—as determined by the number of printed webs and the pages per web. Some folders allow even greater flexibility by enabling one stream to run collect while the other stream runs noncollect.

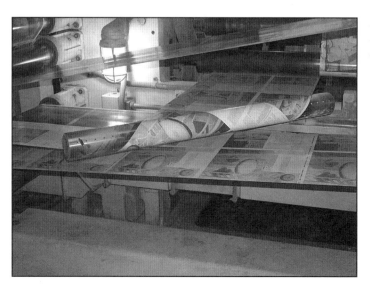

Figure 17-8. A ribbon being turned using an air bar.

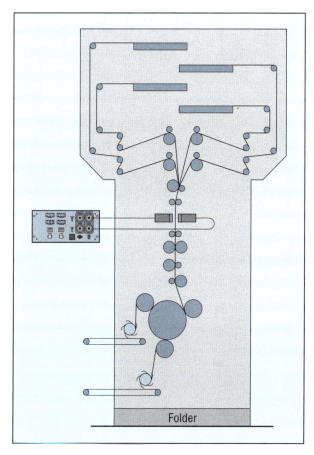

Figure 17-9. *Hurletron ribbon tacking system, which reduces dog-ears in delivered signatures by inducing a controlled electrical charge into the ribbons. This electrical force removes the cushion of air between the ribbons and in effect tacks the ribbons together as they progress through the folder. (Courtesy Hurletron Incorporated)*

Folder

Former Folders

The former folder is generally the simplest of all folders, because the only fold is made on a former. After the web is slit into two-page-wide ribbons, they are led over the former, cut, and delivered.

Virtually all former folders have a least two delivery streams to facilitate delivery without jams or damage to the product. This enables the signature to be slowed down for delivery without the next one catching up and jamming the first signature. Since alternate delivery creates a space equal to the signature length, a maximum slowdown of about 40% is used to allow high-speed, defect-free delivery streams without jamming.

All products coming directly from the former folder are open on three sides, and they are generally used in magazine printing, in which the long-grained product is preferred and flexibility is useful in producing 4-, 8-, 12-, and 16-page signatures to meet regional advertising needs.

A very popular former folder product is the two-up digest for two-up binding lines. The 5½×17-in. (140×432-mm) long-grained signature is often preferred over the same-size short-grained signature produced by double-parallel jaw folders, especially for longer runs.

Double-Former Folders

The double-former folder, also called a form-and-cutoff folder, is found principally in publication and commercial printing plants. It is primarily intended for the high-speed production of a single size of folded product. In this design, the running web is slit in two, with one ribbon guided by angle bars to one former board and the other ribbon to the other. The backbone fold, made by running the paper webs over a former board, is the only fold it produces before the stream is sent to the cutoff knife. Thus, the only result from each fold is a four-page signature with open heads. However, multiple webs can be pulled simultaneously over a single former board, creating an inserting function that increases the number of delivered pages per delivered piece. Although limited to one basic size, the folder delivers up to four streams of signatures simultaneously and in a wide variety of page combinations.

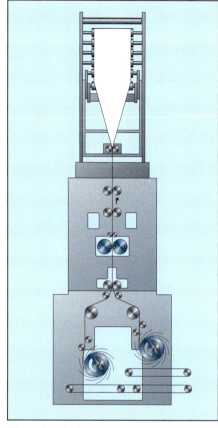

Figure 17-10. *QTI-DFF 2 double-former folder. (Courtesy MAN Roland Druckmaschinen AG)*

Operation principle. The folder has two former boards and four deliveries. The centerlines of the former boards are perpendicular to the press centerline. The formers are slanted with noses pointed downward. A web slitting and angle bar section first slits the one or more printed webs into narrower ribbons of specified equal width. The several ribbons run over turning bars and are brought in line with each other—much in the manner of a ribbon folder—and then pulled over the formers. The ribbons are cut off into individual signatures immediately below the formers by a set of cutoff and delivery cylinders, which transfer them to the delivery conveyors.

Interleaving. The term interleaving refers to the sorting and merging of multiple ribbons to increase folder flexibility. The number of pages in the delivered signatures depends on how the ribbons are webbed over the formers. If all of the ribbons are pulled over one former, two mating deliveries produce signatures containing half the total number of input pages. As an example, two webs totaling thirty-two pages are delivered as two sixteen-page signatures. Four eight-page signatures are delivered if the half-width ribbons are equally divided between the two formers. Sometimes, additional angle bars are used to interleave multicolor and monotone ribbons. This feature provides further flexibility in the placement of color pages within the signatures.

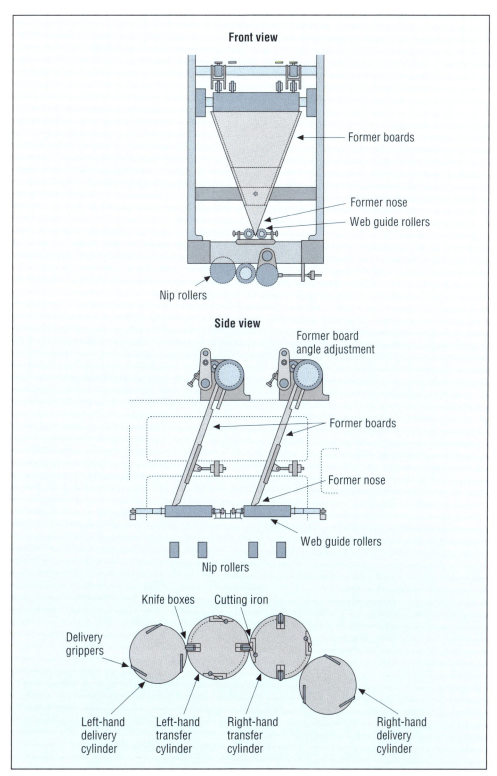

Front view

Former boards

Former nose

Web guide rollers

Nip rollers

Side view

Former board
angle adjustment

Former boards

Former nose

Web guide rollers

Nip rollers

Knife boxes   Cutting iron

Delivery
grippers

Left-hand
delivery
cylinder

Left-hand
transfer
cylinder

Right-hand
transfer
cylinder

Right-hand
delivery
cylinder

Figure 17-11. *A double-former folder. Cutoff sections are alternately delivered to the left and right. This prevents any one station from being overwhelmed with press output. The web-slitting and angle bar section preceding the former is not shown. (Courtesy Heidelberg USA, Inc.)*

Figure 17-12. The double formers and entry rollers of a pinless, high-speed double-former folder for magazine production. The position of the side-by-side formers can be shifted by motors independently of each other. (Courtesy MAN Roland Druckmaschinen AG)

Figure 17-13. The entry section of a double-former folder with self-contained turner bars. (Courtesy MAN Roland Druckmaschinen AG)

Double-Former Products

Unlike a combination folder, there are no options with a double-former folder. The format of the delivered product is determined by page imposition and press webbing in advance of the folder. The following examples show why this is the case.

A six-unit press with an impression length of 22 in. (559 mm) is to run two 34-in. (864-mm) webs. One web is to take four colors and the other, two. Both, of course, are to be perfected. The design of the job, however, requires that full-color illustrations be spread evenly throughout the bound book. Therefore, both webs will be slit in half, and one ribbon from each will run over each former. The two drawings below show the page imposition for each web. Individual page numbers are omitted for simplicity's sake although they are, at this point, fixed.

The impression length of each web carries four four-page signatures. After slitting and mixing, the ribbons cross the former, folding to 8½ in. (216 mm) wide. The top ribbon over the former always forms the outside pages of the signature. Page imposition has required that plate-front signatures from each signature be

delivered together. In this case, signature A must be delivered with signature E, not signature F. Furthermore, the press operator has set up the job to deliver plate-front signatures from both formers to the right and, alternately, plate-back signatures to the left. The results of these considerations can be summarized:

Delivery	Inside Signature	Outside Signature
1	E	A
2	C	G
3	D	H
4	F	B

A general rule applies to planning work for a double former: delivered signatures should contain the maximum possible number of pages to minimize bindery handling. Actually, unless the above job were a combined run, all four ribbons could be run over a single former. Delivery would be of two thirty-two-page signatures: signatures A, E, G, and C (going from outside to inside pages) at delivery 1, and signatures B, F, H, and D at delivery 4. Note that such a change forces a change in page imposition.

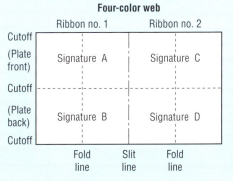

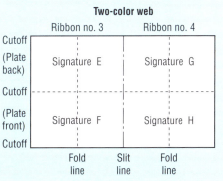

Mechanics of Folders

Folders may be designed to carry the cut signatures through each fold mechanism in one of two ways: with pins or with belts. In the case of pin support, the web is punctured on the lead edge of the fold and transferred through a portion of the folding system gripped by a succession of pin bars. With the belt design (called a pinless folder) a set of belts, one on each side of the cut signature, grips and guides the cut signature through the folding mechanism. Pinless former folders are designed to operate with 16- and 32-page presses. The operational mechanics of each section of the folder are described in the following paragraphs.

Former board mechanics. On former folders, the web first runs over a former board. The former board is triangular in shape, with the apex of the triangle—the former nose—pointing downward. As previously explained, more than one web can be run (one on top of the other) over the former. Directly below the former nose are two web guide rollers that help to control the web passage over the former. Below the web guide rollers are the nipping rollers that, in addition to pulling the web into the folder, press the lengthwise crease into the web.

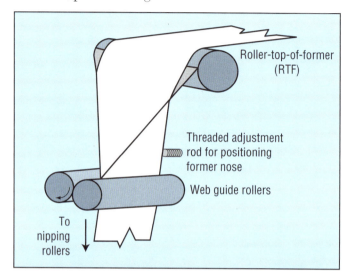

Figure 17-14. The former board and related elements.

Figure 17-15. A closeup of the former board and the guide rollers.

Cutoff cylinder mechanics. From this point on, the basic components of the folder are mounted on rotating cylinders. The press cutoff is located just below the nipping rollers. The cutoff is made by a blade on one cylinder that cuts through the web into a slot lying across the width of a second cylinder. Before the web is cut, sharp pins (called impaling pins) spring up out of the face of the second cylinder and punch through the lead edge of the web just behind the cutoff. These pins pull the lead edge of the web around the cylinder and release just as the jaw fold is made.

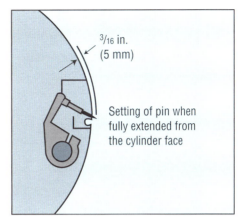

Figure 17-16. Setting of the impaling pins. (Courtesy Heidelberg USA, Inc.)

Jaw fold mechanics. The jaw fold is made across the width of the signature at right angles to the former fold. A tucker blade fixed on the surface of one cylinder swings into a jaw on an opposing cylinder. The paper is between the blade and the jaw. The jaw is mechanically timed to close just as the tucker blade swings out of it. The leading edge of the jaw closes on the paper and creases it. (See the sidebar on the next page.)

Chopper fold mechanics. The chopper folder folds parallel to the former fold. It is often the most difficult part of the folder to operate correctly. Moving tapes carry the paper up to a headstop. An instant after the signature reaches the headstop, the chopper blade descends, forcing the paper between two folding rollers that crease the paper while drawing the signature through. The signature then drops into star (delivery) wheels, which in turn drop it on delivery belts.

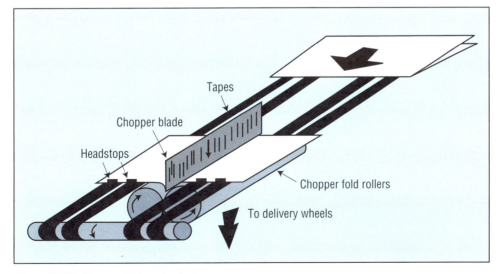

Figure 17-17. Chopper fold mechanism.

How the Jaw Fold Is Made

The movable jaw on this folder is attached to a cam-operated shaft. The jaw opens and closes in the manner indicated by the arrow.

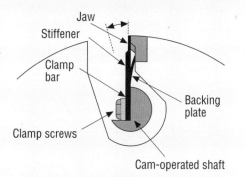

The web is cut off at point A. Note that two cutoffs are being made for each revolution of the folding cylinders. Pins pierce the lead edge of the web (B), maintaining control of the signature. The tucker blade (C) creates a slight protrusion in the web surface.

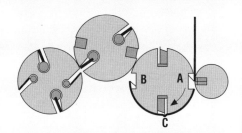

As the cylinder revolves, the jaw on the opposing cylinder closes on the ridge of the signature created by the tucker blade (D). When completely closed, the jaw pinches the fold into the signature. The pins have retracted, releasing the signature from the first cylinder.

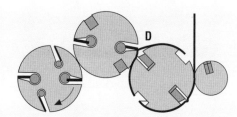

As the cylinders continue turning, the cutoff pins retract, allowing the jaw to pull the signatures from the face of the first cylinder (E). The signature is then carried to point F where the grippers on the third cylinder take control of the signature. The signature is carried around the third cylinder and, on release, dropped onto the delivery tapes at G.

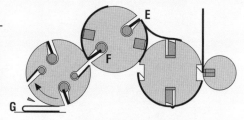

In double-parallel folding, a second set of tucker blades are activated on the second cylinder (H). These will insert the signature into jaws on the third cylinder (J)

Courtesy Goss Graphic Systems, Inc.

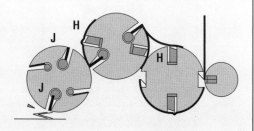

Collect cylinder mechanics. In addition to these components, some folders carry collect cylinders. Assume that a job has been stripped to run with one part of the job on the top half of the plate and the other part on the bottom half. The press cutoff is set to make two cutoffs per revolution of the plate cylinder. The top half of the plate cylinder image is cut off and passed to the collect cylinder, where it is held by the first set of pins. As the lower half of the plate cylinder image is cut off, the second set of collect pins come into position. These pins already hold a top-half cutoff. They accept a bottom-half cutoff, and the two are held together. The cylinder turns, and the third set of pins comes up and takes the new top-half cutoff. Meanwhile, the second set of pins carrying both a top-half and a bottom-half cutoff are releasing them into the jaw fold. A new bottom cutoff is made as the first set of pins comes around, which is still carrying the first top-half cutoff. These accept a bottom-half cutoff and carry both to the jaw folder. In this particular case, the collect cylinder is carrying three sets of pins. Each set of pins is timed to accept a section every time it approaches the cutoff and to release every other time it approaches the jaw fold cylinders. (See the sidebar on the next page.)

All of the folder components are housed on four or five cylinders, with each cylinder carrying out several different operations. For example, the first cylinder often carries the tucker blade for the first jaw fold, as well as the pins and the female cutting iron for cutoff. The next cylinder carries the folding jaws for the first jaw fold and the tucker blade for the second jaw fold. Unlike the other cylinders, collect cylinders serve only the single function of collecting printed sections in proper sequence.

Folder Operation and Maintenance Tips

Though few settings have to be made on a folder, those that are made must be precise. The operating speed of the folder may affect its performance. Sometimes, a setting that works well when the folder is turned over slowly may not work the same when the folder is running at high speed. Accurate and complete records of settings from previous jobs are invaluable aids to efficiently set and operate the folder.

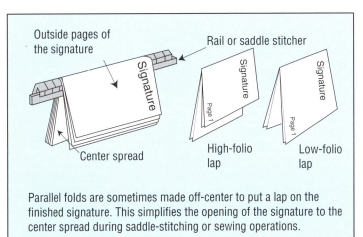

Outside pages of the signature

Rail or saddle stitcher

Signature

Center spread

Signature

Page 1

High-folio lap

Signature

Page 1

Low-folio lap

Parallel folds are sometimes made off-center to put a lap on the finished signature. This simplifies the opening of the signature to the center spread during saddle-stitching or sewing operations.

Figure 17-18. Putting a lap on folded signature to aid in opening the signatures during saddle-stitching or sewing operations. *(Courtesy Heidelberg USA, Inc.)*

How Collect Cylinders Operate

1. The top half of the plate cylinder image is cut off and passes to the collect cylinder, where it is held by the first set of pins. (The circumference of the knife cylinder is one-half the circumference of the plate cylinder, the circumference of the collect cylinder is three times the circumference of the knife cylinder, and the circumference of the jaw cylinder is equal to the circumference of the plate cylinder.)

2. As the lower half of the plate cylinder image is cut off, the second set of collect pins come up; they already hold a top-half cutoff. They accept a bottom-half cutoff, and the top-half and bottom-half cutoffs are held together.

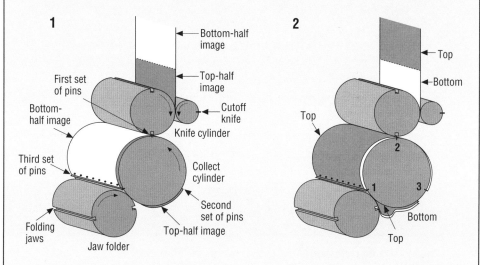

3. The cylinder turns and the third set of pins comes up and takes the new top-half cutoff. While this is happening, the second set of pins carrying both a top-half and bottom-half cutoff are releasing them into the jaw fold.

4. A new bottom-half cutoff is made as the first set of pins come around, which is still carrying the first top-half cutoff. These accept a bottom-half cutoff and carry both to the jaw folder.

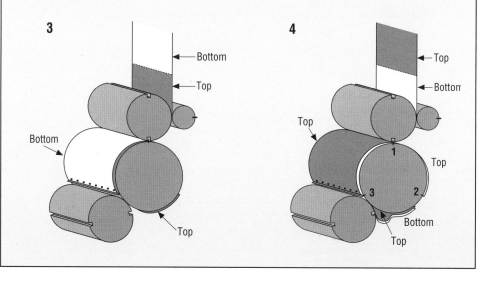

Operating the Folder

1. Identify these parts of the web press folder, and describe their functions:
   - Independent delivery drive
   - Synchronizing mechanism that keeps folder timed to the press
   - Engaging/disengaging mechanism
2. Adjust cutoff mechanism to deliver the sheeted or folded product as desired.
3. Diagram or trace the maximum numbers of pages per product obtainable from a web press, according to number of webs, size of plate, number of forms per plate, number of former plates, and collect action of folding and cutting cylinders.
4. Identify the lubrication points on the folding cylinder.
5. Adjust the spring finger attachment to the thickness of the web, to properly control the inside ribbon.
6. Adjust the holder bar.
7. Adjust the folding and cutting cylinders to the job to be produced.
8. Set the angle of the former board angle.
9. Set the web guide rollers.
10. Set the nipping rollers.
11. Adjust the pins that grip the lead edge of the section after cutoff to the manufacturer's instructions.
12. Set the tucker blades to allow for the thickness of paper.

Stroboscope. One device that aids visual inspection is called a stroboscope or strobe light. This device emits light in extremely short pulses. When synchronized to a repetitive operation, the action seems to stop under the light. Any element in the operation—a folding jaw, for example—can be examined as if in a frozen state under the strobe light while the press is running.

Figure 17-19. Press operator inspecting folded signature leaving the folder on a CROMOMAN web press. (Courtesy MAN Roland Druckmaschinen AG)

Former board settings. The former board angle must be set within a very close tolerance. If set too far forward, the nose may smudge or tear the web. If set too far back, uniform tension is lost and the web wrinkles along the fold. Once the former nose is correctly set, it should not be necessary to change its position.

One method used to determine the correct angle of the former board is to cut or slit the web in half at the top of the former with a cutting wheel. Watch as the split web runs down over the nose of the former. If the two sections begin to overlap one another, the nose of the former is too low. If a gap appears between the two sections, the nose is too high. The webs should run perfectly down the nose of the former side by side when the former is in the correct position. Once the angle of the former board is set, it is usually not necessary to change it except in special circumstances. However, it should be checked periodically for accuracy.

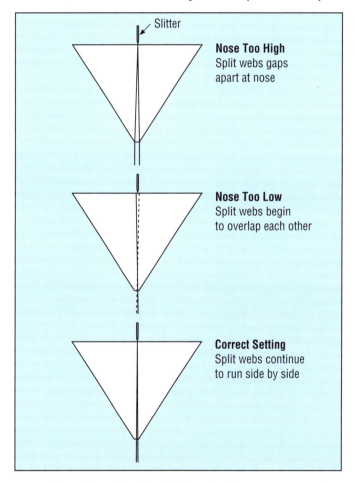

Figure 17-20. Using a split web to determine the correct angle of the former board.

The web guide rollers below the former nose can be set at both ends, close or far apart from each other. The position of the former may create tension differences across the web. The web center may run tight over the former nose while the web edges are running slack, or vice versa. By adjusting the guide rollers, the press operator can equalize tension. The final effect is that all parts of the web enter the nipping rollers under uniform tautness.

Nip roller settings. The nipping rollers immediately below the web guide rollers can also be set closer to or farther away from one another. The nipping rollers on some designs run slightly faster than press speed producing a slippage at this point, which makes pressure settings critical. Setting these rollers properly can be difficult, the critical factor being the caliper and grade of paper. Other designs use cutoff systems that employ no-slip nips. With these systems the nip rollers run at the correct paper speed automatically, adapting to changes in either the caliper or the surface quality of the paper.

Pin and tucker blade settings. The pins that grip the lead edge of the section after cutoff and the jaw fold tucker blades are adjustable on some folders. If they are, their normal setting should be according to the manufacturer's instructions. The pin settings are particularly critical. If they do not spring out far enough, the lead edge will not be gripped sufficiently. If set too high, they may not release in time. Tucker blades can be set on some folders to allow for different thicknesses of paper. On other folders, the tucker blades reciprocate; that is, they retract into the cylinder and come out again as the cylinders turn.

Control element settings. During the various folding operations, several devices including brushes, springs, and rollers control the signature. A tight setting on these elements restricts smooth paper passage; a loose setting defeats their purpose and

Figure 17-21. Papers rolling off delivery of Goss Universal 70 newspaper press. (Courtesy Goss Graphic Systems, Inc.)

may allow the signature to slip. The settings of these components should be uniform, so that no unbalanced forces are applied to the signature as it passes from point to point in the folder.

Paper weight. Folder settings are primarily determined according to basis weight. A 100-lb. coated book stock usually passes through a tabloid fold without much difficulty. Some folder manufacturers suggest 70-lb. coated book stock as the minimum weight for the chopper, even though this fold is made in the direction of the paper grain. A double-parallel fold (second jaw fold) often cracks stocks (along the fold) with a basis weight in excess of 70 lb. Sixty-pound coated book stock should be the maximum when running two webs interleaving. Determine the folder's capability through experience if specific guidelines are not previously provided by the folder manufacturer.

Paper grain. Paper grain is an important factor in folding. Fibers in a web lie in the direction of web travel. Because of the grain direction of the paper, cleaner folds can be made along the length of the web (with the grain) than across the grain. Former folds and chopper folds are made with the grain, while jaw folds are made across the grain.

Running multiple webs on the former board. When multiple webs are being run over the former, several special considerations apply. When a full web and half-width web are run over the same former board, the narrower web should always be on the bottom. This gives better control of the half-web. Whenever more than one web is run, the top web or webs should be made successively looser. Too much tension on the top web wrinkles the bottom web.

Maintenance. Proper maintenance contributes to efficient long-term folder operation. The maintenance schedule should include regular, thorough lubrication of all points suggested by the manufacturer. Expect some wear and tear on moving parts in the folder. The cutoff knives become dull, producing ragged cutoffs. The easiest way to determine their condition is to inspect the last job run. Knives should be mounted according to the manufacturer's instructions. The last nips prior to cutting should be inspected for wear to determine a loss in diameter. Any loss in diameter will contribute to web tension problems.

Sheeters

Sheeters deliver a wide range of fully trimmed sheets or folded products. They significantly reduce bindery costs through operating efficiency at speeds over 45,000 sheets/hr. Sheeters have additional capabilities built into them as well. Gap-cutting is the process of trimming out the nonimage areas between cutoffs. This is achieved by adjustable knife assemblies that cut a 0.125–2.00-in. (3.2–50.8-mm) gap across the web, resulting in sheets. The cutter head is set up using pinned knife holders that

Operating the Sheeter

1. Identify the parts of the sheeter and describe their functions.
2. Identify the lubrication points on the sheeter.
3. Set the sheeter guide, or the positive pull rollers for the size sheet to be delivered.
4. Set the variable-speed slow-down tapes, as related to the speed of the cutting cylinder, to the surface of the web.
5. Identify the parts of the cutting cylinder and the adjustment points.
6. Trace the cutting cylinder drive, relating the cylinder timing with the plate cylinders.
7. Set the backlash gear on the cutting cylinder.
8. Set the blower for various stock thicknesses.

drop into place. Plow-folding stations built into the sheeter further increase product variety. Sheeters are capable of producing fully trimmed sheets, half sheets, four- and eight-page products, and multipage, gatefolded products.

Figure 17-22. INNOTECH high-speed sheeter. (Courtesy INNOTECH Graphic Equipment Corporation)

Figure 17-23. INNOTECH gap-cutting sheeter. This device trims sheets to size, avoiding the need to take a guillotine cut to remove nonprint area or color bars, or to achieve a different length than the press repeat. (Courtesy INNOTECH Graphic Equipment Corporation)

Stackers and Bundlers

The simplest finishing scenario to deal with is for the folder to deliver the same product format all the time. Printers that produce that same product size and format often utilize two former folders with four delivery streams, which delivers into a drum-type stacker, which builds a pile from the bottom. The signatures are usually brick-piled to make a solid pallet load, which is easy to handle and unload.

Books and directories, however, because of their many pages are best handled by bundling. Some folders deliver the signatures on end, building a stack, but more commonly they are fed into a drum stacker, which presses them flat and stacks them on end for manual or semiautomatic strapping using end boards.

Another option is the compensating stacker, which counts groups of signatures and has a rotating table to alternate the position of the fold edge to compensate the pile. Compensating stackers are used widely in binderies and on newspaper and insert products, and some applications are developing in the magazine field.

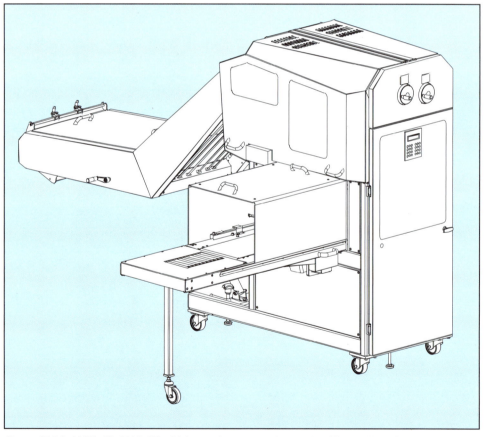

Figure 17-24. AMG's KL 2000-GT, a high-speed compensating stacker. The KL 2000-GT also has a laser counter, lateral jogger at the first gathering stage, and pusher for left or right product delivery. (Courtesy Systems Technology Inc.)

18 In-Line Finishing

In-line finishing equipment works on the running web as it exits the press. A typical in-line finishing package on a commercial web press would include a remoistenable glue applicator, a perforator, plow folders, and a variable cutoff unit.

There are many variations of in-line system configurations, which may be set up in U, L, or T shapes, or in a straight line depending on space limitations in the pressroom. This chapter examines the major characteristics of each broad class of in-line finishing equipment.

Gluers

There are several in-line gluing options available, which have increased product variety and helped to enhance advertisement and promotional capabilities. The main types of gluers are discussed in this section.

Paster wheels. Some folders paste signatures together before delivery. Paster wheels apply paste or glue to the paper. They are driven by the motion of the paper and lay down a line of paste as the wheel turns.

Extrusion gluing. Hypodermic needles may also used for pasting in a process called extrusion gluing, so called because the glue is extruded (forced through) the needle onto a surface. Usually, a 0.03-in. (0.76-mm) head is employed, but the heads can be changed to lay down more or less paste as the job may require. By the use of intermittent feeds, skip-pasting is possible. This allows the glue to be laid down on a predetermined portion of the piece.

Remoistenable pattern gluers. The pattern gluer is eminently suitable for complicated glue patterns, which a paster wheel or an extruded stream of glue in the web direction cannot produce. These types of gluers are used in the production of direct-mail pieces, return-mail envelopes, free-standing brochures, magazine insert cards, and newspaper inserts. Remoistenable pattern gluers are used in the manufacture of these products, which require a dried glue surface pattern on each piece. In addition to glue, these devices can apply a scratch-off or wash-off ink or a liquid fragrance to the printed product. The remoistenable glue is usually applied prior to the dryer so that it can be set. When subsequently remoistened by the end user, this band of glue can seal a return envelope. A pattern gluer may also be used after the chill rolls as a wet gluer.

Pattern gluers stand alone as a unit in-line with the printing units of the press. The pattern gluer unit is composed of the following parts:

- Glue pan to hold the liquid glue.
- Precision-ground rubber feed roller that sits in the glue pan and meters the glue against a transfer roller.
- Transfer roller to transfer the glue film from the pan roller to printing pads on the glue pad cylinder. This may be a water-cooled transfer cylinder in some designs.

Figure 18-1. The WPM RemoistGluer. (Courtesy Western Printing Machinery Co.)

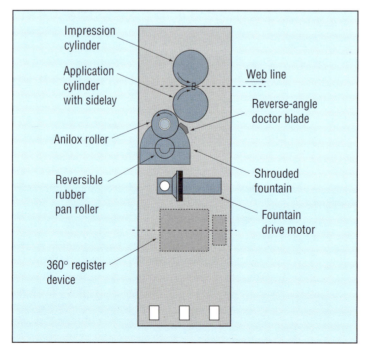

Figure 18-2. The WPM RemoistGluer. (Courtesy Western Printing Machinery Co.)

- Glue pad cylinder that holds rubber pads in the pattern and position for the desired glue pattern.

- Impression cylinder to hold the web to the glue pad cylinder under pressure.

On-the-run independent adjustment of the feed roller helps to provide uniform coverage across the web. A variable-speed electric motor turns the feed roller independently to prevent drying while the press is idle.

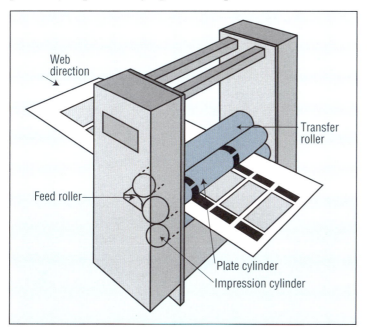

Figure 18-3. Remoisten-able pattern gluer. (Courtesy SPEC)

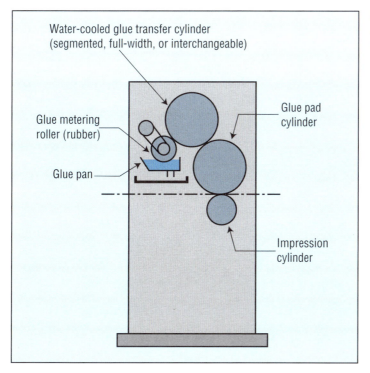

Figure 18-4. Pattern glue unit. (Courtesy Vits-America, Inc.)

The glue pad cylinder is scored for positioning pads. The cylinder is undercut 0.125 in. (3.2 mm) when using sticky-back rubber pads or 0.14 in. (3.6 mm) when using standard rubber plates. The glue pad cylinder can be adjusted laterally or circumferentially for register. The impression cylinder of full press size supports the web while glue is applied, with on-the-run adjustments for printing pressure.

Segmented gluers. Various widths of glue pans and transfer rollers are available to apply glue to only a small portion of the web. Segmented gluers feature the same basic design as the full-web-width remoistenable pattern gluers, with the addition of segmenting capability. The advantage of the segmented gluer is that it more efficiently applies remoistenable glue only where needed. Thus, these gluers reduce waste caused by evaporation, with the added benefit of cutting down on makeready time. An additional benefit is increased flexibility. For example, up to six different segments are possible with the standard 38-in. (965-mm) sixteen-page web offset press. This allows for an extensive range of combinations of fragrances, glues, and removable materials to be applied during a single pressrun.

Envelope pattern gluers. These gluers are designed specifically to seal the edges of envelopes; they can also seal the flap to produce a complete mailing piece. The envelope pattern gluer is normally mounted after a perforator (which punches a pattern of tiny holes in the paper). In addition, combination gluer/perforator or gluer/plow folder units are available.

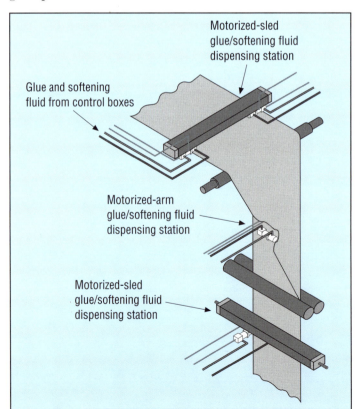

Motorized-sled glue/softening fluid dispensing station

Glue and softening fluid from control boxes

Motorized-arm glue/softening fluid dispensing station

Motorized-sled glue/softening fluid dispensing station

Figure 18-5. RoBond is a microprocessor-controlled application system for precise fold softening and fine-line gluing on web presses and off-line finishing equipment. Used for the production of books and catalogs, RoBond glues signatures together and eliminates saddle stitching. (Courtesy Valco Cincinnati, Inc.)

Spot gluers. Spot gluers apply glue spots in a pattern across the web or in a continuous line around the cylinder. They are designed to operate at high speeds. Spot gluers are used in the production of magazine inserts and gatefolded products used in magazines, catalogs, brochures, and other publications. In the case of a gatefold, the glue spots are needed to hold the short folded flap shut so that it will not complicate bindery equipment operations.

Backbone gluer. The backbone gluer is specifically designed to produce spine-glued books and related products in-line directly from the combination folder of an eight-page, grain-short press. This requires the use of a standard gluer and the combination folder. With these systems, the web is diverted to the gluer with turn bars, the web is glued at the signature backbone, and then the web is guided back to the folder.

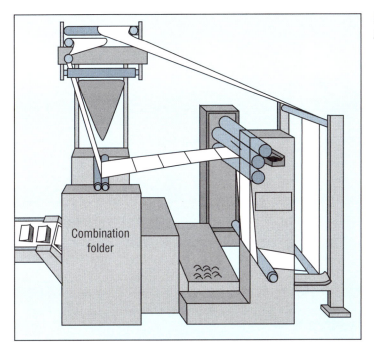

Figure 18-6. Backbone gluer. (Courtesy SPEC)

Auxiliary Folders

Auxiliary folding devices are intended to increase the variety of products that come off the in-line print manufacturing system by enhancing the capability of the existing folder. Using a vertical web path for auxiliary equipment allows easy access for maintenance and conserves space. In a complex in-line finishing system with several pieces of folding, cutting, perforating, and gluing equipment, space considerations may be of paramount importance.

Plow folders. Plow folders work in much the same manner as former boards in that they fold the web continuously in the direction of web travel. They vary in that they can fold anywhere across the web, down to a 1 in. wide (25.4 mm) gatefold. A gatefold is a fold that is only a portion of the paper width, as opposed to folding in half. This folder is an adaptation of a device that is used in the production of kraft-paper bags.

For in-line finishing applications, a plow tower consists of up to four plow heads and a driven paper-pulling nip to fold flaps onto gates in the direction of web travel. Since the plow head is movable and reversible, it offers more flexibility than a former board in location and number of folds. Fold sizes vary from as little as 1 in. (25.4 mm) to as wide as half the web. Installing a number of plow stations in-line permits multiple folds to be made to the web or ribbons. Up to six plow stations have been incorporated into finishing lines to execute complex multiple and sequential folds.

Prefolders. A basic prefolder consists of angle-bar and plow-folding sections that are positioned before the former board. The angle bar can shift the web laterally from the centerline of the press, usually up to 8 in. (200 mm) in either direction. It accommodates odd-numbered page layouts across the web and centers the repositioned web in the appropriate position for the former board after the plow fold is made. The plow-folding section produces either one or two folds in the direction of web travel. After that, the web enters the combination folder.

Insert folders. Insert folders are designed to convert 8.5×11-in. (216×280-mm) signatures into either a delta fold (approximately 3.5×8.5 in., 92×216 mm) or a digest-fold (5.5×8.5 in., 140×216 mm) either for inserting into an envelope or for labeling.

A typical insert folder receives a stream of overlapping signatures that have been plow-folded and converted to the 8.5×11-in. (216×280-mm) format. A bump

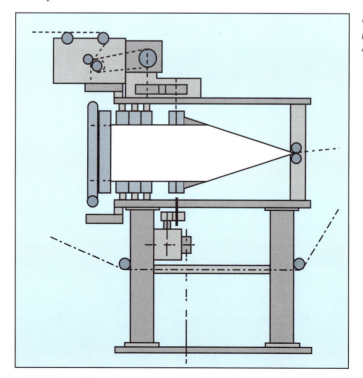

Figure 18-7. Double-former prefolder. (Courtesy Vits-America, Inc.)

Figure 18-8. *INNOTECH prefolder. (Courtesy INNOTECH Graphic Equipment Corporation)*

turn aligns the signatures closed-end-first before making the required folds. Subsequently, the folded products are delivered to a sheeter, after which the printed material is ready to be inserted or labeled.

Perforators

Perforations are patterns of tiny punched holes in the paper. This can function to facilitate the tearing of portions of the printed product, as with coupons. Another purpose of perforations is to enhance the proper folding of signatures in the folder of the web press.

The simplest perforating mechanism is a perforation wheel, mounted against the roller-top-of-former. This roller perforates parallel to the former fold. Perforating across the web width can be done using a set of perforating cylinders mounted below the folder nip rollers. One cylinder carries the perforating teeth and the other a female die.

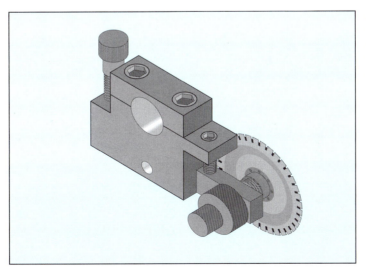

Figure 18-9. *Perforating wheel in an adjustable perforating and slitting wheel holder. (Courtesy RTE)*

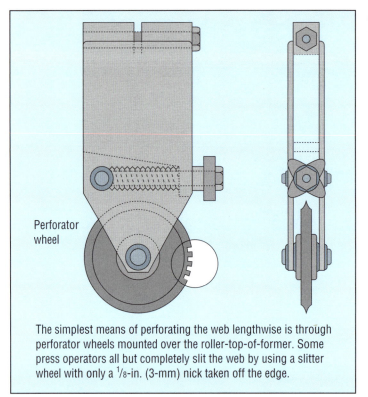

Figure 18-10. Circum-
ferential perforators.
(Courtesy Heidelberg
USA, Inc.)

The simplest means of perforating the web lengthwise is through
perforator wheels mounted over the roller-top-of-former. Some
press operators all but completely slit the web by using a slitter
wheel with only a $^1/_8$-in. (3-mm) nick taken off the edge.

Gusset wrinkles. A common folding problem that can be corrected by perforating is
the gusset wrinkle. A gusset wrinkle forms on the inside pages of a closed-head signature at the corner where the backbone fold meets the head fold. Gusseting occurs
because the outside pages and the inside pages of the signature are subjected to opposite stresses (figure 18-11). When several thicknesses of paper are folded together, the
outside pages tend to be pulled outward by the bend. The inside sheets are not
stretched as much. These differences in stretching would not be a problem if the
pages were not connected at the top by the head fold (which will later be trimmed
away). However, this connection holds the variably stretched paper fast, resulting in
an unwanted wrinkle.

Simple hole perforating along the head of the signature may relieve the strain
to the point where the gusset wrinkle is no longer a problem. In cases where many

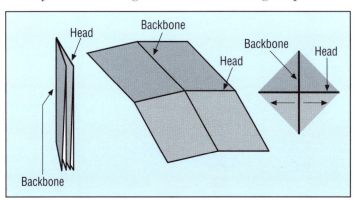

Figure 18-11. The ten-
dency of the inside
sheets to be pushed out
at the lip of the signature.

pages are being folded or the paper is heavier, diagonal-line perforations may be the only way to do away with the gusset wrinkle. Because of the line-segment pattern, this perforation is diecut by a hardened steel cutting head and a female die mounted in two cylinders across the backbone of the signature. The slant of the perforations readily allows movement of the inside sheets away from the backbone; therefore, the stresses that create the gusset wrinkle are eliminated.

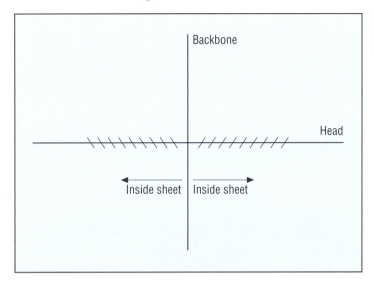

Figure 18-12. *Diagonal perforations to relieve gusset wrinkles.*

Pattern perforating units. The perforation mechanisms so far discussed will result in straight-line vertical or horizontal perforation patterns. Many products are designed to have perforations in geometric patterns, like a rectangle or square. For this purpose, a steel perforating rule can be formed and glued to one of the blanket cylinders. As the paper runs over the blanket, the pressure from the opposing blanket (in the case of a blanket-to-blanket press) or impression cylinder (in the case of an in-line press) forces the perforation rule through the paper.

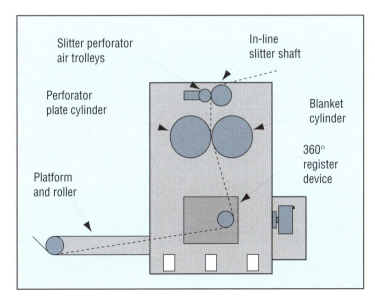

Figure 18-13. *The two-cylinder WPM Pattern Perforator. (Courtesy Western Printing Machinery Co.)*

Because the unit used for perforation cannot also be used to print, most printers who perforate often have separate perforating units that are mounted in the folder. These units have a perforating cylinder and an impression cylinder and are readily adaptable to perform geometric perforations.

Pattern perforating units punch a pattern of tiny holes into the web to allow the end user to tear out sections of the product. Sticky-back perforating strips are mounted on aluminum plates on a plate cylinder, perforating the web against a blanket or backup cylinder. Slitter wheels and numbering units are often added to the perforator to produce coupon products. Pattern perforating is usually carried out after the chill rolls. Theoretically, a pattern perforator could also be installed before the first printing unit; however, piling may occur on subsequent blankets due to paper dust caused by the punching.

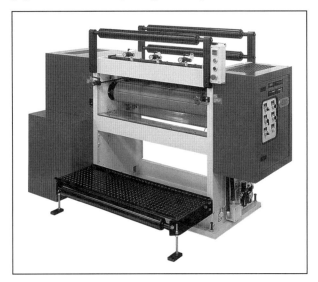

Figure 18-14. The WPM Pattern Perforator, a three-cylinder perforating unit. This perforator can edge-trim, slit the web into ribbons, continuously perforate the web in the running direction, or perforate the web in any pattern around or across the web. (Courtesy Western Printing Machinery Co.)

Imprinters and Numbering Units

Imprinters are often found on web offset presses running jobs where a small amount of copy is going to vary during the run. An example of imprinting would be a catalog prepared for ten different retailers, each set of catalogs requiring a change only in name and address. To accomplish this during the run, names and addresses can be printed on a portion of the catalogs by an imprinter. The imprinter usually consists of a flexographic (rubber or polymer) relief plate, an impression cylinder, and a simple inking system (figure 18-15). Imprinters often carry two plate cylinders, so that one form is running while the other is being made ready, allowing for a rapid changeover.

Numbering can be carried out by the use of indexing heads with raised type or electronically, by inkjet printing. In addition to perforating, the unit can be set up to score the stock, improving the accuracy of folding. Scoring units are comprised of a male and female roller that run together as the web runs through them. The ridge of the male scoring roller disturbs the fibers at the paper surface so that the fold will fall squarely on the score.

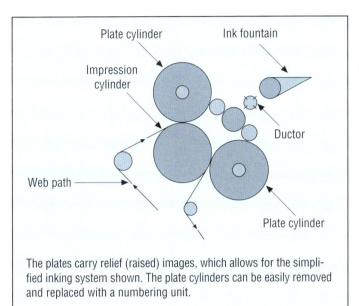

Figure 18-15. *An imprinting unit that carries two plate cylinders on which rubber plates are mounted.*

Plate cylinder

Ink fountain

Impression cylinder

Ductor

Web path

Plate cylinder

The plates carry relief (raised) images, which allows for the simplified inking system shown. The plate cylinders can be easily removed and replaced with a numbering unit.

Variable-Speed Rotary Cutters

Rotary cutters are an essential element in producing the final product off an in-line finishing system. They are used for producing bleed and/or nonbleed products. Bleed products are designed so that ink runs off the edge of the trim. These in-line cutters deliver streams of folded products cut in increments of one, one-half, one-third, or one-fourth of the press repeat, or printing circumference. The cutter can also remove trim scrap between products. Paper waste control is an important consideration in this cutting operation. Some cutters are capable of producing a contoured or diecut style of cutoff (a cut formed to an odd shape) in addition to a single-knife or double-knife straight-bleed cut. This creates paper waste that must be collected and moved from the press at regular intervals.

A typical rotary cutter includes an infeed section that can perform a minifold if needed, an adjustable idler roller to receive folded ribbons, a driven roll with nips, and a driven shear shifter module with male/female cutting heads. The cutting couple consists of a hardened anvil cylinder, a knife cylinder, full-web-width knife bars, and control bars. The variable-speed delivery section has a full-width continuous delivery belt. Options include miniplow folders, driven nip rollers (for high-bulk products), and two-stream capability.

Section IX
Makeready

19 Makeready Procedures

Makeready is the series of operations required to change the press over from one job to the next. Makeready may involve a number of operations including (1) ink system washup, (2) plate bending and changing, (3) blanket changing, (4) roll changing, (5) folder readjustment, (6) presetting the ink fountains, (7) webbing and starting the press, (8) adjusting register, and (9) achieving color tolerances. Whatever the operations required, makeready is considered to be complete when salable signatures are coming off the delivery tapes. A printer can save thousands of dollars per year by increasing efficiency during makeready. This chapter is organized to provide general information on improving the quality and efficiency of makeready for web offset presses.

Safety Precautions for Makeready

- Do not operate equipment unless authorized.
- Check to see that all guards and shields are in place before operating equipment.
- Never release a safe button that someone else has set.
- Do not start a machine that has stopped without an apparent reason.
- Check for persons, tools, or equipment between and around the press before beginning operations.
- Remove all used plates, tools, and equipment from the press area and alert your coworkers before starting the press.
- Wear hearing protection devices when working in areas with high noise levels.
- Do not permit people with jewelry, loose clothing, or long hair near press operations.
- Do not lean or rest hands on the press.
- Do not carry tools in pockets, thus avoiding the possibility of dropping them into the press.
- When making press adjustments, use only recommended tools that are kept in good working condition.
- Keep clear of nips, slitters, and moving parts when operating the press.
- Never reach into the press to make adjustments while it is running.
- Do not attempt to remove hickeys from plates or blankets, or lint or dirt from rollers while the press is in motion.
- Do not wipe down cylinders, plates, rollers, or blankets while the press is in motion.
- To perform any cleaning operation on the press, use rags folded into a pad with no loose, dangling edges.
- Check plates periodically for any loose areas or cracks.
- Follow instructions carefully when mixing or handling chemicals in the pressroom.
- Keep a complete and accessible file of all service manuals, instruction manuals, parts lists, and lubrication manuals or charts for each piece of equipment in the pressroom.

The Press Crew

The press crew's exact titles and duties during makeready differ from plant to plant. A small in-line forms press requires only a single operator, while a full-size, five-unit blanket-to-blanket press may require a press crew of four. A typical press crew might consist of the following job titles and responsibilities:

- *Head press operator.* Responsible for directing the press crew and for communicating all job requirements. He or she assures quality of the printing including register and color and solves any job- or press-related problems that might arise.
- *Assistant press operator.* Responsible for press maintenance, plate changing, blanket changing, blanket washing, and directing the roll tender and materials handler when appropriate.
- *Roll tender.* Responsible for checking and preparing new rolls, mounting and splicing new rolls, and removing cores and paper waste. Also, helps with plate changes and blanket washing.
- *Materials handler.* Responsible for bringing all required materials to the press before the start of makeready, including plates and paper rolls. May also be responsible for helping with plate changes and blanket cleaning.

Importance of Efficient Makeready

Web offset presses, particularly large multi-unit types, are very expensive machines. The hourly rate charged to customers for their operation may be several hundred dollars. The total cost of makeready alone for a sixteen-page press may be in excess of $500,000 per year. For this reason, cutting even small amounts of time from makeready can save the printing company thousands of dollars each year.

Generally speaking, an ideal makeready for a four-color, high-quality job should run no more than 10–12 min. per plate and should generate no more than 1,000 waste signatures. In some plants, inefficient makeready operations may result in plate changes of 30–45 min. with waste in the area of 8,000–10,000 signatures or more. It is difficult to pinpoint proper makeready speed. Most dryers can be activated at about 8,000 or 9,000 impressions per hour (IPH), and in many plants the press is held at that speed until salable signatures are coming out of the delivery.

The following example illustrates the cost of materials and time for one 35-in. (889-mm) web with a 23-in. (584-mm) cutoff. Running a single web, this press consumes approximately 2 lb. of ink per 1,000 signatures. Press time is charged at $425/hr. Makeready averages 2.51 hr. and generates 4,178 waste signatures. Net press hours per year total 5,318. This represents 277 working days, running 24 hr./day. Twenty percent downtime is figured for maintenance, repairs, and delays.

The press in this example prints 79,911,000 impressions per year, averaging 21,800 impressions per hour (iph)—20,000 signatures are good. The average pressrun takes 4.95 hr., thus producing 99,000 good signatures. Makeready and printing take 7.46 hr. Available press time allows 713 jobs to be printed.

Paper cost (e.g., $3,056,000) can be determined by dividing total impressions per year (79,911,000) by 1,000; then, multiply by the weight of 1,000 signatures (e.g., 85 lb.); multiply this value by the cost of the paper being run (e.g., $0.45/lb.). Add 3% for wrapper, core, and strip.

Ink costs $1,200,000/year if every 1,000 signatures consume 2 lb. of ink that costs $7.50/lb.

Working with the previously prescribed parameters, the cost of 1% of waste totals $58,000. Total cost of makeready for this press totals $920,000/year. Therefore, decreasing 713 makereadies by one minute each saves $6,100/press/year. Furthermore, reducing 713 makereadies by 1% saves an additional $9,200/press/year.

Guidelines for Reducing Makeready Time

By following a number of general guidelines, the time required for makeready may be significantly reduced.

Have all materials ready. Good preparation for makeready means having all necessary materials at the press at the time they are needed. If the plates are to be bent at the press, they should be brought to the press and bent while the preceding job is running. Plates should be bent and packed, blankets laid out and ready to be mounted on the press, and packing cut to the appropriate size. If there is to be a color change, ink and washup materials should be waiting at the press.

Procedures for Preparing the Web Offset Press

1. Adjust the press infeed, register system, and press delivery system.
2. Inspect each roll before press mounting for defects and out-of-round conditions.
3. Mount roll stock.
4. Prepare the ink and inking system.
5. Clean the dampening system rollers.
6. Clean plate and blanket cylinder bodies in preparation for packing.
7. Inspect and bend the plates prior to mounting.
8. Mount the plates on each unit, assure proper packing, and change the blankets if needed.
9. Prepare the dryer and chill rolls.
10. Prepare the folder.
11. Start up the printing units and roughly set the ink and water levels.
12. Put the last unit on impression and check the folder and tension setting.
13. Ink the plates after the press and dryer are started.
14. Examine the makeready signatures for position and/or register of image, quality of print, and color of print.
15. Run subsequent makeready signatures and perform any necessary adjustments to plate position and the dampening and inking systems.
16. Choose a good signature to serve as an OK signature and have it approved by the customer or management before running the actual job.
17. Review the Before Webbing the Press and the Before Running the Press checklists to ensure consistent and accurate makeready.

Makeready Checklists

Use the following checklists to ensure consistent and accurate makeready before webbing and running the press.

Before Webbing the Press
- ☐ Engage the clutch on each printing unit to be used.
- ☐ Declutch all printing units not to be used.
- ☐ Declutch the folder.
- ☐ Set the folder angle bars and compensators as required.
- ☐ Wash up the ink trains, leaving the rollers dry.
- ☐ Remove and clean the washup devices.
- ☐ Wash up the ink fountains and fountain roller.
- ☐ Place dividers in the ink fountains where required.
- ☐ Preset the ink ratchet stroke to 75%.
- ☐ Remove the old plates and clean the plate cylinders.
- ☐ Install the new plates.
- ☐ Recenter all plate cylinders, circumferentially, laterally, and cocked.
- ☐ Preset the ink keys.
- ☐ Ink up the ink train rollers and stripe form rollers to plate—reset if necessary.
- ☐ Set up the folder as required.
- ☐ Turn on the dampener supply system automatic mixer and circulators. Check the solution pH.
- ☐ Set the brush dampener water stops to the needs of each plate.
- ☐ Make sure all dampener night latches are in the on-pressure position.
- ☐ Check the blankets. Retorque or replace them as necessary.
- ☐ Check the cooling water supply to the oscillators and chill rolls.
- ☐ Check the lubricant levels in the circulating oil, oil mist, and automatic greasing systems.

Before Running the Press
- ☐ Web up all operational printing units.
- ☐ Set up the roll stand/paster and infeed, making preliminary tension settings.
- ☐ Check that power is on at the control panel.
- ☐ Make sure that dampening solution is flowing through all the dampener pans.
- ☐ Check that the dampener controls on the unit control panels and/or console are preset.
- ☐ Check that the folder is on the timing mark, then engage the folder clutch.
- ☐ Lead the webs into the folder.
- ☐ Check the box tilt web guide settings.
- ☐ Check that the trolleys are set evenly across the RTF (roller-top-of-former).
- ☐ Set the folder nip rollers.
- ☐ Set the web break detectors.
- ☐ Put the air blowers on.
- ☐ Set the folder conveyors as required.
- ☐ Make sure all nip guards and safety shields are in place.
- ☐ Zero the product counters.

Stress efficient teamwork. Crew members that work as a team enhance makeready efficiency. Each member of the crew should be fully occupied during makeready and should know what his or her responsibilities are. For example, two people usually work as a team to change plates and blankets. For efficient makeready a third person should hand the plates, blankets, and packing to the two people working at the units. This allows the people changing plates and blankets to remain at the units, saving time and unnecessary movement.

Provide training. Crew members should be well qualified to perform their assigned duties. For example, the head press operator should be responsible for the makeready on the most difficult side of the web and the second press operator for the other side of the web. Each is responsible for the ink lay, ink/water balance, and register on their side of the web. Thoroughly training each member of the press crew on his or her assigned duties will assure efficient makeready. All tools and procedures should be explained and practiced with before requiring performance.

Keep notes. Keep notes on press settings and other information that might be of value on unusual jobs in the future. Notebooks can provide information that may eliminate an hour or more of trial-and-error experimenting. For multiple-shift plants, the notebook should be kept by the press to benefit each press crew.

Maintain color densitometrically. For many years, the typical printing plant has performed final color correcting on the press by increasing or decreasing ink film thicknesses to match the color proof. To more efficiently match and maintain color on press, the press operator should "run to the numbers" and be mainly concerned with obtaining a predetermined density, trap value, and dot gain value. If the color separations have been adjusted for the press, ink, and paper set to be run on the job, the printed color images will closely match the proof. Ideally, color correcting should take place before the job has reached the pressroom. Customers should approve a proof, and everybody involved in color prepress must ensure that the proofs are matchable on the press.

Use a register system. Several companies offer register systems that ensure that the image is in register with the bends in the plate. All of these systems are capable of ensuring that circumferential register is held very closely on the press, requiring little adjustment. These systems sometimes employ pins attached to the plate cylinder lock-up. By punching slots in the plate corresponding with the pins, side-to-side register is maintained on the press cylinders. Such a system can ensure that the printed images will register to within a few thousandths of an inch every time.

Infeed Makeready

There are a number of important issues relating to infeed makeready. Rolls must be properly inspected and grouped, the roll must be mounted and prepared for the splice properly, and tension controls must be set properly. This section provides information on efficiently organizing and carrying out infeed makeready operations.

Inspect the paper roll. Damaged rolls should not be run on press. The roll tender should closely inspect each roll before mounting it on the press. The following list includes the key factors to examine when inspecting a roll.

- *Roll roundness.* The roll should be round. Out-of-round rolls have a higher incidence of web breaks. An out-of-round roll will wobble on press because it is unbalanced.

Figure 19-1. *A roll with end wrapper removed. End wrappers can be removed some time before the roll is mounted. The tightly wound edges are relatively impervious to atmospheric humidity.*

- ***Condition of wrappers.*** Check for nicked sides and torn wrappers. A torn wrapper, even without damage to the roll underneath, indicates potential trouble. Most wrappers are moisture-proof to preserve the natural moisture content of the web (about 5%). Usually, the first part of the roll to dry out in storage is the surface near the tear. The result is uneven running tension. The best practice is to leave the body wrapper on until just before the roll is pre-pared for splicing.

- ***Condition of roll core.*** The roll tender should check for defective roll cores. A roll with a deformed core wobbles while running, causing problems with all but the most sophisticated infeeds.

Minimizing paper waste. Roll tenders should slab off as little paper from the outside of the new roll as possible. A fairly small depth of paper removed from the outside equates with a good deal of waste left on the core. For example, slabbing ⅛ in. (3.2 mm) from the outside of a 40-in. (1,016-mm) roll generates 1.25% weight loss. Figure 19-2 shows the size of a roll that would result if the same amount of slab waste were wound onto an empty 4-in. (102-mm) core. As can be seen, a ⅛-in. (3.2-mm) depth

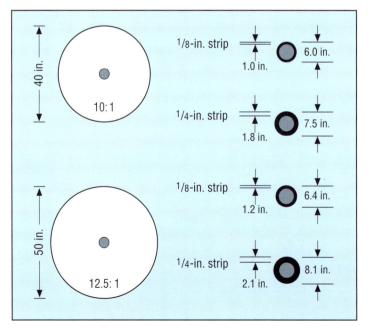

Figure 19-2. *The size of roll that would result if a given amount of slab waste were wound onto an empty core with a 4-in. (102-mm) outside diameter.*

of paper off the outside of a 40-in. (1,016-mm) roll produces a butt roll 6 in. (150 mm) in diameter. The effect is even more striking as the outside diameter of the roll increases. A roll tender should be encouraged to save paper wherever possible and should understand that the greatest savings are realized by conserving paper on the outside of the roll.

Grouping rolls. Press performance is improved by running rolls in a specific order. When paper is manufactured, it comes off the papermaking machine in wide rolls. A 15-ft. (4.6-m) machine roll can be slit into five 36-in. (914-mm) rolls. Paper manufactured on the outside of the 15-ft. (4.6-m) roll may vary slightly in caliper and moisture content than the paper at the center. Because of this, the printer is advised to run all the rolls from the same machine position together. Running rolls according to their machine position minimizes tension and register problems. This improvement is especially apparent during splicing.

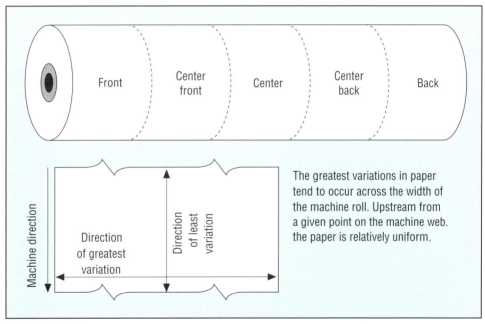

Figure 19-3. *A paper machine roll before slitting, showing the nomenclature for press roll position. Nomenclature varies slightly, depending on the number of rolls slit from the machine roll.*

Some papermakers color-code each roll according to its machine position by either banding or adding color labels. Such color markings make it easy for the printer to store the rolls according to the machine position and bring them out to the press as sets. All manufacturers include roll position information on the roll card that accompanies each roll. Figure 19-4 shows the TAPPI recommended system for sequence roll numbering. In the recommended numbering system, complete information about the roll is recorded. This includes the month of manufacture, the log number, and the reel position. It is the reel position that is used to group the rolls together as sets to minimize disturbances while running.

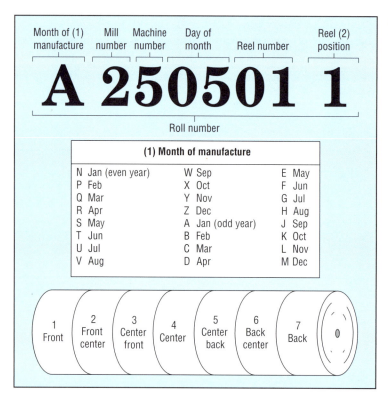

Figure 19-4. *TAPPI recommended roll numbering system.*

Roll stand setup. With a conventional infeed (using a dancer roller connected to the reel brake), a smooth paper feed is achieved with a properly adjusted brake and a tight linkage between brake and dancer. The brake should be adjusted so that the dancer runs (on the average) in the middle of its stroke. This means that the dancer tends to run low on a full roll (more torque generated by the tension in the web requiring more braking) and high (less torque) when the roll is nearly depleted. A brake in poor condition may require adjustment as the roll decreases in size. An uneven feed indicates a faulty linkage, nonuniform braking, or an out-of-round roll.

When mounted, the roll should be aligned exactly with the previous roll to prevent contact with the semidried ink on the edges of the blankets. Misalignment of the new roll will cause it to run in this tacky ink, and a web break may result (figure 19-5). This can occur regardless of whether automatic side-lay controls are in use.

The roll tender should not remove the wrapper or prepare the splice until the running roll is nearly expired. Dust and humidity can adversely affect the paper and reduce the adhesiveness of the splice tape.

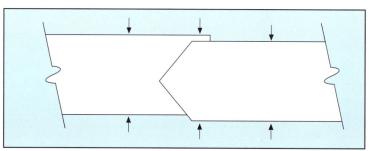

Figure 19-5. *Increase in web width if splice is not registered.*

Preparing the splice. On synchronous (flying) splicers, the lead edge of the new roll should be carefully cut to the proper V-shape. This shape is necessary to the proper functioning of the paster tabs. The splicer manufacturer will prescribe the proper shape for the splice, and each roll should be prepared according to these requirements through use of a template. Adhesive—either tape or glue—should be applied exactly as specified by the manufacturer.

On many flying pasters, the new roll is driven by belts or a roller to bring it up to press speed just prior to splicing. The roll tender should make sure that the new roll rides against the drive mechanisms with little or no slippage. Missed splices can result from improper contact.

On splicers employing photoelectric eyes, the roll tender should routinely check the surface of the eye for accumulated paper lint and dust. If ignored, this material can block out the eye and prevent it from functioning properly.

Washup Issues

Washup solvents are either one-step or multistep. Multistep solvents are used for more time-consuming reconditioning washups. These thorough washups involve the successive applications of two or more separate solutions, plus the use of water, which is essential for removing gum accumulated in the inking system. It is recommended that a reconditioning washup be done at least once a week, and each time an ink color change is made from dark (like black) to light (like yellow).

Simple washups apply to makereadies not involving color changes and usually involve a simple, one-step solvent. The solvent cuts the ink film on the rollers, making it more fluid for easy removal. A properly applied one-step solvent removes any buildup of varnishes and resins on roller and blanket surfaces as well.

To remove the ink and solvent mixture, a washup device is attached to the press. The washup device has a flexible blade that is forced against an oscillating roller, squeegeeing the solvent and ink from the ink train. After use, the blade and pan of the washup device should be thoroughly cleaned. The press operator should always visually inspect the inking system rollers to ensure that they are completely clean. Special attention should be given to the roller ends where ink builds up and dries. If the roller ends are left uncleaned, particles of dried ink will break off into the running ink system, causing hickeys in solid image areas.

Checking rollers. The durometer of rollers should be checked regularly. Information on durometer can be found in chapter 13, "Inking Systems." Rollers that exceed the recommended maximum durometer should be put through a reconditioning washup. Reconditioning hardened rollers may involve thoroughly scrubbing them with the recommended solvents to break through the glaze. Figure 19-6 shows a scrubber pad that is used to remove glaze from blankets and rollers. Products like this effectively scrub the surface of the roller without depositing any troublesome by-products in the inking system. If the durometer is still too high after scrubbing, the roller may be reground. In regrinding, the roller has a surface layer cut off, changing its diameter.

Figure 19-6. Press operator using hand scrubber designed to remove glaze from rollers and blankets.

The best (though most expensive) solution is to have the roller re-covered by the roller manufacturer.

Copperizing rollers. On older presses, the copperizing of steel inking rollers is often made part of the reconditioning washup, with the goal of reducing the possibility of roller stripping. Copperizing solutions deposit a thin layer of copper on the roller. Copper is ink-attracting and makes the roller more resistant to stripping.

Copperizing is not required on newer presses because the surface of modern hard rollers are coated with "Rilsan" (nylon 11), a plastic coating that helps to reduce roller stripping and which eliminates the need to copperize the rollers.

Cleaning conventional dampening rollers. Ordinarily, ink solvents are not used to clean rollers in the dampening system, though they can be used to clean some paper-covered rollers. Cloth covers retain ink solvents and should be cleaned only with suitable detergents, followed by a thorough rinsing.

Changing roller covers. Because changing cloth covers is time-consuming, an extra set of clean rollers should be ready to go on press at all times. The dirty rollers can then be cleaned and changed during the pressrun. Care must be used in mounting cloth covers. The seam should be straight, the cover uniformly tight along the roller length, and there should be equal amounts of overhang at each end. When tied, the ends should be flat, not rounded. (See figure 19-7.)

When removing a covered roller from the press, the press operator must be careful not to reverse the ends of the roller when remounting. A reversed roller rotates in the opposite direction, and a paper sleeve or cloth cover becomes baggy.

Changing dampener covers alters the effective diameter of the roller. When new covers are put on the press, roller settings should be checked and adjusted as necessary.

Preserving plates. Asphaltum gum etch (AGE) is a commonly used one-step finisher that deposits asphaltum on the image areas and gum on the nonimage areas, protecting both areas almost indefinitely. Be sure to use finishers that are recom-

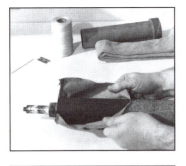
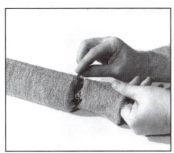

Figure 19-7. When changing a cloth dampener cover, the old cover is first slit and removed. The new cover is pulled over the roller and trimmed to the proper length. The ends are sewn with heavy cord. The cord is pulled tight and tied so that the ends of the cover are snug against the roller ends.

50% overlap spiral

Butted spiral

1/4-in. (6-mm) overlap spiral

Figure 19-8. Winding of paper-strip dampener covers. Variations in the amount of overlap can hamper control of dampening.

mended by the manufacturer. AGE, for example, could cause gum blinding in image areas of certain brands of plates.

To gum a plate on press, first protect the printing image with just enough ink to resist gum blinding. To do this, lift the dampening and ink form rollers and run a few impressions to remove some of the ink and water. This should leave a thin charge of ink on the image areas. If the ink is too thick, it may smear into nonimage areas; if too thin, image areas may be gum-blinded and not accept ink during the subsequent start-up.

Gum the plate using the recommended finisher or an 8° Bé gum. Rub the finisher down thin; then, buff it with clean, dry cheesecloth. Gum left on the image causes gum blinding, and streaks result when the plate is reused. If necessary, the image areas should be lightly rubbed with a clean, slightly damp cloth to remove any overlying gum. The gum, however, must cover all nonimage areas, especially those immediately adjacent to the image. Gum tends to back away from the image, and these areas scum or give the appearance of image gain if left unprotected.

Storing the plate in an area away from water and moist air keeps the gum/asphaltum layer from deteriorating. When rerun, the plate is washed with water to cut through the asphaltum and the gum underneath, loosening them from the surface.

Preparing the New Plate

Generally, it is a good idea to use plates that fit across the entire width of the cylinder body. This helps to prevent water and chemicals from working their way in behind the packing and plate. Following are additional recommendations to properly prepare the plate for mounting:

- After removing an old plate, clean the plate cylinder body and wipe off the bearers. During running, dirt, oil, ink, and various chemicals can collect and build up in both places.

- Inspect the front of the new plate for scratches and other imperfections that may produce printing defects. Also check the back; dirt or platemaking chemicals not removed by the platemaker can cause problems and should be removed. Check the surface for bumps and dents. If any are found, attempt to flatten the plate. If this does not work, get a new plate.

- Properly bend the plate using a plate bending device designed specifically for the press's lockup. Plates are bent to conform to the circumference of the cylinder body as well as the radii at the gap. Should the bends be too far apart or too close together, the plate is likely to crack on the press. Improper bending is a cause of plate cracking. If the plate fits the cylinder properly, it cannot crack. Eventually, it will wear out. (See chapter 10, "Lithographic Plates," for a detailed discussion of plate bending issues and procedures.)

Plate Mounting

Mounting the plate accurately is of utmost importance to an efficient makeready. Any misregister must be corrected as the first impressions are being printed. Significant register moves will take more time and will cause more paper waste and a longer makeready. To assure more effective plate mounting, follow these procedures.

Use protective cover sheets. A cover sheet protects the plate from scratches and contamination. Plateroom personnel should deliver the plate to the press with a cover sheet attached. It should remain attached until plate mounting begins.

Plate packing. Select the proper type and thickness of packing material for the plate and cylinder undercut. See chapter 12, "Packing and Printing Pressures," for a detailed discussion of plate packing issues.

Accurate plate mounting. The platemaker should put center marks at the leading and trailing edges of the plate. The press operator can then match these marks with scribe marks on the cylinder gap, a procedure that helps to ensure accurate register.

The basic procedure for plate lockup is as follows:

1. The lead edge of the plate is tucked into the lead edge of the cylinder lockup. If pins are used, the press operator must assure that the plates are pushed against the pins completely before locking.

2. The press unit is then put on impression and the cylinder slowly turned to the trailing edge. The plate should fit tightly against the cylinder. Any gap between the cylinder body and plate indicates that the plate was improperly bent and will probably crack during the pressrun.

3. The trailing edge of the plate is then inserted into the lockup and tightened, the press unit is taken off impression, and the tightening completed.

4. The press operator then uses a packing gauge to make sure that the plate is packed to the proper height in relation to that cylinder's bearers.

Blanket Concerns

Changing the blanket. Blankets should last for several weeks of heavy use. However, they are sometimes damaged. If a low spot forms, the image will not be transferred properly, requiring a blanket change. Also, the running web will create a slight depression in the blanket where the edge of the paper meets the blanket. If a wider web than usual is to be run, the blanket will need to be replaced. To avoid this problem, schedulers should organize jobs to run the widest webs first, progressing to the narrowest webs.

Press operators should track blanket performance to determine the number of impressions that may be printed by a given blanket. This information and the length of the next pressrun dictate whether the press operator should run used blankets or switch to new ones before start-up.

The general procedure for blanket lockup is as follows:

1. Clean the body of the cylinder of any dirt, oil, ink, or other chemical residue.

2. Insert and lock the lead edge of the blanket into place.

3. Insert the blanket packing underneath the blanket and work it slightly into the cylinder gap; this holds the packing in place and prevents it from creeping while the press is running.

4. Turn the cylinder slowly to the trailing edge, holding the tail edge to maintain tension.

5. Lock the blanket into the trailing lockup and tighten with a torque wrench to achieve consistent tension.

6. At this point, tension on the blanket is probably about 50 lb./sq. in. (350 kg/m^2) of cylinder width. Blankets should stretch a small amount after a short initial run; they should then be re-torqued.

7. Use a packing gauge to make sure that the blanket is packed to the proper height in relation to that cylinder's bearers.

Blanket packing. The blanket packing should be cut so it extends all the way around the cylinder, from leading edge to trailing edge. Many shops cut blanket packing just to web width or slightly narrower, otherwise ink and gum build up on blanket edges beside the running web. (See chapter 12, "Packing and Printing Pressure," for additional discussion of blanket packing.)

Preparing blankets for mounting. Scoring the exact center of the blanket cylinder aids in placing blankets and packing. By marking the exact center of blankets and packing, the press operator can match the cylinder, blanket, and packing marks and accurately center blankets and packing with minimum effort.

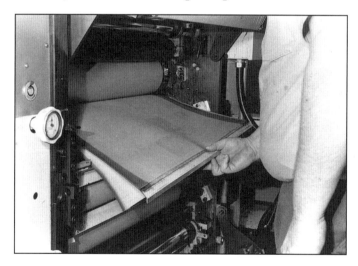

Figure 19-9. Press operator mounting a packed blanket.

Washing blankets. During the washup, ink is removed with a solvent specifically designed for washing blankets. The solvent should readily break down ink and other compounds picked up by the blankets during printing and should have no damaging effects on the resiliency of the rubber. Some press operators will use water to remove gum residue built up on the blanket.

Many blanket washes today are manufactured for low VOC (volatile organic compound) release. This makes the solvents more environmentally friendly but causes them to evaporate more slowly.

Figure 19-10 shows what happens to blankets when exposed to different types of solvents. It can be seen that some solvents have an extreme effect on blankets while others have little or no effect. Inappropriate solvents contain benzol, toluol, or

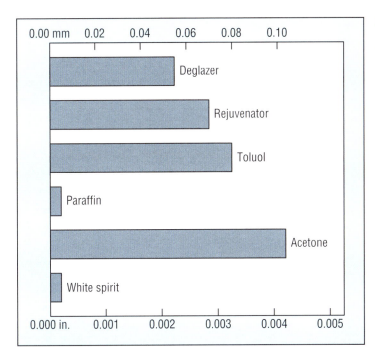

Figure 19-10. *Swelling of a 0.075-in. (1.90-mm) blanket caused by exposure to various solvents for 5 min.*

acetone, which damage natural rubber. Solvents like kerosene and mineral spirits are also not suitable; they have a high boiling point and take a long time to dry. Chlorinated hydrocarbons like perchloroethylene, carbon tetrachloride, and trichloroethylene damage rubber and are highly toxic.

UV ink concerns. The chemistry in UV inks can also cause blanket swelling and can affect certain rubber characteristics. Therefore, the blanket that is ordered must be "UV compatible" so that swelling is minimized.

Dryer and Chill Roll Issues

One crew member should periodically inspect the nozzles in the dryer during makeready. Any contact with wet ink can create a buildup that clogs these openings, causing uneven drying. Some installations require heating experts to periodically clean the dryer box and exhaust stack.

The surface of the chill rolls should be inspected between jobs. Any ink buildup on these rolls can be removed with solvent (always thoroughly remove the cleaning agent).

Dryer and chill roll temperature. Dryer and chill roll temperatures are determined by several factors. The amount of web surface covered by ink directly affects dryer temperature settings. Heavy coverage requires more drying energy than light coverage. A 20-lb. increase in paper basis weight requires a substantial increase in dryer temperature, because the dryer has to heat a larger volume of ink and paper. Finally, dryer temperature is directly related to press speed. Dwell time is decreased with faster moving webs, requiring higher dryer temperatures. Conversely, slower web speeds require lower dryer temperatures.

As a general rule, the dryer should be run at the minimum temperature possible while still removing solvents from the ink but not driving excessive moisture out of the web. Excess heat causes the brittle paper to crack when folded. Also, printed books with below-normal moisture content often gradually pick up atmospheric moisture, which can then result in wavy page edges.

Most operators operate the press in the "web temperature mode," which means that the web temperature is set and monitored by an external infrared pyrometer that reads the temperature of the web as it exits the dryer. The lens of the external infrared pyrometer has a tendency to get dirty, and it must be cleaned periodically. If dirty, the pyrometer gives a false reading of the web exit temperature. As a result, the internal temperature of the dryer can rise significantly, thus causing the web to exit at a much higher temperature than desired.

Folder Makeready

Folder gain is built into most folders. **Gain** means that each successive operation in the folder occurs at higher surface speed, so that the paper is subjected to a steady drawing force as it moves through the folder. Folders can have a gain factor anywhere from zero to 2% or more.

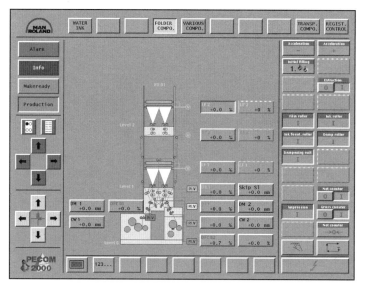

Figure 19-11. Screen mask on the touchscreen for setting the folder from the PECOM central control console. (Courtesy MAN Roland Druckmaschinen AG)

Former board settings. Three or four trolley wheels at the top of the former board are usually spaced across the width of the web. The nip pressure of each roller set must be equal to maintain consistent side-to-side draw. One roller pressing harder than the others will pull the web to the side with the greater nip pressure. The simplest way to check nip pressure is to insert strips of paper under each trolley roller and test them for drag.

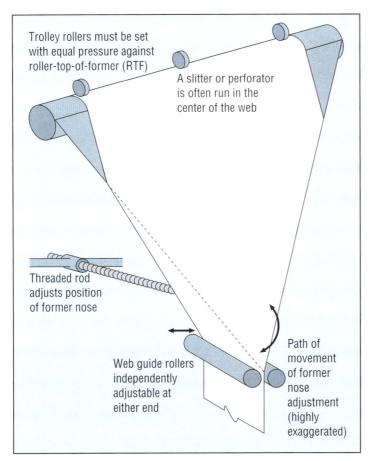

Trolley rollers must be set
with equal pressure against
roller-top-of-former (RTF)

A slitter or perforator
is often run in the
center of the web

Threaded rod
adjusts position
of former nose

Web guide rollers
independently
adjustable at
either end

Path of
movement
of former
nose
adjustment
(highly
exaggerated)

Figure 19-12. Major
operating adjustments
on the former board.
*(Courtesy Solna Web
USA, Inc.)*

Improper positioning of the former board relative to the web guide rollers can cause wrinkles in the fold, smear ink, and may contribute to gusseting. This critical angle is set with a long metal rod located behind the former board. One end of the rod is linked to the folder frame and the other end to the former nose. Both ends are threaded so that when the rod is turned, the nose moves toward or away from the web guide rollers. The initial setting should be marked on the rod, so that the nose can be easily and accurately returned to its original position.

To accurately determine the correct angle of the former board, the press operator can conduct a slit test, which is conducted in the following manner:

1. Set the trolley wheels and the nip rollers to evenly pull the web into the folder.

2. Replace the center trolley wheel with a slitter wheel to cut the full web into two equal sized ribbons.

3. Run the ribbons over the former board.

4. Check the position of the ribbons. If the two ribbons separate too much (more than ⅛ in. [3 mm]), the tip of the nose is too high.

5. If the tip of the nose is too high, adjust the threaded rod to lower the tip of that nose until the two ribbons come closer together, within 1/16 (1.6 mm) to ⅛ in.

6. If the tip of the nose is too low, the two ribbons will overlap each other as they go past the tip of the former nose. In this case, adjust the threaded rod to raise the tip of the nose to get the same $\frac{1}{16}$ or $\frac{1}{8}$ in. of separation between the two ribbons.

7. Lock the nose into position.

At that point, the angle should never need to be adjusted again. For the record, the operator may want to put a protractor on the former board to record the angle at which it was set.

Web guide rollers settings. Accurately set web guide rollers help assure wrinkle-free folds. They can be set farther apart or closer together and are adjustable at both ends. Use the following general procedures to set the web guide rollers:

1. Tear a strip of paper from the roll to be run and fold it in half.

2. Add a strip of paper 0.015 in. (0.38 mm) to this doubled thickness of paper (strips from packing sheets are suitable for this).

3. Insert the entire sandwich between the rollers at the front end, and adjust the rollers until there is a firm drag on the strips.

4. Set the back end of the web guide rollers in the same way, but without the additional 0.015 in. (0.38 mm) strip of material (the web should run fairly tight through the back end of the rollers).

Nip roller settings. The nip rollers perform the vital job of drawing paper into the folder cutoff. Their surface speed is slightly higher than the press to create a positive draw. Changing the pressure between the rollers will affect the feed rate at the nip rollers. Too much pressure will increase the feed rate, producing slack between nipping and cutoff. This could result in cutoff variations. Too little pressure will decrease feed rate, increasing tension and causing the pins to tear out of the web.

Nip rollers are set to one sheet thickness less than the total thickness of the folded paper. For example, if four sheets are to run through the nip, the setting is made with three sheets. The rollers are set so that there is a firm drag on the test thickness (figure 19-13). To ensure adequate paper feed, it is good practice to set the nip rollers tighter rather than too loose. Pressure can always be eased off later.

Setting the back ends of the nip rollers a little tighter than the front holds the web tight against the former board and prevents wrinkling. It is good practice to set the rollers so that more tension is applied to the outside edges of the web going over the former than to the center. The center is where wrinkling, tearing, and smearing can occur. The best approach is to start with uniform pressure on the roller, from front to back, and then experiment to find the settings that produce the least wrinkling.

Cutoff settings. The cutoff knives extend into a slotted, female die on the opposing cylinder at the moment of cutoff. If the female die is filled with a rubber material that absorbs the shock of the cutoff blade, the problem of blade adjustment is less critical. On many folders, however, the female die is a simple slot across the face of

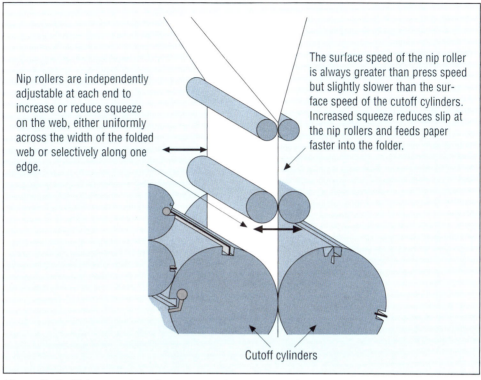

Nip rollers are independently adjustable at each end to increase or reduce squeeze on the web, either uniformly across the width of the folded web or selectively along one edge.

The surface speed of the nip roller is always greater than press speed but slightly slower than the surface speed of the cutoff cylinders. Increased squeeze reduces slip at the nip rollers and feeds paper faster into the folder.

Cutoff cylinders

Figure 19-13. Major operating adjustments on the nip rollers. (Courtesy Solna Web USA, Inc.)

the cylinder. Adjustment becomes much more critical because repeated contact between the knife blade and the sides of the slot will wear out the blade.

Setting the cutoff blade is not required during every makeready, but the setting should be checked when problems arise or when cutting knives are changed. Use the following procedure to check and adjust the cutoff blade:

1. Place tape across the slot at two or three different positions along the length of the blade.
2. Turn the cylinder over slowly and observe where the tape is cut.
3. Set the knife so that it is slightly closer to the leading edge of the slot than to the trailing edge.

This procedure can be 100% effective if the press operator also performs a running check on the cutoff adjustment with the aid of a strobe light.

In the case where the knife box assembly has to be moved, the press operator should ensure that the impaling pins exactly align with their corresponding holes. When adjusting the release of the pins, the timing should be such that the pins retract and release the sheet exactly as the folding jaw closes.

When the press is running at idling speed, it is normal for alternate signatures to tear out of the pins. This occurs because the web tension increases in the folder at very slow speeds, and tension on the alternate signatures is excessively high. This does not indicate improper adjustment on the folder, unless the problem continues at running speeds.

Jaw fold settings. The key to clean, accurate, parallel folds is the proper adjustment and timing of the tucker blades and folding jaws. The tucker blade protrudes from the surface of the cylinder, usually by about ⅛ in. (3 mm). The height of the blade and its position on the cylinder is adjustable to some extent. The height of the blade should be adjusted while observing the entry of the tucker blade into the fully opened jaw. If set too high, blade contact with the folding jaws will cause damage.

The blade should be set to clear both sides of the jaw. Allow for the paper thickness that will pass between the blade and the jaw. At the point of its deepest penetration into the jaw, the blade should be set slightly closer to the fixed trailing face of the jaw than to the movable jaw. Close settings against the movable leading face will probably result in loose folds, cracked folds, and wipeout. Wipeout is the removal of the inside fold on the face of the tabloid signature. This is caused by the tucker blade not being set at the correct position.

Many press operators cover the tucker blade with the thinnest available Teflon tape to allow slippage between the tucker blade and the fold. This assures that the inside fold will not pull out with the tucker blade as it continues on to the next stage of folding.

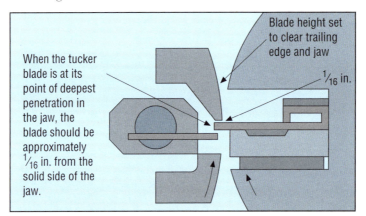

When the tucker blade is at its point of deepest penetration in the jaw, the blade should be approximately ¹⁄₁₆ in. from the solid side of the jaw.

Blade height set to clear trailing edge and jaw

¹⁄₁₆ in.

Figure 19-14. Major operating adjustments on the folding jaws and tucker blades. Note: On some folders, the tucker blade and folding jaw assemblies can be moved slightly on the cylinder face to put a lap on the signature.

Chopper settings. When adjusting the chopper, set the folding rollers (located under the chopper table) first. To make the setting, use a large piece of the stock to be run. Fold the sheet to one sheet thickness less than the thickness that will go through the chopper. For example, if four-sheet thicknesses are to go through the chopper, use a setup piece three sheets thick. When testing the two rollers for drag, adjust for uniformity of tension from end to end on the rollers.

Adjusting the timing of the chopper blade requires that the headstops have approximately a ⁹⁄₁₆-in. (14-mm) gap between the signature head as the chopper blade touches the quarter fold. The chopper blade inserts the backbone of the signature into the center of the knurled position on the nip roller. Setting the chopper blade too low can cause the blade to pull the signature out from between the rollers as the blade retracts.

Scoring the signature along the intended fold line can improve folding accuracy. Also, moisture applied to the chopper fold lines softens the web and improves the accuracy of chopper folds. A water-glycerin mixture can also help make the chopper

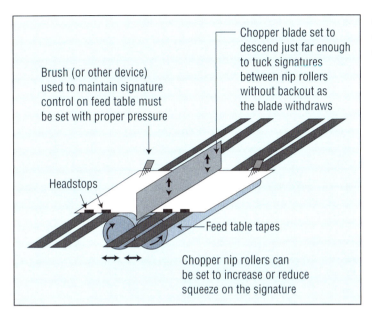

Figure 19-15. *Major operating adjustments of the chopper folder.*

fold more accurate, but it must be used with caution. Such a mixture can permanently soften the paper. Scoring, unlike perforation, is generally used for jobs that are to be sewn or saddle-stitched.

Perforator settings. Perforators are used in folders either to improve the fold and eliminate gusseting or to create a tear-out. There are two basic perforator designs. Circumferential perforators slit the paper in the direction of web travel. Usually, circumferential perforating is performed by a wheel mounted at the top of the former board. Cross perforators, on the other hand, perforate the web across its width. This perforation is executed by a set of rotating cylinders below the nip rollers. One cylinder carries the perforating teeth and the other acts as a die for the teeth.

A circumferential perforator wheel set at the top of the former perforates the web lengthwise, usually along the former fold line. Once set and centered, this perforator should not be moved. If the web begins weaving to either side as it crosses the former board, the web guide rollers should guide the perforation back into proper position. The perforator wheel can be replaced by a slitter wheel to slit the web into ribbons.

Running Makeready

Running makeready begins with the press start-up and ends with the delivery of approved, saleable signatures. During this period, work tempo picks up because the paper running through the press is being wasted. The aim should be to keep this waste down by completing the running makeready as soon as possible. The press operators' chief concerns at this time are attaining a proper ink/water balance, setting colors to a specified density and maintaining it, adequately drying the web, and achieving and maintaining register. During running makeready, the press operators must decide whether a given problem requires a press shutdown. Each moment of hesitation increases paper waste.

Setting ink and dampening levels. Prior to webbing the press and before running makeready begins, ink and dampening levels are roughly set. It is good practice to keep the ink level fairly low before starting to print. This allows the press operator to increase ink feed to the minimum needed for good color. Dampening levels should also be set to the minimum required to keep the plate clean.

Initial tension test. When ready to print, the press is put on impression and given a short burst at running speed to test for proper folder and tension settings. This test is usually not necessary if the previous job ran successfully with a similar web-up and comparable paper. The paper used during this test is wasted, so, if settings are found to be far off, it is best to stop the press to make adjustments. Under most conditions, tension is adequate if the web runs fairly tight without weaving from side to side in the press. Additional minor adjustments will be required during the pressrun.

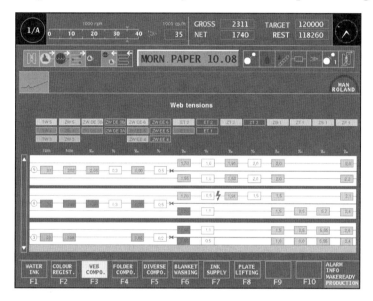

Figure 19-16. Press web tension as shown on the monitor of the PECOM central control console. (Courtesy MAN Roland Druckmaschinen AG)

Makeready running speed. Running makeready is usually accomplished at a relatively low speed. However, the speed should be high enough to fire up the dryer. Otherwise, wet ink will track on the chill rolls, the former board, and the folding cylinders.

Placing the press on impression. The plates should be inked after the press and dryer have started. The press is put on impression. Too much ink or too high dampening levels can lead to web breaks at this point. The impression pressure may be gradually inched on to avoid web breaks and picking, particularly with lightweight stocks. Easing up to full impression allows the web to pull the half-set tacky ink from the blanket little by little rather than all at once.

Folder concerns. As the speed of the press slowly increases, the web may tear out of the pins in the folder cutoff. One remedy is to increase the infeed speed or the chill-roll speed, which will decrease web tension in the folder. When the paper no longer tears, the infeed speed or chill roll speed can be reduced to bring web tension back up.

A second method used to decrease folder tension is to increase the nip roller pressure, forcing more paper into the folder. This setting, too, can be backed off once the press is running properly. When making this adjustment, it is usually necessary to increase the squeeze only at the front or the back of the nip rollers.

Check for proper tension at cutoff by inspecting the holes left by the pins. There should be enough tension on the web to slightly tear the paper at the pins. Too much tension causes the pins to tear out of the paper, while too little tension shows little or no tear behind the holes.

This check is meaningful only if the pins have been properly timed. If they retract too slowly, the pins hold the paper and take it in the opposite direction of the folding jaw. The result looks as if the web is under too much tension because the pins tear out of the sheet. Likewise, if the pins retract too fast, the paper runs free in the folder, causing erratic folding. Also, pins set too high do not fully retract from the paper and tear out, giving the appearance of too much web tension, improper timing, or both. Most manufacturers supply gauges to set the folder pins and tucker blades to the correct height.

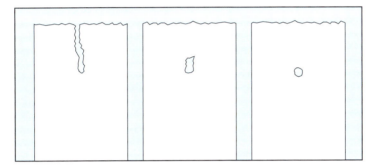

Figure 19-17. Holes left by the cutoff pins, indicating the amount of tension present at the cutoff. Round holes indicate insufficient tension. Long slits indicate excessive tension. Proper tension is evidenced by slightly elongated holes.

Achieving proper ink density. As a general rule, the press should always be set to run with minimum ink and minimum water while achieving proper ink film density. Ink/water balance and color are rarely up to standard at the beginning of the pressrun.

Procedures for Bringing the Press up to Color

1. Prior to webbing the press, start up printing units and roughly set ink and water levels, depending upon plate demands.
2. Turn on dryer to avoid wet ink tracks on chill rolls, former board, and folding cylinders.
3. On start-up, and after water and ink are on the plates in their relative, light amounts, slowly turn on impression and run press at approximately 400 ft. (122 m) per minute.
4. Adjust ink and water by cutting back on the water until the plate begins to catch-up and is carrying as much ink as the rollers deliver.
5. Add just enough water to stop the catch-up and allow the press to run for a while.
6. Add more ink followed by just enough water to stop catch-up. (If color is still strong enough, repeat this process. Press should always be running an ink-water balance with minimum ink and minimum water.)
7. Gradually increase the running makeready speed from about 400 ft. per minute to the highest speed possible, keeping register and color under control.
8. Check for proper tension.

Because presses tend to start with too much water on the plate, a good practice is for the press operator to initially cut back on the dampening solution. The ink density will print too lightly at this point if the press operator started with a light ink film. Alternate between increasing the ink level and increasing the dampening level to maintain balance until proper density is attained. Check the densities with a densitometer as the signatures come off the tapes.

Starting the Run

Strategy for moving to running speed. During running makeready, a typical press will run at approximately 400 ft. (122 m) per min. Once the signatures are approved, makeready officially ends and the press operator eases the press up to running speed. The increase from makeready to running speed should be gradual with the press raised to running speed in a series of incremental steps. At each speed increase, the ink/water balance is checked and restored (if necessary), the tension readjusted, and finer settings are made on the folder. If something goes wrong after any one of these increases in speed, the press should be slowed to the previous level to solve the problem. Once the problem is corrected, increase to the next increment toward running speed. The objective is to continue to produce salable signatures at all times. Increasing speed in relatively small increments keeps the press under control and thus avoids waste production.

Register problems. Besides getting up to color, poor register of images on the plates will lengthen makeready significantly. Improper fit and/or misregister require the press operator to spend time making adjustments. Consistently poorly registered plates indicate problems in the prepress department, and these problems should be solved there—not in the pressroom.

Procedures for Achieving Register

1. Punch plates prior to plate bending.
2. Mount plates carefully on the pins in the bending fixture.
3. Bend plates so that the two bends will correctly match the distance around the plate cylinder body from gap to gap.
4. Check to assure parallelism between the bends from side to side across the plate.
5. Cut packing material ½ in. (12 mm) narrower than the plate on each side.
6. Apply adhesive or oil on packing sheets at the lead edge of the plate to make it easier to simultaneously insert and position both plate and packing on the press.
7. Observe the scribe marks on the cylinders and the image marks on the plates to position the plates on the cylinders.
8. Put press on impression and turn cylinder slowly to the trailing edge.
9. Insert trailing edge of plate into the lockup and tighten.
10. Take press off impression and complete plate tightening operation.
11. Run the press and adjust the movement of each plate cylinder to bring the image into register.
12. Bring the press up to speed to correct misregister.

High-Speed Makeready

If the quality of the job and the fit required are not high, a high-speed makeready is appropriate and more economical. With this system, the press is quickly taken up to running speed and a few hundred signatures are run. The press is then shut down and adjusted to the actual printed signatures. After this, the press is brought up to running speed again, with salable signatures coming out of the delivery.

20 Understanding Quality Control

Test images are quality control devices that, when printed, allow the press operator to evaluate print quality and press performance. These devices are specifically designed to indicate variations between an original image and the final printed image. Numerous factors contribute to print quality, and many test images may be used singly or in combination to assess press performance.

Control images are typically placed in the trim area of printed materials to help monitor press performance and provide information regarding the consistency and reproduction capability of the production equipment. Unlike control images, test images are usually too large to be included in the trim area of an actual job; therefore, they are printed separately as press tests.

Densitometry

Densitometry, colorimetry, and spectrophotometry are the three methods used to measure color in the graphic arts. The densitometer measures optical density. It is essentially colorblind, measuring only the level of reflectance of a print or the transmittance of a film. Density varies with ink-film thickness, but data can also be used to determine dot area. Colorimeters define color numerically in various color spaces, but they do not measure density or dot area. Colorimeters are useful to determine color difference in an accurate metric called ΔE (delta E). Spectrophotometry yields the highest level of color information; it is the most accurate and versatile. Spectral analysis yields color curves and permits calculation of density, dot area, and colorimetry.

Of the three, the densitometer is the most widely used device in the pressroom for measuring and evaluating color testing and control images. Its primary use in the pressroom is to analyze test forms and color bars to determine how the press is performing and to monitor the pressrun, providing feedback on a number of factors including ink density, dot gain, trapping, and print contrast. The densitometer can also be used to inspect the quality of supplied off-press proofs and press proofs and determine their conformance to standards and specifications. It can also be used to analyze the incoming materials, such as ink and paper, used in the pressroom.

A densitometer indirectly determines the amount of light absorbed by a surface by measuring the proportion of light reflected from or transmitted through a measured surface. It is assumed that—for practical purposes—the amount of light absorbed is equal to the amount of incident light (the supplied light) minus the amount of light reflected. The densitometer then calculates density using an equation. (A reflectometer records or reads the percent reflectance of a beam of light compared to the standard beam, while a densitometer converts the numbers into the familiar logarithmic density units.) Transmission densitometers compare transmitted light, and reflection densitometers compare reflected light. Because different makes of densitometers may vary in spectral response, the printer should choose one brand of densitometer as the standard and use it throughout the company.

The densitometer provides numerical data in the form of percentages or logarithms. Many digital densitometers will apply the formulas discussed in this chapter automatically to derive many important evaluation numbers.

A densitometer consists essentially of a light source, a lens and filter assembly, and a photocell. The light source emits a measured amount of light to the paper surface where some portion of the light is absorbed and the remainder of the light is reflected through the lens and filter to the photocell. The photocell responds to the light, and a ratio or percentage is derived. The percentage describes the value of the intensity of the light source divided by the light reaching the photocell.

$$Reflectance = \frac{Value\ of\ light\ at\ the\ photocell}{Value\ of\ light\ at\ the\ light\ source}$$

Converting the percentage of light to optical opacity and optical density requires a simple algorithm to be built into the densitometer's circuitry. Optical opacity is the reciprocal of percent reflectance:

$$Optical\ opacity = \frac{1}{Reflectance}$$

Finally, the optical density is the logarithm of optical opacity. A logarithm is the expression of a number as an exponent to a base of ten.

$$Optical\ density = Log_{10}\ Optical\ opacity$$

The following example shows how optical density is derived using the densitometer. Assume that a densitometer's light source has a measured value of 100 units of light. The densitometer is placed on a printed control patch and 10 units of light are reflected to the photocell. This means that the 90 units of light not reflected to the photocell were absorbed by the control patch. Determining the reflectance of this patch:

$$Reflectance = \frac{Value\ of\ light\ at\ the\ photocell}{Value\ of\ light\ at\ the\ light\ source}$$

$$= \frac{10}{100}$$

$$= 0.10$$

Given a 0.10 light reflectance value, optical opacity is determined as follows:

$$Optical\ opacity = \frac{1}{Reflectance}$$

$$= \frac{1}{0.10}$$

$$= 10$$

Finally, since optical density is the logarithm of the optical opacity, the density is equal to 1.

$$Optical\ density = Log_{10}\ Optical\ opacity$$

$$= Log_{10}\ 10$$

$$= 1$$

Color filters. The densitometer can measure the process colors of yellow, magenta, and cyan with the aid of color filters. On many modern digital densitometers, the filtration of the colors happens automatically, without the press operator's intervention. However, older densitometer models require the press operator to dial in the proper filter for the density measurement.

Densitometers are usually equipped with three color filters—red, green, and blue. The blue filter measures the reflected light in a range between about 400 nm and 500 nm and reads it as the blue density. Similarly, the green filter measures the reflected light in a range between about 500 nm and 600 nm, and the red filter measures between about 600 nm and 700 nm. The exact range of wavelengths measured depends on the specifications of the filters used with the densitometers. For example, so-called wide-band filters typically measure a range of 50–75 nm, while the "narrow-band" filters generally measure a range of about 20 nm. With both types of filters, the density reading is an average of the densities over the range of wavelengths measured, not the density at a specific wavelength.

The purpose of each filter is to render the process color as if it where black. Yellow when viewed through a blue filter will appear to be black. The green filter will render the color magenta as black, and the red filter will render the color cyan as black. Using the correct filter allows the densitometer to single out a reading for each process color, even if the colored ink films are overlapping.

Scanning densitometer. A scanning densitometer is a computerized quality control device that measures and analyzes printed color bars using a densitometer. There are in-line scanning densitometers, which are mounted directly on the press to read density information on the fly. More commonly in use are off-line scanning densitometers, which require the printed sheet to be manually positioned for scanning.

The densitometer provides an indication of printed ink film thickness by measuring the ink densities on color bars printed on the sheet. The ink patches to be read are spaced within each ink key zone of the printed sheet. The results are compared

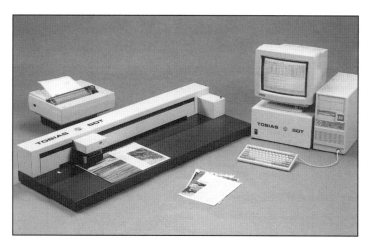

Figure 20-1. *Tobias SDT scanning densitometer. (Courtesy Tobias Associates, Inc.)*

with desired densities, and a computer monitor displays the degree of variation. Ink fountain keys and fountain rollers can be then be adjusted if necessary. Scanning densitometers may also provide information on dot gain, print contrast, trapping, hue error, grayness, slurring, and doubling.

Densitometric Color and Image Analysis

Solid ink film measurement. To maintain image quality during the pressrun, the operator is concerned with the ink film thickness being printed on the paper. When ink films are too thin, the images will appear to be washed out. When ink films are printed too thickly, many problems can result including poor drying, halftone plugging, and scumming or toning. The densitometer is used to provide numerical feedback on ink film thickness. These optical density readings are taken from solid color patches printed in the trim area of the form. Generally, the thicker the ink film, the higher the optical density reading. Published standards for the printing industry, like the Specifications for Web Offset Publications (SWOP), specify density values and tolerances for solid ink film.

Hue error and grayness measurements. The densitometer can be used to evaluate the character of each of the process colors used to print full-color images. Ink color quality may vary from manufacturer to manufacturer. For example, a cyan ink from one manufacturer may not match the cyan ink from another manufacturer. Without consistency of process ink color, preparation and editing of color images in the prepress stage cannot be accurate.

- *Determining hue error.* The process colors of yellow, magenta, and cyan are not completely pure. For example, a theoretically pure magenta would reflect only its own color and no cyan or yellow when measured with a densitometer. Because magenta is impure, it acts as though it is reflecting small amounts of yellow and cyan. Hue error, expressed as a percentage, describes the deviation of the process color from a pure state. For example, a magenta ink film might have a hue error of 40%, while a yellow might have a hue error of 3%.

These values indicate that yellow is the purer of the two colors and that magenta has a higher hue error. To measure hue error, each process color is measured with the densitometer through each of the three filters: red, green, and blue. For example, when magenta is measured through a blue filter, the density reading shows how much yellow is reflecting from the magenta. When read through a red filter, the density of the cyan contamination in the magenta is determined. Once all readings have been taken, hue error values are determined by applying the following formula:

$$Hue\ error = \frac{D_M - D_L}{D_H - D_L} \times 100$$

where D_H is the highest density value, D_M is the middle density value, and D_L is the lowest density value.

- **Determining grayness.** Equal amounts of yellow, magenta, and cyan will produce gray. Process color inks are slightly grayish, each color behaving as though it were contaminated with slight amounts of the other two. The grayness of the ink film, expressed as a percentage, can be easily determined with a densitometer. Each process color must be measured through each of the three filters—red, green, and blue. Once all measurements have been recorded, grayness is determined by applying the following formula:

$$Grayness = \frac{D_L}{D_H} \times 100$$

where D_H is the highest density value, and D_L is the lowest density value.

Trapping measurements. Trapping values express how well one process color ink film is adhering to the previously printed ink film. Because the adhesion characteristics of plain paper are different from those of a wet ink film, not all of the ink film density that prints on paper may also be printed on a previously printed ink film. Because secondary colors are created by printing overlapping cyan, magenta, and yellow ink films, trapping values will give insight into the relative effectiveness of the blues, reds, and greens. Before applying the trapping formula, each process color ink film and each secondary overlap color must be measured through each filter of the densitometer. Also, the order that the ink films were printed must be known.

$$Trapping = \frac{D_{OP} - D_1}{D_2} \times 100$$

where D_{OP} is the density of the overprint, D_1 is the density of the first color printed, and D_2 is the density of the second color printed. **Note:** All three densities are measured using the color filter for the second color printed.

- **Interpreting trapping values.** A trap of 100% is considered to be perfect. This means that the second ink film down is adhering with the same ink film thickness on the paper as it is on the previously printed ink film. Normally, this does not happen. Rather, typical trapping values range from 80–90%, meaning that some adhesion is lost on the second color down. Very low values, like 60% trap, indicates that the secondary color is very weak and the colors in the printed image corresponding with that color will not be rich.

- *Factors affecting trap values.* Trapping capacity is directly related to the tack of the ink (as well as other factors). Tack is a measure of the resistance of the ink film to splitting. The inkometer is a device used to measure the tack of ink. When asked, ink manufacturers will provide the printer with tack values. When printing four colors of ink in sequence on a web press, the tack of the inks should be graded to yield good trap percentages. The first ink film printed should have the highest tack value, followed by successively lower tack values. This assures that the ink films will adhere well. If the second ink film down had a higher tack value than the first, the ink on the blanket would "stick" more effectively to the blanket than it would to the already printed ink film, resulting in poor transfer.

Print contrast measurements. Print contrast is a densitometric measurement that indicates how well the three-quarter tone to shadow areas of an image are reproducing on press. To make the measurement, a 75% patch (three-quarter tone) and a solid for the color in question must be measured with the densitometer and the following formula applied:

$$Print\ contrast = \frac{D_S - D_{75}}{D_S} \times 100$$

where D_S is the density of the solid and D_{75} is the density of the 75% tint patch.

- *Interpreting print contrast.* The print contrast value will increase when the difference between the solid ink film and the 75% halftone dot is highest. When the value is relatively high, better shadow detail will reproduce in the image. When the value is low, the image will produce with flat shadow detail. An ink film that is too light will yield a low print contrast value. The light solid will not contrast well with the 75% dot. Similarly, an ink film that is too thick will also cause print contrast to drop. This is because of excessive dot gain in the 75% dot area, making contrast between the 75% dot and the solid weak.

Dot gain. Dot percentages on the printing plate will grow when transferred to the printed sheet. This occurs because the pressure between the plate and blanket causes the ink film covering the dots on plate to spread outward. More spreading occurs as the dots are transferred to the paper. The densitometer is used to measure dot gain, usually from the 50% dot. The Murray-Davies dot gain formula is as follows:

$$Dot\ gain = \frac{1 - 10^{-D_T}}{1 - 10^{-D_S}} \times 100$$

where D_T is the density of the tint, and D_S is the density of the solid.

	Tone Value Increase		Print Contrast Range
	Target Value	Range	
Yellow	18%	15–24%	20–30%
Magenta	20%	17–26%	30–40%
Cyan	20%	17–26%	30–40%
Black	22%	19–28%	35–45%

Table 20-1. SWOP tone value increase (total dot gain) and print contrast target values and tolerances. (From the ninth edition of SWOP, Specifications for Web Offset Publications)

Process Optimization Devices for Web Offset

Newspaper test forms. The two-page GATF/SNAP Test Form, developed with the SNAP Committee and the technical staff of the Newspaper Association of America, provides greater measuring capabilities and compatibility with today's digital imaging systems. The image size of this 85-lpi test form is 20×11.5 in. (521×292 mm).

The GATF/SNAP Test Form, which comes in both film-based and digital formats, consists of several test images that help nonheatset web offset presses to be evaluated and adjusted to achieve high-quality color printing. The combination of quality control devices on the form allows press operators to diagnose press, platemaking, and image assembly problems. Target values for various printing parameters, such as tone reproduction, gray balance, dot gain, print contrast, and ink trapping, are provided.

The first page measures the characteristics of the newspaper printing system. It can be used to calibrate the output of the prepress system to achieve optimal color reproduction on the press. It can also be used to establish meaningful process control aim points for the pressroom. The second page contains a color chart to help the user get predictable colors from the newspaper printing system.

The GATF/SNAP Newspaper Test Form includes the following elements:

- Four-color single-tier and four-tier color control bars
- A type resolution target
- Transfer grids
- Four photographs: female portrait, fleshtones photograph, memory colors photograph, and high-key photograph
- Maximum ink coverage target, tone scales, and gray balance chart
- Hexagon target
- IT8.7/3 basic data set
- A plate control target
- Unidirectional register track

GATF web test forms for heatset printing. Three versions of a two-sided web press test are available from GATF: 17×22-in. (432×559 mm) four-color form, 17×24-in. (432×610 mm) six-color form, and 22×34-in. (559×864 mm) six-color form. These test forms were designed to help heatset web offset printers achieve higher quality

Figure 20-2. The first page of the GATF/SNAP Newspaper Test Form.

Figure 20-3. *The second page of the GATF/SNAP Newspaper Test Form.*

color printing with less waste and reduced makeready time. They enable the printer to diagnose, calibrate, and characterize the printing conditions (press, ink, fountain solution, paper, plate, blanket, etc.). These two-sided forms print on both sides of the web, taking advantage of the two-sided blanket-to-blanket press.

The test forms incorporate numerous quality control devices to measure a wide variety of printing variables, such as dot gain, trapping, and print contrast, along with readily indicating whether printing problems like slur, doubling, or ink/water imbalance are occurring. For example, the 22×34-in. (559×864-mm) six-color form includes the following elements:

- Unidirectional register track, three-color gray bar, and transfer grids
- Minimum/maximum dot size target, gray balance chart, and ladder targets
- Proof comparator, color correction target, and IT8.7/3 basic data set
- Six-color control bar, mottle patches, and dot size comparator
- Ink coverage target, ghosting indicator, and plate control target
- Paper fluting, show-through, and star targets

Figure 20-4. The 22×34-in., six-color, two-sided GATF Web Press Test Form.

SWOP calibration kit. The kit includes a set of four-color 30×40-in. (762×1016-mm) films or digital files of the SWOP Calibration Test Form, which contains a variety of quality control devices and photographic images taken from the ISO Standard Color Image Data (SCID) set. It is designed to calibrate the color reproduction system to conform to SWOP (Specifications for Web Offset Publications). Measurements including ink film density, dot gain, trapping, print contrast, hue error, and grayness can be measured and compared to standards.

GCA/GATF Proof Comparator. Ad agencies, publishers, separators, and printers frequently use off-press proofs to communicate color reproduction requirements. A proof comparator, such as the GCA/GATF Proof Comparator, is an ideal visually oriented tool needed to evaluate the accuracy and consistency of off-press proofs. The GCA/GATF Proof Comparator allows the user to measure solid density, print contrast, hue error, gray error, gray balance, dot gain, and resolution.

The GCA/GATF Proof Comparator III is a film target that evaluates the accuracy and consistency of off-press proofs. This visually oriented device permits users to confirm that a proof has been properly made and accurately represents the proofing system. The comparator consists of a set of separation films. A proof of these films becomes a "reference comparator" against which the proof comparator imaged on production proofs are evaluated.

The GCA/GATF Digital Proof Comparator, a native PostScript file, also contains targets that measure the exposure, resolution, and directional effects of the imaging system in addition to its reproduction characteristics. A unique aspect of the digital proof comparator is that it carries on a two-way dialogue with output devices by changing its dimensional tolerances and element sizing in response to the resolution of the raster image processor. This interactive target reports several key imaging attributes as well as provides the highest resolution measurements that can be obtained for a given imaging device.

Figure 20-5. GCA/GATF Digital Proof Comparator.

Plate control targets. Plate control targets are highly precise test images for diagnosing, calibrating, and monitoring imaging steps in the graphic reproduction process. As the name implies, plate control targets are designed for platemaking. However, they are also useful for other graphic arts processes that utilize contact printing and digital imaging. GATF plate control targets are produced for both traditional film-based platemaking and digital computer-to-plate workflows.

The film-based GATF Plate Control Target consists of seven elements:

- Calibrated continuous-tone step wedge that is used to monitor the exposure levels given to plates or film.
- Continuously variable microline resolution target to measure the resolution of plates and for monitoring exposure level.
- Calibrated halftone wedge—a wedge that contains patches (150 lines/in.) of uniform tone value.
- Fine-screen midtone patches used to measure dot gain differences due to screen ruling.
- Highlight and shadow dot control patches—the highlight (0.5–5%) and shadow (95–99.5%) section contains a range of tone values.
- Three Dot Gain Scale-IIs© for measuring mechanical dot gain during platemaking. The screen rulings are 150, 175, and 200 lines/in.
- Patented GATF Frequency Modulated Acutance Guide that quantitatively measures the apparent sharpness of an image printed on a substrate.

The continuing development of computer-to-plate technology provided the impetus for the development of the GATF/Systems of Merritt Digital Plate Control Target. It is a native PostScript file that allows the user to monitor the output of digital systems, particularly computer-generated printing plates. The file generates a target that contains a wide variety of elements that measure the exposure, resolution, and directional effects of the imaging system in addition to its reproduction characteristics, including the following elements:

- Dialog box that reports variable information from the RIP
- Horizontal and vertical microlines in negative and positive patches
- 1-, 2-, 3-, and 4-pixel checkerboard patches
- Negative and positive curved microlines
- Star targets
- Solid patch
- 50% reference tint patches at 150 and 200 lpi
- Corrected and uncorrected highlight and shadow tints
- Corrected and uncorrected tone scales

Also available from GATF is the UGRA Plate Control Wedge, which contains a sensitivity guide, positive and negative microlines from 4 to 70 microns, halftone dots

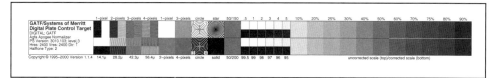

Figure 20-6. GATF/Systems of Merritt Digital Plate Control Target.

from 0.5% to 99.5%, and a slur target. It can be used for plotting plate reproduction curves, determining reproduction characteristics of plates at different exposures, detecting slur or doubling, and determining dot gain or loss on plates and prints.

Ladder target. The ladder target, incorporated in many GATF test forms, is a test image that visually shows variations in the amount of slur and doubling along its length—when printed around the cylinder. Slur and doubling may also be determined using a densitometer. The ladder target also detects gear streaks and wash marks. The ladder target is one of the most effective diagnostic test targets for troubleshooting the printing units.

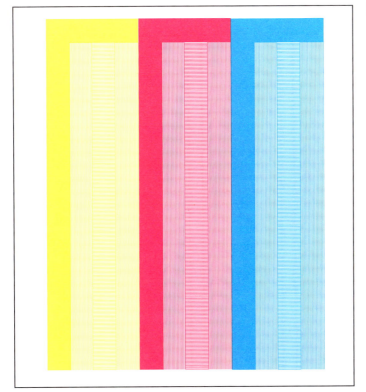

Figure 20-7. Ladder target.

Gray balance chart. A gray balance chart (figure 20-8), also incorporated in various GATF test forms, allows the image editor to determine the dot percentage combinations of magenta, cyan, and yellow to achieve a neutral gray. Gray balance values change depending upon the ink, paper, and plate combinations used under production conditions. The chart establishes empirically determined tone correction for gray balance in color reproduction.

Production Control

Test images for production control show print quality variations during a pressrun. They are usually small enough to fit in a trim area.

Cyan		Yellow					
	7	6	5	4	3	2	1
Magenta 6							
5							
4							
3							
2							
1							

30	28	26	24	22	20	18	16
28							
26							
24							
22							
20							
18							
16							

60	58	56	54	52	50	48	46
58							
56							
54							
52							
50							
48							
46							

80	78	76	74	72	70	68	66
78							
76							
74							
72							
70							
68							
66							

Figure 20-8. Gray balance chart.

Star Target. The Star Target, which is incorporated in several GATF quality control devices, consists of many tiny wedges that converge in the center of the target. It quickly indicates ink spread, slur, and doubling in screen rulings as fine as 2,200 lines/in. The sensitivity of this device magnifies ink spread approximately 23 times, thus press operators can easily detect any variation.

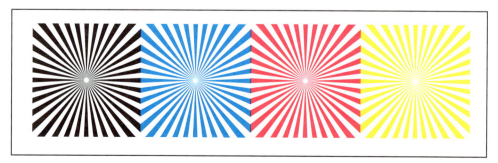

Figure 20-9. GATF Star Targets, enlarged 6×.

Color control bars. GATF provides the web printer with a variety of color control bars to monitor the pressrun. The typical color control bar consists of solid patches, tint patches, Star Targets, and Dot Gain Scale-IIs. These bars enable the press operator to monitor ink density, dot gain, print contrast, ink trapping, slurring, and doubling during the web pressrun.

The GATF/SWOP Production Control Bar was designed in cooperation with the Specifications for Web Offset Publications (SWOP) committee. This color control bar permits ink density, print contrast, and trapping measurements. It also indicates slur, doubling, and dot gain. Solid color overprints, solid ink patches, tint patches, Star Targets, and Dot Gain Scale-IIs comprise this test image, which easily fits into trim areas.

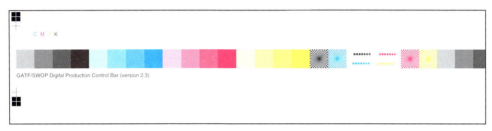

Figure 20-10. A portion of the GATF/SWOP Production Control Bar.

The GATF/SWOP Production Control Bar is just one of many color control bars available from GATF. GATF color control bars are available as film images or as digital files. Some are designed for use with scanning densitometers. Contact the GATF Order Department for a copy of the GATF Process Controls Catalog, which describes and illustrates color control bars and other quality control devices.

Dot Gain Scale-II©. The GATF Dot Gain Scale-II, incorporated in many GATF color control bars, allows quick visual inspection of dot gain in seven midtone increments: 1%, 2%, 5%, 10%, 15%, 20%, and 30%. A pattern of round dots and squares display dot gain when the round dots increase in size and touch the squares. The corresponding percentage that is printed below the image indicates the degree of dot gain.

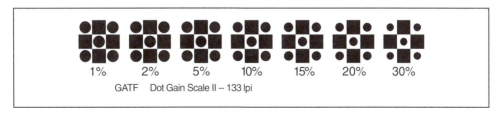

Figure 20-11. GATF Dot Gain Scale-II, enlarged approximately 20×.

Right Register Guide. The Right Register Guide, incorporated in some GATF test forms, indicates misregister between colors (in thousandths of an inch or hundredths of a millimeter). Upon visual inspection, the press operator can determine which way to move a specific plate cylinder to register the color printed by that plate.

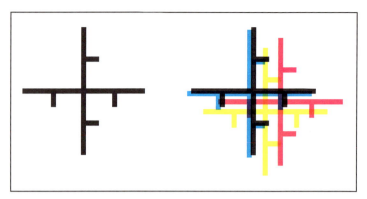

Figure 20-12. GATF Right Register Guide, enlarged approximately 20×.

QC Ink/Water Balance Strip. The "QC" strip, incorporated in some GATF quality control devices, is a test image that indicates variations in ink-film thickness, ink/water balance, and dot quality. The press operator compares inspection sheets to the OK sheet to detect any differences in the size of the image and nonimage areas of the printed strip.

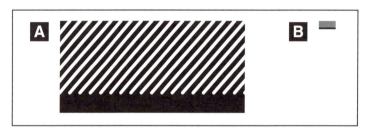

Figure 20-13. GATF QC ink/water balance strip. Image A shows a "QC" strip enlarged approximately 12×, and image B shows the strip at its actual size.

GATF Newsdot. The GATF Newsdot, incorporated in some GATF color control bars, is a test image that allows the printer to monitor dot gain during newspaper production. The minute size of the Newsdot makes it applicable to jobs that have no trim area for color bars, because it is not readily apparent to newspaper readers. Dot gain is visually assessed using a hand magnifier.

Figure 20-14. GATF Newsdot, enlarged approximately 20×.

Section X
Press Production

21 Control of Color and Register

Web offset presses may measure well over 100 ft. (30.5 m) from infeed to delivery. A press of great size has to be run with the same register accuracy as a 15-ft. (4.6-m) web offset forms press. Multicolor images must print within an acceptable tolerance—impression after impression. Technological advancements and sophisticated precision equipment permit press operators to produce quality printing on a press of any size.

As has been explained, paper can shrink and stretch on the press. From the unwinding roll to the press cutoff, the web runs through the press under tension.

Safety Precautions

- Do not operate equipment unless authorized.
- Check that all guards and shields are in place before operating the press.
- Never release a safe button that someone else has set.
- Do not start a machine that has stopped without an apparent reason.
- Before operating, check for persons, tools, or pieces of equipment between units and around the press.
- Remove all used plates, tools, and equipment from the press area and alert coworkers before starting the press.
- Wear hearing protection devices when working in areas with high noise levels.
- Do not permit people with jewelry, loose clothing, or long hair in the pressroom.
- Do not lean or rest hands on the equipment.
- Do not carry tools in pockets, thus avoiding the possibility of dropping them into the press or other hazardous locations.
- When making press adjustments, use only recommended tools that are kept in good, clean working condition.
- Keep clear of nips, slitters, and moving parts when operating the press.
- Never reach into the press to make adjustments when it is running.
- Do not attempt to remove hickeys from plates or blankets, or lint and dirt from rollers while the press is in motion.
- Do not wipe down cylinders, plates, rollers, or blankets while the press is running.
- Use only folded rags with no loose edges dangling when performing any cleaning process on press.
- Carefully follow instructions when mixing or handling chemicals.
- Keep a complete and accessible file of all service manuals, instruction manuals, parts lists, and lubrication manuals or charts for each piece of equipment in the pressroom.

Consult the manufacturer's press manual for information about specific safety devices, their proper identification, and operation.

When put under such a stress, the paper stretches. The dampening solution picked up by the web, dryer heat, variations in moisture content within the paper, and the composition of the paper all contribute to dimensional changes.

These factors, singly or in combination, change the length of the paper or cause the web to weave from side to side in the press. This causes images to print in and out of register as they change position on the web from unit to unit. The press operator adjusts the longitudinal and lateral position of the images printed by each unit during the pressrun. This is accomplished by maneuvering the web, adjusting the position of the plate cylinders, or both.

In addition to press and paper factors, prepress operations can also lead to registration problems, mainly in the areas of film assembly and platemaking. A range of image and web control problems and solutions relating to paper, press, and prepress are discussed in this chapter.

Register

Register, or *registration,* is the precise imposition of image elements—usually multicolor—relative to each other, so that they align exactly in final printed form. Each color printed must exactly align with the preceding colors printed, and the full-color printed image must be positioned exactly on the trimmed and folded product. Register is a consideration in prepress stages before presswork. Prepress procedures include the digitization of hard copy images (like photographs), page layout, film output (in conventional workflows), image assembly (either manually by positioning films or electronically by imposing image elements on computer), and platemaking. With the advent of modern digital prepress equipment, misregistration in prepress is less of a problem than ever before. However, the traditional prepress operations of film assembly and platemaking are potential sources of misregister, which means that a job on press could be out of register before printing begins. Prepress procedures and equipment must be scrupulously monitored to ensure accuracy.

Platemaking Considerations

Plates for the press may be made in one of two general ways: (1) digitally with a platesetter or (2) photomechanically with a vacuum frame and films. Though misregistration with platesetters is rarely a problem, there are some considerations. Many more variables must be controlled with traditional film-based platemaking procedures.

Platesetting. Platesetters are devices that rasterize (turn into tiny dots) the image information sent from the computer workstation. The platesetter then uses a laser or, in some cases, a light-emitting diode (LED) to expose the plate, exposing tiny dots in a microscopic grid pattern on the plate surface to form the image.

There are various designs of platesetters (these platesetter designs are presented in more detail in the chapter on plates). The design differences do not affect registration accuracy of plate images, but rather the speed of production and the plate types that can be handled.

When a platesetter is being used, unexposed plates are loaded in various ways. Some have automatic loading, which means that the platesetter mechanism picks up a single plate from a stack and moves and positions the plate onto the imaging drum or bed. Manual designs require the operator to load the plate by positioning it against head stops, which are small fingers to rest the plate against. If the plate is mispositioned when loading, the images will be out of register. The plate is held fast on the imaging drum or bed by vacuum pressure. This assures that no movement will occur during imaging.

Data that is sent from the computer workstation to the platesetter may contain errors that cause images to misregister. Normal practice is for the platesetter operator to send out the images to a proofing device to check for position. However, image elements are sometimes not positioned properly on the plates. The press operator must be alert to check the initial printed images carefully through a magnifier before continuing the pressrun.

Photomechanical platemaking. Misregister frequently originates in the vacuum frame, where plates are exposed while in contact with film. Misregister in the vacuum frame is one of the most common, troublesome, and expensive problems encountered by web offset printers that are not using a computer-to-plate workflow.

Conventional prepress procedure is to assemble film flats that are carefully registered using a pin register system. These flats are then used in a vacuum frame to expose the four or more plates needed for process color printing. The plates are punched with a die (a plate punch) and are also aligned on the die-shaped pins.

Even though register pins are used during film assembly and platemaking, each plate may not register perfectly with the others. This is rarely visually apparent and may not be discovered until the plates are used on press. When misregister is discovered on press, it is a common practice to check the film assembler's work by overlaying the flats on the register pins over a light table to check register. The films may register on the pins; however, this does not necessarily indicate that the plates register properly or that misregister is attributable to press settings. Often, problems that occur in the vacuum frame cause serious misregister. Three common causes of misregister in the vacuum frame are:

1. The film sticks to the glass and buckles or distorts during vacuum drawdown.
2. The platemaker uses improper pins in the vacuum frame.
3. The film expands or contracts between exposures.

Drawdown problems during photomechanical platemaking. As the rubber blanket on the vacuum frame is drawn against the glass, the film (or flat) contacts the glass and tends to adhere. As the vacuum drawdown continues and air is removed, the punched edge of the film and pins are drawn tightly against the glass. The vacuum holds the film flat and plate tightly between the blanket and the glass; therefore, a buckle may develop near the pins as the film is not stiff enough to shift with the frictional force between it and the cover glass. Often, this distortion causes misregister, as shown in the following illustration. The buckle in the film would not cause misregister if films on subsequent exposures were drawn to the glass in exactly the same manner; however, film randomly responds to vacuum pressure. Thus, the film first contacts the glass and adheres in different areas on different exposures to cause misregister.

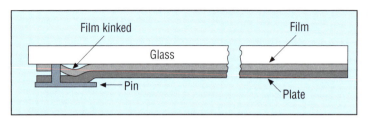

Figure 21-1. Film distorting near register pins, a cause of misregister.

Using the GATF Register Test Grid. Uneven vacuum, static electricity, or vacuum frame leaks contribute to unpredictable and uncontrollable drawdown. Follow this procedure to test the vacuum frame for repeat register capabilities.

1. Adhere register pins to the vacuum frame base in a position that corresponds to the holes in the film flats and plates.

2. Place a plate on the register pins.

3. Position a GATF Register Test Grid over the plate.

4. Make four separate exposures on the same plate; between each exposure, open the vacuum frame, remove the grid, and then replace it on the pins. After vacuum drawdown, make the next exposure.

5. After the first exposure, place four pieces of opaque tape vertically on top of the glass, one in each quadrant.

6. After the second exposure, place pieces of opaque tape horizontally across the vertical pieces on the vacuum frame glass.

7. After the third exposure, add pieces of tape diagonally across the previously taped areas. (Unmasked areas of the plate have received four exposures.)

8. Compare the image areas under the tapes to the background areas. Differences in density or line width, or multiple images of the grid pattern of the GATF Register Test Grid indicate misregister.

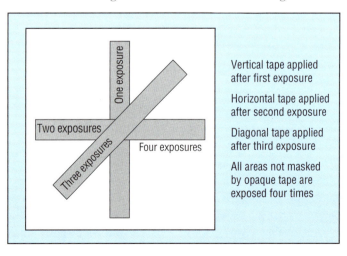

Figure 21-2. Tape pattern used to test the vacuum frame for repeat capabilities.

Vertical tape applied after first exposure

Horizontal tape applied after second exposure

Diagonal tape applied after third exposure

All areas not masked by opaque tape are exposed four times

Maintaining register in the vacuum frame. Some recommended practices to follow for achieving good register in the vacuum frame are as follows:

1. Be sure to use pins that fit tightly in the holes in both the plate and film. Ideally, the holes in the lead edge should be as far apart as practical.

2. Do not use pins that are tapered or beveled. The film can draw against the glass and move around in the area that is tapered.

3. Make sure that the pins are of the proper height, so that the film cannot move up and down significantly on them.

4. Punch films far enough away from image areas to prevent halation. Halation occurs when there is poor contact between the film and plate upon exposure. When this occurs, the image becomes distorted.

5. Be sure to maintain the vacuum pump according to the manufacturer's instructions. Most vacuum pumps should be flushed out every two months. All vacuum hoses and inlets should be unobstructed and in good condition.

6. Use a vacuum frame that is several inches larger on all sides than the plates.

7. Use antistatic devices to eliminate static charges.

8. Maintain the vacuum frame seals. Blanket rejuvenators can be used to clean the seals and keep them soft. Damaged seals should be replaced.

9. Use 0.007-in. (0.178-mm) polyester base sheets to resist distortion and achieve complete drawdown.

10. Draw down vacuum in the vacuum frame in two stages. If the vacuum is kept below 10 psi (pounds per square inch) for about 1 min. after the film is drawn against the glass, the film will slide against the glass more easily and give better register. Full vacuum (20–25 psi) should be reached before exposing.

11. Where register is critical or there is a chronic register problem, use a tail pin to prevent tail-end whip. The hole for this pin should be oblong and positioned in the center near the tail edge of the plate. Both the plate and flat should be punched with the same punch. The same system should be used when stripping flats.

12. Keep the vacuum frame glass and the film or flat clean and free of any sticky material. Depleted fixing solution may leave a deposit on the film that causes it to adhere to the glass.

13. Place a clear plastic overlay on top of the film to increase the number of sliding surfaces. This may improve register slightly.

14. Because misregister on some vacuum frames can be caused by excessive heat buildup in the frame, be sure that the exhaust fan is operating properly and that the vent or exhaust is not clogged on flip-top units. The top can also be left open between exposures for faster cooling.

Folding and Cutoff Register

Most printers use multipurpose register marks on the form to obtain backup register and folding accuracy. Usually, the register marks are short lines in the trim areas of

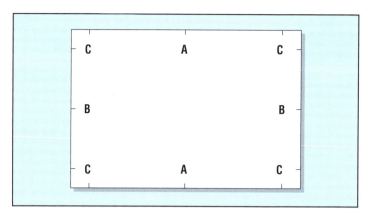

Figure 21-3. Register marks.

the signature. If the press operator is going to rely on these marks during the press-run, the image must be square on the plate before the plate is bent and mounted.

The "A" marks in figure 21-3 are helpful in locking up the plate on the cylinder squarely. When the "A" mark on the leading edge is lined up with the "A" mark on the trailing edge just across the gap, the plate is assured of being mounted with the image square to the cylinder. To center the plate from side to side on the cylinder, it is useful to scribe a mark at the center of the tail edge of the plate cylinders. The "A" mark is then aligned with the scribe mark.

Usually the "A-A" line is the line of the former fold. It is good practice to check the actual fold against these marks to measure former fold accuracy. The first jaw fold is made along the "B-B" line. Misalignment of the first jaw fold will result in an error in cutoff. Thus, the "B" marks serve as a guide to adjust the cutoff compensator, which is a device that retards or extends the fold.

Many shops put the "C" marks on all four-color process work. These marks provide a simple visual reference that indicates which colors are printing long or short. More information on this concept is presented in the section on relative print size in chapter 8.

Backup

Backup describes the proper positioning of the image on the top side of the web relative to the image on the bottom side. Establishing backup is a cylinder timing issue, which was described in detail in chapter 9. To reiterate, initial backup is established between the top and bottom couples of one key unit, which becomes the reference unit for all other printing couples. Once this unit is in register and backup is acceptable between the top and bottom couples of this unit, all other couples of the press are adjusted in reference to it. This will ensure proper backup at all times.

Plate Bending

A high-quality plate bending device (bending fixture) and proper plate bending procedures are critical to accurate register. Poor bending produces plates that print skewed images, because the plates will not mount squarely on the cylinder.

Procedures for Press Start-up and Checking Instructions from Previous Shift

1. Web the press, roll stand, or splicer and set up the folder.
2. Check that all components are ready for the run.
3. Make start-up checks for the infeed.
4. Make start-up checks for the active printing units.
5. Check and set up the dryer according to the manufacturer's instructions.
6. Make checks for the chill rolls.
7. Make checks for the folder.
8. Develop a checklist (to be used by the press crew for the start-up) of the press configuration and accessories of the web press being used.
9. Review job ticket or production work order with outgoing shift supervisor.
10. Review press record form with outgoing shift operators regarding any problems that should be anticipated. (See the sample form, figure 6-4, in chapter 6, "Controlling Paper Waste.")

Unless the plate or the plate cylinder can be counter-skewed (as it can be on some presses), no amount of circumferential and lateral movement will bring the image into register. To make matters worse, skewed plates or skewed cylinders may lead to plate cracking or squeeze differentials. The only acceptable solution is the proper positioning of the plate images and proper plate bending.

Color Register

Full-color reproduction of an image requires the use of four (sometimes only three and sometimes as many as six) specially matched inks. Pinpoint register of the successive colors is necessary to reproduce detailed images. Color register is achieved by fitting all colors to a key color or by printing to register marks in trim areas of the form. The blanket-to-blanket press operator always has two sets of colors to register for each multicolor web run: one on the top and one on the bottom.

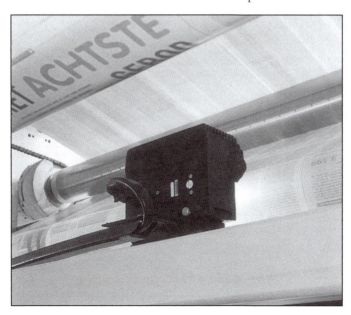

Figure 21-4. QTI's Register Guidance System V scanner in operation. (Courtesy QTI)

Color register is achieved by moving the plate cylinders in each unit. The plate cylinders can be moved in two directions: circumferentially (forward and back) and laterally (toward the operator side or toward the gear side). As discussed above, some plate cylinders may also be skewed. Any of these movements can be made independently of the others, and when carried out in combination can bring any point into register (within a variance of about 0.125 in., or 3 mm).

On older presses, these adjustments are made manually at the unit. This requires the press operator to unlock and shift the unit the proper amount. On newer presses equipped with remote control consoles, register adjustments are made through small motors mounted on each printing couple. On these presses, each printing couple is controlled remotely by a separate circumferential and lateral control on the master console.

Controlling Color

The problem of controlling and maintaining color during the pressrun involves controlling the thickness of the ink film, the ink trap in overprints, halftone tint values, and dot gain. The ink film thickness is controlled by observing color changes and density differences that occur as the thickness varies and making the necessary inking adjustments.

The most useful tool to control color is the densitometer (see detailed discussion in chapter 20), which provides numerical feedback to changes in density and halftone tint values. Another useful tool is a dot gain scale, typically incorporated into the color control bar, that visually indicates changes in dot gain and, usually, slurring and doubling.

A densitometer equipped with red, green, and blue filters is necessary to measure the reflection densities of the magenta, cyan, and yellow inks and their overprints. In addition to density measurement, the use of a densitometer allows trapping, dot gain, and print contrast to be calculated and monitored during the pressrun. (See chapter 20 for the equations used to make these calculations.)

Procedures for Maintaining Consistent Quality and Color

1. Position multicolor register marks on the form in order to obtain backup register and folding accuracy.
2. Position the image on the top side of the web in register to the image on the bottom side.
3. Wash blankets regularly in order to maintain image quality.
4. Maintain minimum ink and minimum water at all times during the pressrun.
5. Remove a signature from the delivery every few minutes.
6. Compare the delivered signatures to the okayed signature.
7. Use a magnifier to double-check image register and make the necessary adjustments to the press registration devices.
8. Use a densitometer to measure ink densities in solids and dot gain in screen tints; adjust ink feed if necessary.
9. Verify folds as indicated by the marks and make adjustment or realign the web as needed.

Figure 21-5. *The control room, or "quiet room" for a Goss Universal newspaper press. (Courtesy Goss Graphic Systems, Inc.)*

Causes of color variation. Variations in ink feed is an apparent cause of color variation. Another leading cause is uncontrolled dot gain. When dot size changes on press as a result of variations in ink feed or printing pressure, it produces inconsistent color.

A third cause of color variation is trapping. As explained in chapter 20, trapping values express how well one process color ink film is adhering to the previously printed ink film. The trap equation given in chapter 20 indicates apparent trap, because effects such as changes in gloss between single- and two-layer ink films could influence the percentage.

In printing, the amount of ink film transferred to a previously printed ink film can be more than, equal to, or less than that transferred to paper. The amount of transfer is referred to as *trap.* In printing, wet trap seldom exceeds 90% in red and blue, although trap exceeding 90% occurs often with green; 80–90% trap values are most common. Wet trap values less than 75% on a sheetfed press are considered unacceptable. Although a high trap value is important, it is more important that the ink trap on production sheets match that of the OK sheet. Any change in ink trap causes changes in hue, saturation, and lightness of overprints. Such changes can be detected in the overprint portion of the color control bar when compared to those on the OK sheet. The following recommendations help to minimize trap problems:

- Use only inks that are balanced correctly for color strength. Trapping is improved if the ink film thickness increases slightly from one unit to the next. In order to increase ink film thicknesses in such a manner, the color strengths of the ink must be adjusted accordingly.

Figure 21-6. *Omnicolor color control console, designed to be used with the Heidelberg M-600 web offset press. (Courtesy Heidelberger Druckmaschinen AG)*

- Use tack-rated inks for wet-on-wet printing; that is, purchase a set of inks so that the highest tack-rated ink is printed on the first printing unit, the second highest tack-rated ink on the second unit, and so on. The difference in tack of consecutively printed inks should be only one or two tack units, as measured using an Inkometer.
- Make sure that the ink/water balance is correct and that the inking system is operating properly.

Specifications for Web Offset Publications

The Specifications for Web Offset Publications (SWOP) was first published in 1975 for magazines and catalogs printed by web offset lithography and gravure. The ninth edition of SWOP was published in 2001. The impetus for SWOP was the increasing popularity of web offset printing of publications in the 1960s and early 1970s. At that time materials supplied to printers (proofs and film) varied considerably, making it difficult for printers to obtain consistent color. In late 1974, an industry group met informally to discuss the writing of specifications for material supplied to web printers. Among other specifications, SWOP includes screen ruling, screen angles, gray balance, and total ink coverage for separation films; inks, substrate, and dot gain for proofs; and inks, substrate, density, and dot gain for production printing.

Adhering to SWOP helps is to ensure that printers receive uniform input, better enabling the printer to produce quality reproductions that meet the expectations of the advertisers. The printer is responsible for visually matching the supplied proofs, provided they have been made to SWOP specifications.

Following are just a few of the requirements as stated in the ninth edition of the SWOP booklet:

- A content proof made from the supplied digital file must be furnished to the prepress service supplier with all supplied digital files.
- Thin lines, fine serifs, and medium or small lettering should be restricted to one color.
- Use dominant color (usually 70% or more) for shape of letters and, where practical and not detrimental to appearance of job, make lettering in subordinate colors slightly larger to reduce register problems. Small letters with fine serifs should not be used.
- 133 lines/inch is the recommended nominal screen ruling.
- In general, for four-color wet printing the sum percentages of tone values (total area coverage) should not exceed 300%.
- All proofs must contain a color control bar in order to be considered an acceptable SWOP proof and be clearly marked with available job information and proofing system identification.
- Ink density and color in proofing should be controlled by the use of the SWOP Hi-Lo Color References.

Other topics include image trapping, vignette or fadeaway edges, screen angles, gray balance, film properties, image orientation on film, register marks, proofing, color bars, press proofing, print contrast, tone value increase tolerances. For additional information, please refer to the ninth edition of the SWOP booklet, available from SWOP, Incorporated as well as the GATF/PIA Bookstore, or visit www.swop.org.

Standard Viewing Conditions

Proper lighting and viewing conditions are important in the pressroom. Light levels must be sufficiently high around the press to permit the press crew to install plates, blankets, and packing easily and to make the necessary press adjustments without eye strain. Having a light shining between each pair of press units is recommended. However, the lights should not be placed directly above the press, but off to one side. Aisles around the press should also be illuminated for safety reasons.

In addition to lighting around the press, the booths or tables where press sheets are viewed must not only have sufficient lighting but must have lighting that conforms to recommended standard viewing conditions. To avoid misunderstandings, printer, supplier, and customer must agree on the illumination under which the print is to be viewed. Standard viewing conditions, consisting of standard lighting and surround conditions, are necessary to communicate the desired results and to ensure accuracy and consistency in color reproduction. Viewing conditions for the graphic arts industry are specified by the major standards organizations, the American National Standards Institute (ANSI) in the United States and the International Organization for Standardization (ISO) for international standards. ISO 3664:2000, *Viewing Conditions for Graphic Technology and Photography,* is the currently recognized international standard.

Following is a summary of the standard viewing conditions for photomechanical reproductions:

- Correlated color temperature of lighting—5,000 K, which closely represents average white light.
- Spectral power distribution (the relative output of the light source across all the wavelengths of the visible spectrum)—D_{50}. This light source consists of essentially equal amounts of energy in the red, green, and blue portions of the visible spectrum.
- Color rendering index—at least 90. The color rendering index provides a measure of the degree to which the perceived color of objects illuminated by the source conform to those of the same object illuminated by a reference illuminant.
- Light intensity of print illumination—2,000±500 lux.
- Uniformity of print illumination—the intensity at the edges of the viewing plane should be at least 60% of the intensity at the center.
- Surround—matte, neutral gray of 10–60% reflectance, 60% being equivalent to Munsell N8/.

Several companies manufacture viewing booths in which the lighting conforms to industry standards for illumination of the press sheet.

Section XI
Preventive Maintenance

22 Maintenance and Mechanics

Maintenance of printing plants and equipment is essential to a graphic communications organization. The delays in production that result from equipment failure can create problems because printing is a service industry.

Web offset presses are complex mechanical systems that are required to run with precision at high speeds. Each component of a web offset press performs a specialized function, interrelated to the functions of other press parts. Press manufacturers prescribe operating procedures that should efficiently produce optimum results.

The initial investment in a web offset press is extremely high; therefore, the equipment is expected to run continuously for a long time. Printers must implement the manufacturer's established preventive maintenance program to keep equipment in top running condition. The pressroom, too, must be kept as clean as possible, and regular cleaning should be scheduled. A reliable press is productive and profit-generating; regular maintenance contributes to dependability. Income is not realized during downtime; however, a breakdown may require much longer downtime than a scheduled shutdown for maintenance. Excessive repair costs and overtime pay may also be incurred as the result of a breakdown. Furthermore, jobs fall behind schedule and fewer are printed as a result of lost press time.

Identifying and servicing press components that require regular attention help to eliminate major repairs. Timely, continuous maintenance is essential because the pieces of equipment need to be serviced at different intervals. Stocking spare parts contributes to more efficient operation by reducing downtime. Management should compile a central file of press manufacturers' manuals. These manuals can be used to develop a preventive maintenance checklist from the manufacturers' recommendations. A malfunction report allows the press crew to itemize problems that require service. Furthermore, the job planning board should be used to post scheduled maintenance.

This chapter briefly discusses the primary types of maintenance, explains how to establish a preventive maintenance program, focuses on equipment lubrication (probably the single most important aspect of equipment maintenance), and provides the reader with an understanding of the basics of mechanical, electrical, pneumatic, and hydraulic systems.

Special note: Much of the material in this chapter has been adapted from *Maintaining Printing Equipment* (Apfelberg, Herschel L., Pittsburgh: Graphic Arts Technical Foundation, 1984).

Types of Maintenance

An effective equipment maintenance program is made up of four elements: (1) restoration maintenance; (2) preventive maintenance; (3) prediction maintenance; and (4) safety. The degree of how much each maintenance element is applied may vary, depending on the nature and technology of the printer's equipment.

Restoration maintenance. *Restoration maintenance,* or *repair and corrective maintenance,* is the most common maintenance performed. Restoration maintenance consists of repairing broken or damaged equipment to restore necessary operating conditions. Another part of restoration maintenance includes replacing abnormal or worn parts that cause materials and product to be out of specification. Restoration maintenance is basically a fix-it-when-it-breaks function, addressing sporadic and sudden equipment losses that are actually unscheduled equipment downtime.

There are two types of restoration maintenance activities: emergency restoration and scheduled restoration. *Emergency restoration* refers to addressing a sudden and total shutdown of the equipment or the failure of any of the equipment's safety elements, such as safe-stop buttons and interlock guards. If the equipment's component failure does not shut down the machine or compromise safety or quality, *scheduled restoration* is performed at a designated time, usually on weekends when the equipment is not scheduled for production.

Preventive maintenance. Preventive maintenance is a program approach to preventing the sporadic and sudden failures that will totally shut down equipment. The main things that are required for a quality preventive maintenance program include knowledge of the equipment's operating components, structured scheduling, and the discipline to adhere to standards and procedures.

Preventive maintenance activities are assigned on a periodic schedule. Normally these maintenance activities will be segregated into daily, weekly, monthly, quarterly, semiannual, and annual schedules.

Common Checkpoints of Preventive Maintenance

- Safety buttons and interlock guards
- Ink and dampening rollers
- Ink and dampening roller bearings
- Transfer gripper bar cam followers
- Fountain solution recirculation tank pumps
- Double-sheet detectors
- Web turn-bar bearings
- Blanket tension bars
- Sheetfed delivery chains and gripper bars
- Press ink fountain, cleaning and maintenance
- Spray powder unit, cleaning and maintenance
- Plate/blanket cylinder bearer ring contacts and condition
- Oil in gear compartments
- Fountain solution recirculation tanks and hoses

Typical activities of a preventive maintenance program will include periodic cleaning, inspection, lubrication, adjustment, and replacement of equipment components that are subject to high degrees of friction and repetitive operations. Common checkpoints of preventive maintenance for press equipment are shown below.

Prediction maintenance. Prediction maintenance takes preventive maintenance to a higher level. ***Prediction maintenance*** utilizes more state-of-the-art technology to predict when equipment components will need maintenance before they fail. Major maintenance and overhaul intervals are now being determined by scientific methods and accurate data analysis. Prediction maintenance requires monitoring specific elements that could cause catastrophic failure of the equipment. Prediction maintenance monitoring activities include lubrication analysis, vibration analysis, thermal analysis, crack detection, and noise monitoring.

Maintenance safety. Safety is the most important issue for consideration when addressing printing-plant maintenance effectiveness. Establishing the proper safety elements of a maintenance program should include education in safety and required protective gear, training in the use of MSDS sheets and lockout/tagout procedures, developing the proper lifting techniques, and establishing documented checklists. The safety checklists should cover extension cords, appliances, drive motors, electrical control panels, maintenance shop power tools and equipment, portable electrical tools, hand tools, ladders, compressors, compressed gases, and solvent safety containers. Poor safety practices can cause serious, possibly fatal, personal injuries, and damage to equipment. (See safety precautions on next page.)

Establishing a Preventive Maintenance Program

Preventive maintenance is simply an organized program that will keep facilities and equipment in the best possible condition to suit the needs of production. Not all machinery is given the same time and effort when preventive maintenance programs are being established. Some equipment is more valuable, or more central to the heart of production, or more sensitive to losing adjustments and must be considered first in a well thought-out program.

Preventive maintenance should be looked upon as a necessary production center, without which deadlines will not be met and quality will suffer. There are several reasons for this increased emphasis on the maintenance function:

- Increased mechanization and automation
- Increased complexity of equipment
- Increased parts and supplies inventories
- Tighter control over production
- Tighter delivery schedules
- Increased quality requirements
- Rising costs in terms of parts and labor

Safety Precautions

- Exercise extreme care and precaution when working on any printing press.
- Observe and practice all safety rules, regulations, and advice given in the press manual.
- Never make repairs and adjustments or perform maintenance and cleaning jobs when the machine is running. Any work on the press, under the infeed mechanism, around the folder, on the motor, on the main control panel, or on other electrical or moving parts of the equipment must be performed only if the main switches on the control panel are in the OFF position, and locked or tagged out.
- Heed all verbal and written instructions before performing maintenance or operating the press.
- Wear protective gear for eyes, ears, head, and feet where necessary to protect against injury.
- Wear clothing that will not become entangled in any part of the press equipment.
- Stand clear of the press immediately when the run signal is given.
- Make sure the press is completely stopped before touching any of its operating parts.
- Check all safety devices on the press every day to ensure that they are reliable and working.
- Never switch off safety devices or remove or otherwise bypass guards.
- Before working on the press, check to make sure it has been put on "safe" using the "stop (security)" push button and follow the proper lockout/tagout procedures where necessary to ensure the safety of everyone involved in press maintenance.
- Check that all guards, covers, and swiveling footrests are securely fastened or completely locked in place before performing maintenance or operating the press.
- Check that stairs, footrests, running boards, gangways, platforms, and other equipment surfaces are clean and free of grease. Do not place tools and supplies on these surfaces.
- Grasp handrails securely when ascending the platforms, standing on the platforms, and before leaving the platforms.
- Only clean the ink fountains while the press is stopped to avoid injury and press damage.
- Do not work on moving rollers with rags, tools, etc., because of the high risk of accident and damage.
- Reinstall guards immediately after removing the press washup devices.
- Use the "reverse" button on the press only for plate removal—not for cleaning or gumming cylinders, etc.
- Do not operate equipment unless authorized.
- Check that all guards and shields are in place before operating the press.
- Never release a safe button that someone else has set.
- Do not start a press that has stopped without an apparent reason.
- Check for persons, tools, or equipment between and around the press before starting it.
- Remove all used plates, tools, and equipment from the press area and alert your co-workers before starting the press.
- Wear a hearing protection device when working in areas with high noise levels.
- Do not permit people with jewelry, loose clothing, or long hair in the pressroom.
- Do not lean or rest hands on the press.
- Do not carry tools in pockets to avoid the possibility of dropping them into the press or other hazardous locations.
- When making press adjustments, use only recommended tools that are kept in good working condition.
- Keep clear of nips, slitters, and moving parts when performing maintenance or operating the press.
- To perform any cleaning process, use rags folded into a pad with no loose edges dangling.
- Keep a complete and accessible file of all service manuals, instruction manuals, parts lists, and lubrication manuals or charts for each piece of equipment in the pressroom.

Consult the press manual for information about specific safety devices, their proper identification, and operation.

Implementing a Preventive Maintenance Program

1. Establish and maintain equipment records.
2. Maintain a central file of all press manufacturers' manuals.
3. Develop a preventive maintenance checklist from the manufacturers' checklists.
4. Develop a checklist of possible malfunctions so that the press crew can list problems that require attention.
5. Schedule regular preventive maintenance as you would a normal production job.
6. Complete a machine inspection form and perform periodic inspection of the equipment.

Although the primary objective of preventive maintenance is to keep the printing equipment in top operating conditions and prevent breakdowns, there are also several subobjectives:

- Less downtime
- Fewer large-scale repairs
- Fewer repetitive repairs
- Less standby equipment needed
- Reduced maintenance costs and lower repair costs
- Identification of high-maintenance items
- Shift from inefficient "breakdown" maintenance to less-costly preventive maintenance
- Better work control
- Better spare-parts control
- Safer working conditions
- Improved cost control
- Increased employee morale
- Lower unit cost
- Longer equipment life
- Maximum equipment availability
- Minimum pollution
- Reduced maintenance inventory

A preventive maintenance program has four areas of concern: (1) lubrication, (2) an inspection system, (3) cleaning, and (4) parts adjustment and repair.

Lubrication is discussed extensively in the next section. In regard to preventive maintenance, the primary lubrication-related concerns are type of lubricant, lubrication device or system to use, general effectiveness of the lubrication program, lubrication schedule, and personnel to perform lubrication.

Inspection is vital to the proper maintenance of all equipment. The following factors help to determine the equipment inspection schedule: age, condition, value, severity of service, safety requirements, hours of operation, susceptibility to wear, susceptibility to damage, susceptibility to losing adjustment, and operator experience.

Cleaning is essential to preventive maintenance because it allows one to inspect for proper lubrication, wear factors, proper adjustment, and malfunctions. It is also important in developing an attitude of respect for the proper operation of printing equipment. A clean machine gives all concerned the feeling of producing a high quality printing product, and in fact if one has an attitude of respect for the equipment the end product will be more satisfactory.

Parts adjustment and repair must be performed on a regular basis if an acceptable printed product is to be the end-product. Some organizations replace certain parts regularly, determining the effective life of a part and then replacing it just before it would wear out. This is an extremely successful approach if the effective life of a part is known. However, in the printing industry there are very few parts that have a predetermined life expectancy and, therefore, inspection programs help to determine when to replace parts. Parts replacement will decrease if machine adjustments are checked on a regular basis. Regular inspection of machine adjustments ensures consistent end-product printing quality as well as cost savings in that wear factors will be decreased.

The preventive maintenance program should be structured according to the frequency at which specific items require service.

- *Daily* maintenance operations must be performed every day.
- *Weekly* functions may be divided among the days of the week but should be performed on designated days.
- *Monthly* service includes those items that must be checked after every 500 hours of operation. The frequency at which these operations are performed varies, depending upon the number of shifts that a printing press is run each day. A two-shift schedule accumulates 500 operating hours every 8–9 weeks; a single-shift operation, 16–18 weeks; and a three-shift operation, 4–5 weeks. Using an hour meter to record running time assures accurate and timely maintenance.
- *Semiannual* procedures are to be performed after every 1,500 operating hours; add them to the 500-hour list every third time.
- *Annual* maintenance is performed after every 3,000 hours in addition to all other procedures. The frequency at which the semiannual and annual operations are performed also varies, depending upon the number of shifts that a press is run each day.

Maintenance checklists help to ensure that all equipment is serviced as required. The person who performs a procedure should initial and date the appropriate checklist accordingly.

To enhance the efficiency of the preventive maintenance program, some press manufacturers color-code the lubrication points that need to be lubricated at the same interval. Furthermore, special lubricants for specific functions should be clearly identified, and the location for their use should be clearly marked. Preventive maintenance should be treated like a job—i.e., placed in the schedule and completed on time.

Lubrication

Lubrication Principles

Simply defined, **lubrication** is the adequate application of a lubricant to the surface of a part to coat it, to maintain or reduce friction. Proper lubrication increases printing equipment life, facilitates quality reproduction, and reduces downtime and maintenance. Lubrication of moving equipment components and parts helps to prevent friction, and in turn to prevent deterioration of the parts. The primary functions of lubricants are to minimize friction, control wear, maintain normal temperatures, dampen shock, and remove contaminants. Lubricants minimize friction with higher-viscosity film on the moving parts, preventing energy loss from friction. The right lubricant will cause the equipment to run smoother and faster, and will prevent component wear and deterioration. Lubricants can also provide seal formation.

Control of friction. Lubricants control friction under fluid film conditions through the effect of their viscosity on film thickness and energy losses. These energy losses (resulting from friction) are important for the printer to keep in mind as they can result in machinery slowdowns and also the need for larger motors to compensate. Increased friction also results in unnecessary wear.

Control of wear. Wear can occur in lubricated systems by abrasion, corrosion, and metal-to-metal contact. The lubricant plays an important role in combating each type of wear.

The excessive wear that results from a poorly chosen lubricant will cause particles of a part to break off and grind, or abrade, into a moving part. Also, if a lubricant can't absorb or wash out offset spray, paper dust, and ink particles, abrasion will result.

Corrosion, the deterioration—e.g., rusting—of a metal, is caused by oxidation, when the air comes in contact with unlubricated metal parts, or by the contact of developers and fountain solutions with moving printing equipment that has not been protected by a lubricant. The role of a lubricant is, therefore, twofold. When machinery is idle, the lubricant may act as a preservative. When the machinery is in actual use, the lubricant controls corrosion by coating lubricated parts with a protective film (which may or may not contain additives) to neutralize corrosive materials and to wet the surface preferentially.

Metal-to-metal contact is caused by the use of an improper lubricant or no lubricant at all. This contact is very dangerous because it can create "hot spots" that can literally weld parts together when exposed to the heat and pressure of the day-to-day printing operation.

Control of temperature. Temperature control is an important function of lubricants. Lubricants can conduct the heat caused by friction, thus cooling the rotating parts. The effectiveness of control depends upon the amount and type of lubricant supplied, the ambient temperature, and the provision for external cooling such as fans, air conditioning, and circulating systems.

Dampen shock. Lubricants function as shock-dampening fluids, used in shock absorbers, which may be found on dancer rollers of web presses and elevators. These fluids are often vegetable oils, such as peanut oil.

Removal of contaminants. Lubricants—oils as well as greases—are called upon to remove contaminants in many systems. This treatment requires that an oversupply of grease constantly be fed to the lubricated parts on a once-through basis. Handy in cleaning out press bearings, this process can be considered a system of purging.

Forming a seal. A special function that can be performed by lubricating grease is that of forming a seal. Seal formation is important in keeping dirt out of bearing surfaces, thereby avoiding abrasion and undue parts replacement. Be aware that a grease seal may sometimes be mistaken for the overapplication of grease.

Types of Lubricants

Four types of lubricants encountered in maintenance are petroleum oils, animal and vegetable oils, grease, and graphite.

Petroleum is a complex mixture of hydrocarbons of varying volatility together with small quantities of substances containing oxygen, nitrogen, sulfur, and ash (derived from the vegetable and animal organisms from which the petroleum was formed). Fractional distillation and other processing of petroleum yields a number of products, including various grades of oil. Many petroleum fractions have to be treated with special additives to give them the desired properties. For example, machine oils are processed to be nonresinous, pale in color, odorless, and oxidation-resistant.

Animal and vegetable oils—fatty or greasy liquid substances derived from animals and plants—were the original substances used as lubricants. However, they are sensitive to exposure to air and therefore oxidize readily. This oxidation process causes the oil to become rancid, creating fatty acids that can pit metal parts. Oxidation also causes oil to gum up, another detriment to the lubrication process. Yet another disadvantage of these oils is that they decompose when subjected to high temperatures, thus losing their ability to lubricate. Last, they support bacterial growth, which can become a health problem. The vegetable oil that is an exception is olive oil. With regard to lubrication purposes, pure olive oil is probably the purest and most perfect vegetable oil known. Olive oil is therefore often used to lubricate delicate instruments where oxidation and bacterial growth would be harmful.

Grease is oil that has been mixed with soap to thicken it so it will stand up under greater heat and pressure. Greases are classified broadly according to the type of soap used in their manufacture. The general-purpose greases, such as calcium-(lime-) base, are used for lubricating surfaces that have temperatures up to a 160°F (71°C); sodium-base greases are usable at temperatures up to 250°F (120°C); and lithium-base greases are usable at temperatures up to 275°F (135°C). Some special greases are made for temperatures that exceed 275°F.

Dropping, or *melting, point* is the temperature at which a grease will soften. Generally, the higher the temperature the better the grease is suited for high temperature operation.

Soft carbon, which has a metallic luster and a slippery texture, is called *graphite* or black lead. Graphite, a dry lubricant, is used in areas where exposure to dust and other airborne solids would cause gumming of liquid lubricants. Graphite is also used as a polishing agent. Another successful use is the lubrication of door lock keyways, which need lubrication but would be impossible to keep clean if a liquid lubricant were used. Graphite can be applied directly or as a spray. Graphite, however, should not be used in conjunction with liquids, as it will become sticky and lose its ability to lubricate.

Characteristics of Lubricants

Wetting ability is probably the most important characteristic of a lubricating film. In short, how well the lubricating film actually wets or clings to the surfaces of the parts in motion relates directly to the friction taking place. A simple test is to coat the surface that is to be lubricated with the oil you are testing and then dip it in water. With a satisfactory lubricant the surface will retain the film of oil when dipped into the water and the water clearly drains off when it is removed.

Surface tension is related to a lubricant's ability to be cohesive and is directly related to its viscosity, temperature, and emulsion-forming tendency.

Viscosity is the relative strength of a lubrication film, determined by the relationship of tack and flow. The higher the viscosity index of a lubricant, the less it will be affected by temperature. As temperature increases, the surface tension of all oils decreases because of the proportionate drop in viscosity. The lubricant will look thinner and generally will not perform as well.

Adhesion is related to the wetting ability of a lubricating film. Generally, a lubricant with good wetting ability has good adhesion. The two main factors that affect adhesion and wetting ability are the smoothness or polish of the surface to be lubricated and the refinement of the lubricant.

Additives and Their Importance

Additives are used with lubricants to improve their performance. Additives are available in many varieties and serve many uses in the industry, as described below.

An *antioxidant* is an additive that inhibits oxidation. Oxidation in grease and oils causes gum, lacquer, and sludge buildup, which results in expensive repairs and excessive cleaning of equipment. Most lubricants have antioxidants added.

A *foam depressant* is an additive that prevents or retards the formation of foam when oil and air are agitated together. Foam, which is the addition of air, reduces effective lubrication.

Anticorrosion additives and *rust preventives,* very important in circulating oils and some types of greases, retard corrosion and rusting of metal surfaces exposed to moist air or water. The petroleum industry includes these additives in most of its premium grade lubricants to increase the wetting ability and to neutralize any acidity that could cause pitting.

Extreme-pressure additives, widely used in situations where meshing gears are not only sliding but rolling at the point of contact, are additives intended to reduce friction. Their reduction of friction is twofold: (1) they resist being wiped off at high rubbing speeds, and (2) they resist being squeezed out under high pressures at high temperatures. These additives are very important in printing equipment utilizing hypoid gears.

Establishing a Lubrication Program

The first step in establishing a lubrication program is selecting or determining the proper lubricants to used. Equipment manufacturers specify the precise lubricant to be used on each designated lubrication point. Often these recommendations are specific to the point of telling the printer the particular manufacturer—i.e., Shell Darina Number 2 E.P. Often the terms of the machine warranty require that these recommendations be followed. Never substitute one lubricant for another, unless the manufacturer has approved such a substitution. Improper lubrication may result in serious damage to the equipment. (See figure 22-1 and the accompanying three-page sidebar.)

Performing an equipment survey to determine lubrication needs of equipment is helpful in establishing an effective lubrication program. The survey should indicate what lubricants are being used, the frequency of application, the method of application, the condition of the machinery, and any other general observations pertaining to the equipment. The lists showing lubrication requirements should be compiled from manufacturers' manuals and from experience.

Important to any lubrication system is the standardization of lubricants through the development and maintenance of standard lubrication specifications. Each lubrication specification is identified with a code number that states the requirements of the lubricant for oils in terms of such qualities as viscosity, flash and fire point, specific gravity, pour point, and oxidation stability. For greases, the requirements include penetration, type of soaps, dropping point, viscosity, and extreme-pressure (E.P.) additives. All lubricant specifications should be subjected to constant revision and testing, which will allow for substantial savings, both in terms of purchase cost and in improvements to machine efficiency.

After you have selected the correct lubricant, only about one-third of the job is finished. The most important task is to set up and organize a lubrication system and schedule. Printing plants should use lubrication charts to implement the proper lubricant program. The use of these charts cannot be emphasized enough, as a chart is the only way to be sure that the proper types of lubricants have been used at the proper intervals with the proper methods of application and that different parts of a particular piece of equipment have been lubricated properly. The person who lubricates the equipment should date and initial the chart.

Maintenance personnel must be trained in the proper application of lubricant and informed continually as to uses and specifications of all lubricants being brought

(Text continues on page 379)

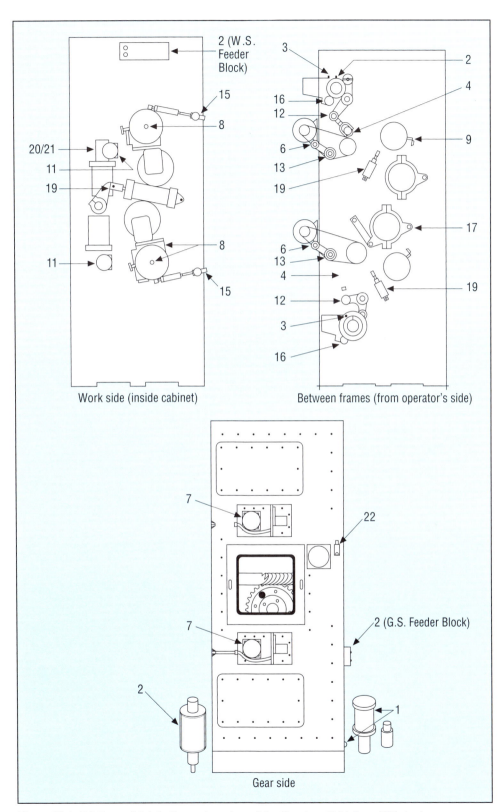

Figure 22-1. Printing unit lubrication and service points. (Courtesy Heidelberg USA, Inc.)

LUBRICANT MANUFACTURER RECOMMENDATIONS

LUBRICANT LETTER	A*	B*	C	D
	Mobil SHC 630 (Synthetic Oil)	Mobiletemp SHC-32 (Synthetic Grease)	600W Cylinder Oil	DTE Light
	Mobilegear 630 (Natural Oil)			

PRINTING UNIT LUBRICATION AND SERVICE POINTS

This chart covers the printing units only. Most of the points are shown in the illustration on page 375.

Item	Component	Procedure	Lubricant
		DAILY	
1	Circulating oil system behind pump, fill as necessary	Check sump oil level in window	A
2	Trabon automatic greasing system	Check reservoir level, fill as necessary	B
3	Ink fountain shoes	Grease	B
4	Inker cams (4 per unit)	Oil	A
		WEEKLY	
6	Dampener brush roller drive belts	Check tension and wear	—
7, 8	Circumferential (7) and lateral (8) register systems	Check for freedom of movement	—
9	Bearer wiper felts	Remove, check for wear and dirt, replace as necessary	—
10	DUO-TROL pan and slip roller gears	Grease	B
11	DUO-TROL pan roller drive brake	Clean by squirting with alcohol	—
		EVERY TWO WEEKS	
12	Ink ductor linkage (1 fitting on each side)	Grease	B
13	Brush dampener drive belt idler	Grease	B
		MONTHLY	
1	Circulating oil system	Replace external filter	—
8	Lateral register (4 fittings each)	Grease	B
15	Plate cylinder cocking devices (operator's side)	Oil	—
16	Fountain pivots (2 per fountain)	Grease	B

Item	Component	Procedure	Lubricant
		MONTHLY (continued)	
17	Impression linkage (4 fittings each side)	Grease	B
18	Plate and blanket cylinder eccentric boxes (4 remote fittings on edge of operator's side frame, 2 inside gear side frame)	Grease	B
19	Air cylinder pivots (all)	Oil	—
20	Brush dampener pan roller drive gearboxes	Remove level plug on side, fill as necessary	C
21	DUO-TROL roller drive gearboxes	Remove level plug on side, fill as necessary	C
22	Phaser	Grease	B
		EVERY SIX MONTHS	
1	Circulating oil system	Drain, remove, and clean strainer replace external filter, refill, clean pump cooling fan	A
20	Brush dampener pan roller drive gearboxes	Drain and refill to level plug on side	C
21	DUO-TROL roller drive gearboxes	Drain and refill to level plug on side	C
23	Dropdown drive bearings and tensioner on last unit	Grease	B
		YEARLY	
24	Check unit timing	See Section II	—
25	Check bearer pressures	See Section II	—

PRESS SYSTEM LUBRICATION AND SERVICE POINTS

This chart generally covers all components of the printing system except the printing unit. Most, but not all, of the points are shown in the illustration on page 378.

Item	Component	Procedure	Lubricant
		DAILY	
1	Automatic greasing systems	Check level, fill as necessary	*
2	Oil mist lubricating systems	Check level, fill as necessary	*
3	Circulating oil systems	Check sump oil level, fill as necessary	*
4	Air line lubricators	Check, fill as necessary	D
5	Dampener circulator filter	Clean, replace as necessary	—
6	Photoelectric web guide or break detectors	Clean senders and receivers, check lamps	—

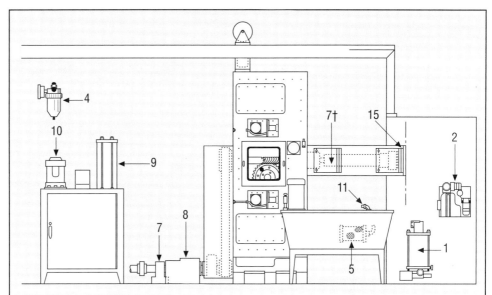

Press system lubrication and service points

Item	Component	Procedure	Lubricant
		EVERY TWO WEEKS	
7†	Pillow block bearings of drive shaft	Grease	B
8	Folder clutch	Spray lubricant sparingly onto hub splines	NEVER-SEEZ® or equivalent high-temp. antiseize lubricant
9	Dampener solution concentration reservoir	Remove and clean	—
10	Dampener solution mixer motors	Clean cooling fan	—
11	Dampener system	Flush	—
15	Unit separation clutch	Spray lubricant sparingly onto hub splines	NEVER-SEEZ®
		EVERY SIX MONTHS	
3	Circulating oil systems	Drain, remove, and clean sump strainer, replace external filter, refill	*
12	Chill roller cylinders	Flush out	—
13	Main drive motor	Cleaning cooling fan filters	—
		YEARLY	
14	Infeed, chill roller, and folder web pressure and nip rollers	Check condition, replace as necessary	—

*See Operator's Manual for recommended lubricant   †Not on all units.

into the plant so that they know the use and the specification requirement of each lubricant and all manufacturer's recommendations for each piece of equipment.

Lubricating Systems and Devices

The final step in lubrication, after choosing the right lubricant and training for its application, is the actual application. Over the years, systems and devices have been developed that have become standards in the printing industry. These lubrication systems and devices fall into three categories: manual, semiautomatic, and automatic. The manual devices are those to which lubricants are applied with an oil can or grease gun. These are oil holes or grease fittings (figure 22-2).

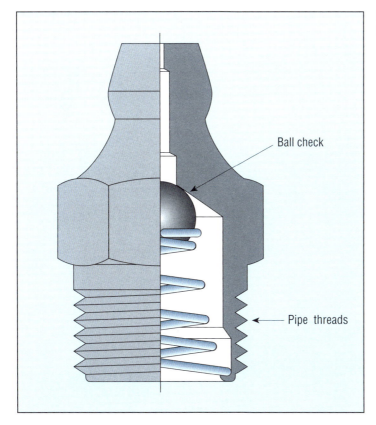

Figure 22-2. Lubrication fitting for grease gun.

Ball check

Pipe threads

With semiautomatic devices the lubricant is applied by some sort of reservoir and fed to the spot of lubrication by some action on the part of the operator. A few examples are sight feed oiler, bottom feed wick, bottle oiler, grease cup, siphon-type oiler, and pad oiler (figure 22-3).

Automatic devices, which are becoming more popular, use pressure to apply the lubricant. These systems are varied and complex. In addition to being designed to apply the lubricant, they are timed to give the right amount at the right time. Some systems are timed to feed oil or grease on a regular schedule. Figures 22-4 and 22-5 show a gravity-feed lubrication system and a force-feed lubrication system and point out various components in a system.

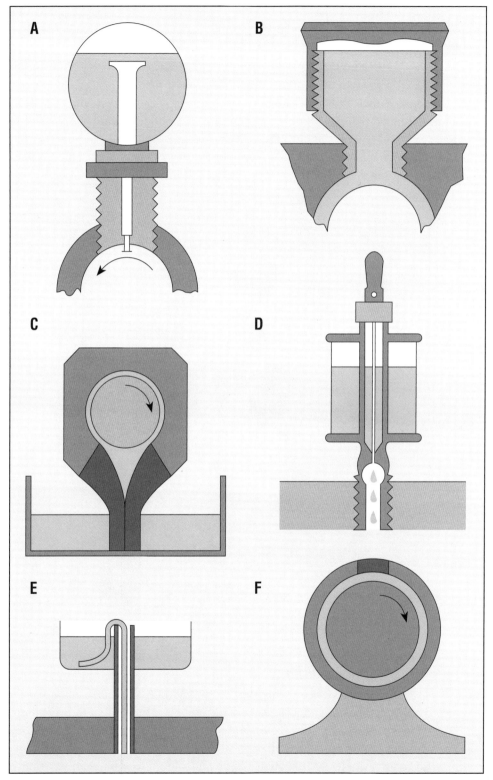

Figure 22-3. Six semiautomatic lubricating devices: (A) bottle, or needle, oiler, (B) capped grease cup for gravity feed, (C) bottom-wick oiler, (D) sight gravity-feed oiler, (E) siphon-type oiler, and (F) pad oiler.

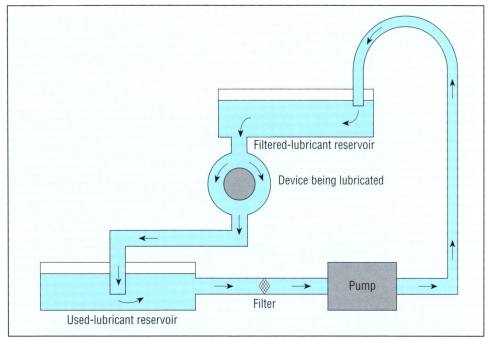

Figure 22-4. Gravity-feed lubrication system.

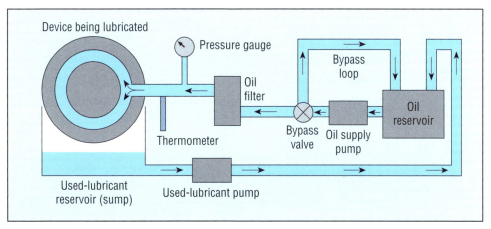

Figure 22-5. Force-feed lubrication system.

Mechanical Systems Maintenance

Mechanical drive systems are mechanisms that convert and transmit physical power to printing equipment. The most common mechanical drive mechanisms include chains, sprockets, gears, cams, pulleys, and belts. These mechanical drive mechanisms transmit drive power, convert and reduce speed, and physically convey and elevate materials. Virtually all printing equipment uses mechanical drive mechanisms.

Proper maintenance of these mechanisms will make the difference between maximizing equipment effectiveness or mechanical failure of the equipment. These mechanisms must be properly lubricated and inspected periodically to determine if replacement is necessary.

Chains and Sprockets

A *chain drive* consists of an endless series of chain links that mesh with toothed wheels, called *sprockets.* Sprockets are usually keyed to the shafts of the driving and/or driven mechanism. Almost without exception, when chains are used, they are used in connection with sprockets.

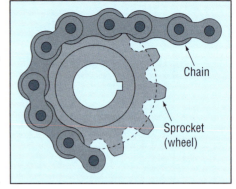

Figure 22-6. *Chain and sprocket.*

As mechanical drives, chains and sprockets are extensively employed throughout the printing industry for power transmission, speed conversion, and elevating and conveying systems. Examples in the web offset pressroom include roll hoisting mechanisms at the roll stand.

The following terms are commonly used with chain drives:

- *Chain,* a flexible series of joined links, usually metal
- *Sprocket,* a toothed wheel engineered to engage the chain
- *Pitch,* the distance from the centerline of one chain pin to the centerline of another chain pin, which determines chain size
- *Width,* the normal width of the link or the length of the pin or rivet measured within the side plates
- *Diameter,* the outside diameter of a roller, approximately 5/8 of the pitch
- *Chain guard,* a guard of sheet metal around a chain drive
- *Load classification,* a classification of drive loads based on the amount of shock imposed on the drive

Chain drives/sprocket systems have several advantages over other types of mechanical drive systems, including the following:

- Very high drive efficiency
- Lower loads on bearings
- Drive speed remains uniform, even when there is some loss of chain tension
- Drive chain and sprocket types are standardized throughout industry, so replacements are rather easy to find
- The entire chain does not require replacement; individual links can be quickly replaced on the equipment
- Chains have a long life due to low wear and very little deterioration

Among the disadvantages of chain drives are the following:

- Can be noisy, and are more cumbersome than cams and gears
- Run at lower speeds than cams and gears
- Require frequent lubrication and cleaning
- More subject to failure, due to the many working parts

Types of Chains

There are many types of chains, including the roller chain, the silent or inverted-tooth chain, the Ewart chain, the block chain, and the bead chain.

Roller chain. A *roller chain,* the most common type of chain, is made up of side plates with alternating roller/bushing and pin-link assemblies. It is used for accurate and high-speed chain drives. This type of chain is adaptable to widely varying needs, from small-strand drives for light applications to large multiple-strand chains for heavy-duty industrial applications. The following are the key components of a roller chain mechanism:

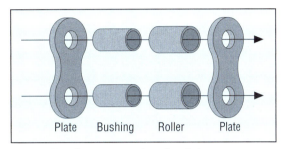

Figure 22-7. Components of a roller chain.

- *Roller chain,* which consists of a finished steel roller, alternate assemblies roller, and pin links
- *Link,* which is a heat-treated stripstock steel that has been perforated and blanked
- *Pin,* which is a hardened alloy steel that has been ground to specific tolerances
- *Bushing,* which is a case-hardened core that prevents metal from rubbing metal
- *Rollers,* made of heat-treated, high-carbon steel

A *roller link,* a single unit of a roller chain, consists of two links (plates), two bushings, and two rollers. The roller links are joined together by a pin link at each end to form a continuous chain. The pin link is fastened by one of the following pins: rivet, colter, or spirol. A series of roller links and pin links form the roller chain, and the ends are joined together by a common coupler link, the *master link,* to form a loop. The master link is an important part of roller chains that is often overlooked. If a chain breaks or has to be repaired, it requires a master link of the proper kind to fit the chain.

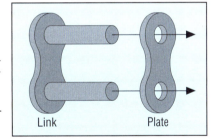

Figure 22-8. Pin link, used to join roller links at each end to form a continuous chain.

There are several types of roller chains. One is the *light steel roller chain,* generally used for light industrial work where the running speed of the chain is not likely to exceed 700 ft./min. (200 m/min). Its rollers are constructed of either malleable iron or hardened steel, depending on service requirements. The side bars are generally flared for easier engagement with the sprockets, which are constructed of either cast-tooth steel or cast iron. The *rough-finished roller chain* has rollers made of case-hardened steel, which has high strength and the ability to resist shocks.

The pins are forged and accurately machine-hardened. This chain operates on cast-tooth steel or cast-iron sprockets and is used mainly for heavy applications. The **twist-type roller** is a specialty chain. With this chain, specially machined angled bushings allow for ease of operation when sprockets are not in parallel alignment.

Silent chain. The **silent chain,** also known as an **inverted-tooth chain,** consists of a series of inverted-tooth links held together by joint pins to which washers have been riveted. The silent chain, while not exactly silent, is much quieter in operation than other transmission chains.

Unlike a roller chain, the silent chain has a tooth engagement with gradual sliding action. This chain "rolls" on its sprockets rather than riding on them. The straight sides of the sprocket teeth mesh with the straight-sided working jaws of the chain links to move the chain along. The chain is constructed of precision-formed leaf links that are perforated and blanked from strip steel. Several different types of guide rails can be used to prevent lateral movement of silent chains.

Because of the high speed and whipping action of a silent chain drive, the chain must be well oiled. The viscosity of the oil depends somewhat upon temperature conditions and the chain speed, but an oil of about SAE 30 viscosity will suffice.

Ewart chain. The **Ewart chain** is principally used for conveying and elevating equipment. The open-link type of Ewart chain uses a one-piece cast link, unmachined, with no bushings, permitting ease of assembly and disassembly. A second type uses a pin to form a closed link, or pintle, which improves chain life, performance, and strength, but tends to be noisy. A third type of Ewart chain

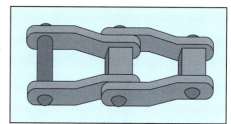

Figure 22-9. Ewart chain.

contains a bushing for quieter service and hollow-cored cylinders with offset side bars. This type is for use under more severe conditions.

Block chain. The **block chain** consists of solid or laminated blocks connected by side plates and pins. The sprocket teeth engage the blocks to move the chain along. For a low-speed, light-load drive, the block chain is preferred to the Ewart type.

Bead chain. A **bead chain** consists of beads that swivel and turn on metal links or pins. Because the beads are relatively delicate, the bead chain is used for light service. Sometimes special situations such as misaligned sprockets, skewed shafts, or non-parallel planes call for the use of this type of chain.

Figure 22-10. Bead chain (cutaway view).

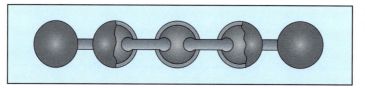

Sprockets

The American National Standards Institute (ANSI) specifies dimensions, tolerances, and tooth forms for sprockets—toothed wheels engineered to engage (drive) the chain. There are four sprocket types included in the standard: Type A is a plain sprocket without the hub; Type B has the hub on one side only; Type C has the hub on both sides; and Type D has a detachable hub.

Between the hub and the rim, a sprocket may take one of several forms according to size. The smaller sprockets, under 15 in. (380 mm) in diameter, are usually solid. Medium-size sprockets, over 15 in. (380 mm) in diameter, have cored holes between the hub and rim to reduce the weight. The larger sprockets are usually cast in the form of spoked wheels to obtain the maximum possible reduction in weight. A split sprocket is mounted to permit installation or removal without disturbing the shafting.

Shear-pin and slip-clutch mechanisms are designed to prevent damage to the drive or to other equipment caused by overloads or stalling. A *shear-pin sprocket* consists of a shear pin, a hub keyed to the shaft, and a sprocket that is free to rotate either on the shaft or on the hub when stresses cause the shear pin to break. The shear pin is made of a material with a known shearing strength. The breaking point of the shear pin can be established

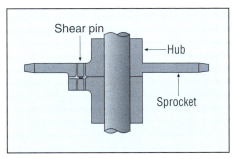

Figure 22-11. *Shear-pin hub sprocket.*

with reasonable accuracy by machining it with a groove of a calculated diameter at the shear plane. The assembly can be returned to working condition by inserting a new shear pin.

Maintenance and Lubrication of Chain Drives

Four factors assure long life and efficiency of a chain drive:

- Alignment of chain and sprocket
- Adjustment of chain tension and sprocket
- Constant lubrication for most types
- Cleanliness—oil casings being the best assurance

Lubrication of chain drives is extremely important for long life. Application of oil is determined by the drive shaft location and the chain's speed.

- Power lubrication from oil pumps is used for high-horsepower drives
- Oil bath or splash lubrication systems require sealed enclosed housings. For moderate-speed motors, oil must be changed periodically
- Drip lubrication, a semiautomatic system, applies a small amount of oil over a predetermined time span
- Manual lubrication (oil-can application) is usually performed daily

Gears

A *gear* is a mechanism that transmits mechanical rotary power from one shaft to another or from a shaft to a slide (rack) by means of successively engaging projections of tightly fitted teeth. A gear drive is one of the oldest methods but still is the best when constant speed ratios are necessary.

Some basic uses of gears are changing speed, substituting for frictional drive, changing torque, changing direction of a shaft, and changing direction of rotation.

Gears have the following advantages over alternative mechanical devices:

- *Economy of operation.* They have very efficient power transmission and will last, with a minimal amount of maintenance, through many years of service.
- *Adaptability.* They require small mounting space and will operate accurately under adverse conditions.
- *Safety.* With proper installation and the use of safety guards, gears are very safe. However, turning gears that are not guarded can be extremely dangerous.

Gear Terminology

The following terms are used when discussing gears:

- *Pitch circle,* the line of contact of two cylinders that would have the same speed ratios as the gears
- *Addendum,* the portion of the tooth between the pitch circle and the outside diameter
- *Dedendum,* the portion of the tooth between the pitch circle and the bottom of the clearance
- *Pitch diameter,* the diameter of the gear at the pitch circle
- *Tooth face,* the working surface of the tooth above the pitch circle
- *Flank,* the working surface of the tooth below the pitch circle
- *Circular pitch,* the distance from the center of one tooth to the center of the next when measured on an arc of the pitch circle

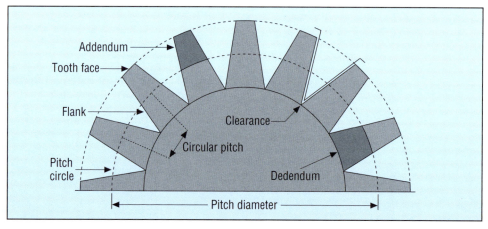

Figure 22-12. Gear terminology.

- **Clearance,** the space between the bottom tooth and the tip of a tooth fully meshed into the tooth space

- **Backlash,** the free movement of the pitch circle of one gear in the direction of the circumference; or, simply the "play" between teeth

- **Rack,** the teeth cut on a flat surface

- **Undercut,** the process of cutting the base of the gear slightly below the teeth, protecting the gear teeth from breaking if an unusual amount of stress is placed upon the gear

Gears Used for Printing Equipment

Spur gear. A *spur gear* has teeth that are cut parallel to the axis of the shaft, and for this reason spur gears can be fitted only to drives with parallel shafts. They are used in a wide range of printing equipment, including folders and the drive systems for the inking and dampening systems.

The ability of a pair of well-made spur gears to facilitate a smooth, regular, and positive drive is of the greatest importance in many engineering designs. Spur gears offer the added advantage of simplicity. Spur gears are one of the best means of transmitting a positive motion between two parallel shafts.

The following are the main characteristics of spur gears:

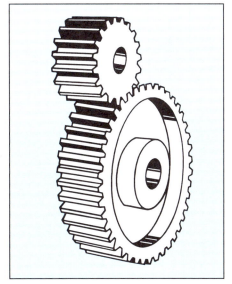

Figure 22-13. *Spur gears.*

- Transmit power between parallel shafts

- Commonly used in drives of moderate speeds

- Straight teeth parallel to gear axis

- Low cost

- Absence of end thrust

- Economical to maintain

- Most commonly used gears

Helical gear. A *helical gear,* also called a *spiral gear* or *skew gear,* has teeth that form a helixical angle around the center; i.e., the teeth are cut so that they lie along a helix. The helix angle is the angle between the helix and a pitch cylinder element parallel to the gear shaft. Helical gears are about the most accurate, strongest, and quietest gear used in printing equipment today.

Helical gears are stronger and quieter than spur gears because the contact between mating teeth increases gradually. With helical gearing, the teeth mesh gradually, and at no point is the full width of any tooth completely immerged. This gear meshing eliminates most of the shock and jarring associated with straight teeth when gears are operating under heavy loads.

The following are several characteristics of helical gears:

- Teeth are cut on an oblique angle across the face of the gear
- Several teeth mesh at any given time, resulting in increased load-carrying capacity, transmission constant velocity, and reduced noise and vibration
- Accommodation for end thrust through the use of bearings
- Higher operating speeds

Figure 22-14. Helical gears.

Spiral gear. A *spiral gear* has teeth that are cut in a spiral. Spiral gears are used to connect nonintersecting and nonparallel shafts. Their advantage is in the precision instrument field. Spiral gears transmit motion between nonparallel shafts that are in different planes, and also transmit motion between parallel shafts. As the teeth mesh, the loading edge first engages the mating gear and the balance of the tooth slides into mesh as the gears rotate. This is less efficient and more power-consuming than spur and helical gears.

Bevel gear. A *bevel gear* has teeth cut on an angular face for transmitting motion between power shafts that are at an angle to each other; it can be thought of as a spur gear cut an an angle. The face angles of the gears vary widely because of the varying shaft angle and gear diameters. Where there is a large variance in size, the small gear is usually called a *pinion* (as is also the case with spur gears).

The bevel gear is one of the oldest methods of transmitting power at an angle. Although bevel gears are not as precise as spur gears, bevel gears are the

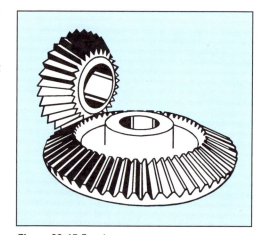

Figure 22-15 Bevel gears.

necessary choice where right-angle (or an angle) drives are unavoidable along with significant loading. They can be designed to transmit almost as much power as the most powerful helicals. Bevel gears not readily interchangeable, because bevel gears are developed in pairs to improve precision.

Several types of bevel gears are used in a variety of industrial applications. These include the straight bevel, the spiral bevel, and the hypoid bevel. Both the spiral and the hypoid bevel gears offer quiet operation due to the gradual engagement of the teeth while meshing. The straight bevel, while less expensive, is noisier and of lesser quality.

Worm gear. *Worm gears* combine the use of a screw and *worm wheel,* a helical gear with a concave face to match the outside of the screw worm. A worm gear can be thought of as a shaft that has been machined into a long spiral gear. A common application of worm gears is to control the amount of impression between a plate and blanket cylinder on a nonpneumatically operated press.

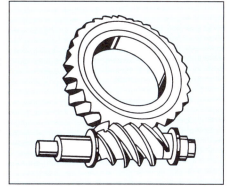

Figure 22-16. Worm gears.

Worm gears are the most versatile gears used to connect shafts in parallel planes. This versatility is due to the wide range of ratios that can be employed in gears meshing at the same center distances. Worm gears are used for high-precision, high-quality work because of their simple straight-side tooth form. They can handle high load capacity and are suitable for right-angle transmission.

The following are some characteristics of worm gears:

• Large ratio of reduction in a small space
• Operation at high speeds
• Quiet and vibration-free operation
• Good load-carrying capacity
• Transmission of power at right angles
• Greater heat generation and reduced mechanical efficiency at higher speeds because of sliding action

Herringbone gear. A *herringbone gear,* or *double-helical gear,* is a gear in which there are opposite-angle helixes on the two sides the gear. They are sometimes made of two opposite-angle helical gears fastened together. Herringbone gears are becoming increasingly popular where it is necessary to transmit heavy loads with a minimum of noise and side thrust. This thrust is balanced by placing the gears at opposite angles.

Rack and pinion. The rack has teeth cut on a flat surface. As there is no curvature on the pitch line, teeth cut to a basic involute tooth form have straight sides. The teeth may be cut at right angles to the direction of travel (as for a spur gear) or, if inclined, they may be cut to mesh with a helical gear. The pinion rides in a fixed length of track, and a spur or helical gear is used.

Cams

A *cam* can be defined as a machine part that either rotates or slides back and forth to produce a prescribed motion in a contacting element known as a *follower.* The shape of the contacting surface determines the prescribed motion and the shape of the follower, which is usually circular or flat. Cam-follower systems are particularly useful when the uniform revolving or reciprocating motion of one part of a machine is converted to any kind of nonuniform alternating, elliptical, or rectilinear movement, which may also include periods of rest known as dwells. The combination of movements and dwells desired in a part determines which type of cam should be used.

Types of Cam Mechanisms

The following is a description of some of the different types of cams.

Disk cam. The most common type of cam is the *disk cam,* which is a flat rotating plate with a curved contour that usually rotates with a constant velocity to impart a desired, constrained motion through the movement of a follower device riding on the cam.

Translation cam. A *translation cam* produces a vertical motion in the follower when the cam is moved horizontally. In some translation cams the follower travels over the surface of a stationary cam.

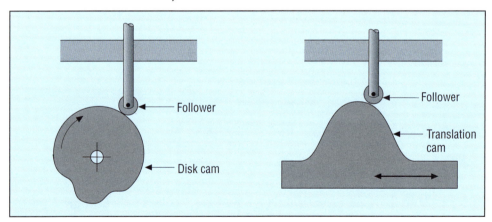

Figure 22-17. Disk cam (left) and translation cam.

Groove-plate cam. A *groove-plate cam* has the follower restrained within a groove cut in the cam plate. It is used primarily for high-speed applications. With this design, springs are not required to keep the follower in contact with the cam face, as the groove guides the follower along the path accurately.

Cylindrical cam. A *cylindrical cam* is a cylinder into which a contour has been grooved. The follower is guided through a contour, which usually pushes or pulls the arm in a vertical or horizontal motion to perform the work. A cylindrical cam is a very sturdy, accurate design.

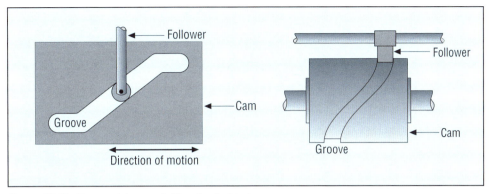

Figure 22-18. *Groove plate cam (left) and cylindrical cam.*

Eccentric cam. An *eccentric cam* is a circular cam in which the rotating shaft is off-center.

Tow-and-wipe cam. A *tow-and-wipe cam* is a cam that rotates and imparts movement that lifts or lowers the follower, depending on the direction of rotation.

Types of Followers

A *follower* is a device that follows the contoured cam surface, producing a prescribed motion of output. As with cams, there are several different designs for followers. The desired performance and the work to be done determine which follower has the best design for a specific application. Various followers are described below.

Roller follower. A *roller follower,* the most common type, is a roller mechanism that rides on the cam face to impart linear motion. Its advantage is the reduction of sliding friction as the follower travels over the cam face. The disadvantages of the roller follower are that weak points are created at the pin axis and that steep cam contours may jam the translating roller follower. The follower shown in figure 22-17 is a roller follower.

Knife-edged follower. A *knife-edged follower* is a pointed device that rides on the cam face to impart linear motion. Because the sharp edge of the follower causes excessive wear on the cam face, its practical use is limited to precision, light-load mechanisms.

Flat-faced followers. A *flat-faced,* or *"mushroom," follower,* the type used in automotive engines, is a flat-faced spherical-surface mechanism that rides on the cam face to impart linear motion. There are two types of flat-faced followers: plain face and spherical face. Both of these designs perform well with steep cam contours. The plain-face design, which has a flat surface in contact with the cam, may give high surface stress to the cam because of deflection and misalignment. The spherical-face design alleviates this condition because of its rounded surface face. (See figure 22-19.)

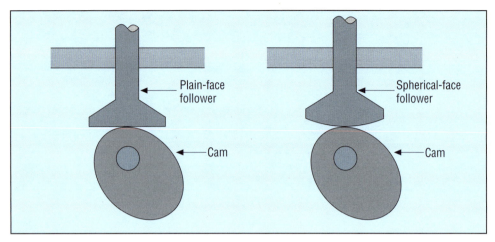

Figure 22-19. *Two flat-faced followers.*

Motion of Follower

The ***stroke,*** or ***throw,*** of a follower is the maximum movement away from the cam shaft from its lowest position. Because cams are nearly always drawn with the follower at the top, this movement is also called the ***rise*** or ***lift,*** although the follower might be moving in some other direction on the machine. The opposite movement is called the ***return. Dwell*** is the condition that occurs when the follower remains stationary for some period of rotation of the cam.

When considering follower motion, there are three basic cam classifications:

- ***Dwell-rise-dwell cam,*** in which part the cam action consists of a zero-displacement portion (dwell), a rise contour, and then another dwell period. This cam is the most common type found in machinery. Actually, the complete cam cycle is ***dwell-rise-dwell-fall-dwell.***

- ***Dwell-rise-return-dwell cam,*** in which rise and return are preceded and followed by dwells. It has some application in machinery.

- ***Rise-return-rise cam,*** in which the cam contour has no dwells. This cycle suggests an eccentric mechanism since it fulfills the follower action more accurately than a cam contour.

Belts and Pulleys

A ***belt*** is a continuous band of flexible material used to transmit power or motion between two parallel shafts. These parallel shafts are connected to a wheel or pulley. A ***pulley*** is a wheel that turns or is turned by a belt so as to transmit or apply power. Belts and pulleys are used in a variety of ways in the printing industry. For example, belts and pulleys are normally the main power transmission from the electrical drive motor to large equipment.

A ***belt drive,*** in its simplest form, is a drive consisting of an endless belt fitted tightly over two pulleys (driving and driven) transmitting motion from the driving pulley to the driven pulley by frictional resistance between the belt and the pulley.

With a belt, the inside portion of the belt that takes the compression strain as it is forced against and around the pulley is called the **high-compression area.** The outside portion of the belt that takes the tension strain as it is pulled and stretched around the pulley is called the **high-tension area.** The portion of the pulley actually in contact with the belt at any one time is referred to as the **arc of contact.**

Belts

Belts are classified according to their cross section, design, and composition. The cross section is the most important factor in the design of pulleys and of the entire system. The major types of belts include:

- Rope (hemp, steel, leather)
- Flat (leather, rubber, woven cotton, woven wool)
- V-belt (cord fabric, cord wire, cord wire toothed)
- Cog or timing belts (positive drive belt—no slippage, used where timing is essential)

Many flat belts are used for conveyor belts rather than power transmission; however, flat belts can function as infinitely variable speed reducers if conical pulleys are used. This type of belt can provide a great speed-reduction ratio and operate at greater speeds than other belts. V-belts are used on short center-to-center drives at high reduction ratios. They are able to transmit higher torque with less width and tension than flat belts. V-ribbed belts have the strength and simplicity of the flat belt with the positive tracking of the V-belt. The pulleys for these belts must be precisely aligned. Timing belts are used when precise synchronization between drives and driven elements is required. The positive tooth and groove engagement of this belt prevents slipping, creeping, and speed variations.

There are three major types of belt joints: cemented joint (leather or rubber), laced joint (flat cloth), and hinged joint (flat fiber or leather belts).

Maintenance of Belts

V-belts. The maintenance of V-belts includes the following procedure:
- Replace all belts in a matched set at the same time.
- Check tension and keep belts tight.
- Never attempt to correct belt slippage by using a belt dressing or lubricant.
- If belts slip when properly tensioned, check for an overload (which might be caused by worn grooves or oil and grease on belts).
- Never pry a V-belt or force it into the sheave groove. Loosen the tightener before installing the belt.
- Store belts in a dry, cool place. Relieve the tension if stored on the machine.
- Never attempt to check or adjust belts while the machine is running.

Flat belts. In addition to the previous procedure , the maintenance of flat belts includes the following:
- Dress leather or canvas flat belts to prevent deterioration
- Make sure the tension of flat belts is adequate
- Check pulley alignment because flat belts can run off of the pulley easier than V-belts
- Never force a belt off of a moving pulley

Belt slippage, the main disadvantage of belt drives, causes a loss of power and belt destruction. The following factors must be considered to avoid this problem:

- Suitability of type and thickness of belt for the pulley diameter and speed
- Overloading and underloading
- Sufficient belt tension
- Use of dressing to avoid dry, glazed, and hard belts
- Formation of lumps on the pulley face due to resinous dressings
- Atmospheric effects
- Proper alignment

Pulleys

For efficient, economical power transmission by belting, the type and design of pulley used must be proper for the existing operating mechanical and atmospheric conditions. The four main classifications of belts have corresponding pulley types:

- **Flat belt pulleys.** These pulleys are made mainly of three materials: cast iron, fabricated steel, and pressed steel. To keep a flat belt riding true on a pulley, a set of flanges or a crown on the pulley must be provided. Crowning is the preferred method, but it must be done in moderation to prevent undue stress concentration in the middle of the belt.

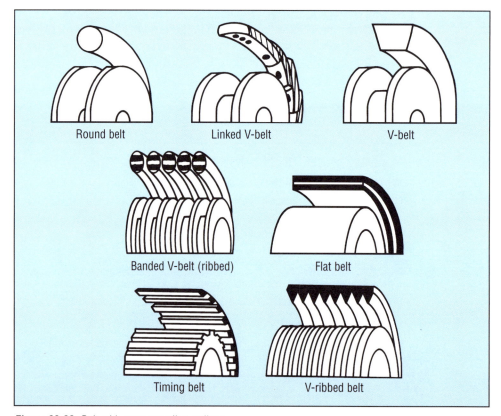

Figure 22-20. Belt with corresponding pulleys.

- **V-belt pulleys.** V-belt pulleys are made of pressed steel and grooved cast iron.
- **V-ribbed pulleys.** V-ribbed pulleys, usually made of cast iron, are used for drives that range from fractional horsepower to 10 h.p. Their sizes range from 3–40 in. (76–1000 mm) for multi-horsepower drives.
- **Timing belt pulleys.** These pulleys are similar to uncrowned flat belt pulleys, except that grooves are cut in their faces parallel to their axes. Steel and cast iron are used for this type of pulley.

Advantages of Belt Drives

The following are the principle advantages of belt drives when compared to chain drives:

- Minimum vibration
- Minimum noise
- Cushioning of motor bearings—due to the slight give of the belt—against fluctuations in the load
- Practical where long center-to-center distances occur
- No need for lubrication
- Simple and quick replacement of belts
- Clutch action obtainable with flat belt by shifting the belt from a loose pulley to a tight one
- An economical means for changing the velocity ratio obtainable by employing step pulleys

Disadvantages of Belt Drives

The following are the principle disadvantages of belt drives when compared to chain drives:

- Slippage between the belt and the pulley
- Damage to the belt due to oil, grease, and heat
- Higher load transmitted to the bearing because of the belt drive's dependency on tension to maintain friction
- Not usable when exact timing or speed is required unless a special timing belt is used
- Wearing out of belts due to fatigue failure

Bearings

Bushings and *bearings* are machined elements in which another part (e.g., axle or shaft) turns or slides and which reduce friction between moving and stationary parts of machinery. This reduction is achieved by either radial or axial (thrust) motion. *Radial motion* is circular, as in a rotating printing cylinder, whereas *axial motion,* or thrust, is the upward/downward or forward/backward pressure. Bearings are classified according to how the rotating part moves within the bearing.

Sliding Bearings

A *roller-contact,* or *antifriction, bearing,* which is discussed later, is a bearing in which the rotating part rolls. Both ball bearings and roller bearings are of this type. If the rotating part slides within the bearing, that bearing is known as a **sliding,** or **plain bearing,** or **plain bushing.** Sliding bearings include journal, thrust, and guide bearings.

Journal bearing. A *journal,* or *sleeve, bearing,* the most common type of sliding bearing, is made up of a journal (a rotating shaft), a bearing (a bushing made of a special metal, alloy, or plastic) and a housing that holds the bearing. The journal and bearing are separated with a small clearance by a thin film of lubricant, which keeps the two surfaces apart and thereby minimizes wear, reduces friction, and dissipates heat.

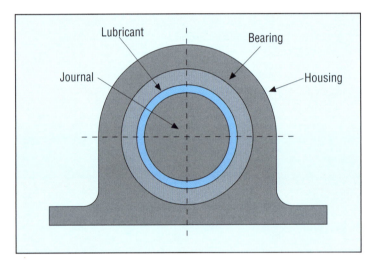

Figure 22-21. *Journal bearing.*

Hydrodynamic bearing. When an oil film is interposed between a shaft and the outer member surrounding the shaft (as with journal bearings, above), an important phenomenon occurs as the shaft begins to turn. The phenomenon, known as the **wedge effect,** is the self-generation of pressure sufficiently high enough to support the load of the shaft. A bearing that operates by this principle is called a **hydrodynamic bearing** because pressure is generated by the flow of a fluid (the lubricant) through a wedge-shaped passage.

Hydrodynamically lubricated bearings rely heavily on the design of the fluid system to provide the pressure needed to support the load. In addition to the physical nature of the fluid system, other important factors are the operating speed, the viscosity of the fluid, and the temperature.

Thrust (axial) bearing. The simplest type of *thrust bearing* consists of two plane surfaces but can support only a slight load at ordinary operating speed, which is 100 ft./sec. (30 m/sec) or less. The load capacity of the plain thrust bearing may be increased by altering one of the bearing surfaces so that an oil film of gradually decreasing thickness is provided in the direction of relative motion. The exact angle to be given the tapered portion of this bearing depends upon the load, speed, dimensions of the bearing, and viscosity of the lubricant.

Thrust bearings can also be of the antifriction type, which is discussed in the following section.

Antifriction Bearings

An antifriction bearing, used to separate the rotating shaft from the stationary parts of the machine, must carry the load imposed by the moving parts. The load is radial (in the running direction of the shaft and the axis of the bearing) and/or thrust (running perpendicular to the axis and bearing).

The three basic types of antifriction bearings are ball bearings, roller bearings, and needle bearings. They can be self-aligning or non–self-aligning, and they are used for radial applications, thrust applications, or a combination of radial and axial use.

The major advantage of antifriction bearings over bushings is the reduction of the frictional loss associated with bushings. Since wear is significantly reduced, there is less need for an elaborate lubrication system. Some antifriction bearings contain a sealed-in lubricant, which in most cases lasts for the life of the machine; others require replenishment of the lubricant after a certain number of years or hours, depending upon the machine and the type of work done with it. Designed to operate at much higher speeds than sliding bearings, antifriction bearings can also carry much heavier loads.

The greatest disadvantages of these bearings are their high initial cost and the fact that they take up more radial space than bushings. Though they must be mounted much more accurately than bushings, their precise action will produce better performance from the machine.

Ball bearings. A *ball bearing* is a bearing in which a *journal* (a shaft, axle, roll, or spindle) turns upon loose, hardened steel balls that roll easily in a *race,* a track or channel in which something rolls. There are many kinds of ball bearings, the most common being radial ball bearings, including deep-groove, self-aligning, angular contact, and ball thrust. The *plain single-row ball bearing,* or the *Conrad bearing,* most prevalent in simple machines, is an assembly in which the inner and outer races (or rings) are placed together so they form a crescent-shaped space on the opposite side, into which the balls are placed. When the greatest possible number of balls has been inserted, they are spaced, and the two parts of the retainer are joined together, usually by riveting. The construction of the Conrad bearing limits the amount of balls that can be inserted, and this limitation restricts the radial load capacity of the bearing.

The high noise level associated with high-speed ball and roller bearings results partly from their nonhydrodynamic operation and partly from the windage loss due to the retainer. The inner ring, outer ring, balls, and rollers are the basic components of a rolling-contact bearing. In addition, nearly all bearings have a *cage,* a device that keeps the rolling elements properly spaced and prevents adjoining rolling elements from touching.

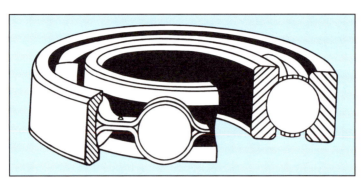

Figure 22-22. Deep-groove ball bearing.

Radial roller bearings. A *roller bearing* is a bearing in which a journal rotates in contact with a number of rollers, usually in a cage. Roller bearings have a greater load-carrying capacity than ball bearings, because they have line contact rather than point contact. There are many types of roller bearings, both cylindrical and tapered, each designed to meet a specific need.

A *spherical-roller bearing* is a two-row, self-aligning bearing (somewhat similar in appearance to the self-aligning ball bearing) in which spherical rollers provide line contact on the inner race and point contact on the outer race. Skewing is eliminated and the retainers must merely space the rollers rather than guide them.

A *tapered-roller bearing* is a bearing that uses rollers in the shape of truncated cones in such a way that thrust as well as radial loads can be supported. These bearings are constructed so that lines drawn coincident with working surfaces of the rollers and races meet at a common point on the bearing axis. The cage of the bearing is relieved of maintaining the alignment of the rollers and is used simply to space them.

Needle bearing. A *needle bearing,* or a *cageless needle bearing,* is essentially a cylindrical roller bearing without a cage. The needle bearing, however, is characterized by long, thin rollers of a length-to-diameter ratio of 3–10. Since the needles (rollers) are small in diameter, more can fit into the bearing. This provides for greater load-carrying capabilities. However, the relatively great length of the needles makes them sensitive to shaft misalignment, resulting in unequal distribution of load along the needles. Needle bearings also have a larger coefficient of friction than roller bearings. The main advantage of

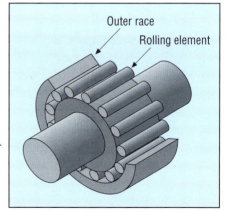

Figure 22-23. Needle bearing.

needle bearings over roller or ball bearings is the conservation of space due to the small diameter of the needles.

Lubrication of antifriction bearings. Although friction values are lower for ball, roller, and needle bearings than for sliding bearings, lubrication is still needed. Point and line contact cut down on friction, but the pressure applied on the contact point is very high. For this reason many antifriction bearings require extreme-pressure lubricants. Lack of lubrication will usually cause destruction of the bearing, which in turn could result in major damage to the machine of which it is a part. Lubrication is also essential to the bearings for the following reasons:

- To reduce friction at all sliding points
- To control the heat caused by friction
- To prevent corrosion
- To aid in protecting the bearing from contamination due to dirt and dust

Other Types of Bearings

A *porous bearing* is a bearing made from powdered metal that is compressed and molded but which is 25% porous (i.e., 25% of the bearing consists of pores or air spaces). These air spaces are filled with a nongumming lubricating oil. When the shaft rotates, oil is drawn out of the pores, thereby lubricating the shaft. When the shaft stops turning, the oil is drawn back into the pores by means of capillary action. All parts of the shaft are lubricated. A porous bearing is desirable for light bearing needs but not for heavy loads because it is not as strong as a 100%-metal bearing.

Although a porous bearing with a replenishing oil supply is sometimes referred to as an "oilless bearing," a true *oilless bearing* is a bearing that uses no oil but may use some type of solid lubricant. Some oilless bearings have sleeves impregnated with slugs of graphite or other solid lubricant. For other light duties and particularly in noiseless situations, nylon bearings, another type of oilless bearing, are used.

Bearing Failure

Ink and dampening roller companies have found that bearing failures are a common cause of streaks and dampening control problems. Following are some possible causes of bearing failure:

- Bearing seats of shafts are defective
- Misaligned journals, bearings, and shafts
- Improper procedures followed in the mounting of interchangeable parts (such as variable-size printing cylinders)
- Incorrect bearing being used
- Incorrect type of lubricant being used
- Leaking bearings due to loose seals
- Excessive lubrication of bearing when not in use
- Metal fatigue due to electric current (arcing), which happens often with motor bearings

Electrical Components Maintenance

The majority of equipment in the printing plant is electrically operated. Only qualified electrical technicians should be allowed to work on the electrical systems, in order to ensure safety and help eliminate electrical component failure.

Common Problems with Electricity

Voltage fluctuation. One of the most frequently overlooked items in most plants is voltage fluctuation, which can cause major problems in many areas of a printing plant. Perhaps the most obvious example is changes in exposure intensity of a timed exposure due to voltage fluctuation. Voltage fluctuation is also a problem in platemaking and can be detrimental to computers and other solid-state equipment. Voltage fluctuation can be overcome by the rather simple combination of a voltage meter and an audio-transformer. Computer equipment should also be equipped with uninterruptible power systems (UPS).

Proper installation. A critical factor leading to efficient electrical operation is the correct installation of electrical components, which requires timely consultation with the equipment's manufacturer prior to and during the installation process. Poor installation of an electrical component could result in safety hazards, the component's failure, and negate any manufacturer guarantees and warranties. Correct wiring and grounding of equipment cannot be stressed too much.

Power requirements. When a printer is looking for a new building, one of the biggest considerations should be the power available in the building (or if and when the utility company will bring in the power needed). Never assume that power is available. Always check with your utility representative and make sure you have the right power and, if not, when the utility will be able to schedule the installation. Never sign a lease until this is clearly established in writing.

When buying new equipment, always check with the manufacturer as to its electrical power requirements. Sometimes, the purchaser of a new piece of equipment finds that the voltage requirement to operate it is not the same as the voltage level presently in the building. This is not an infrequent occurrence. It can cost thousands of dollars to bring the proper power lines to your plant—if the power company can even do it. The printing equipment manufacturer and the equipment purchaser should never assume that there will be no problem with operating voltages. Power requirements should be one of the first items checked before purchasing equipment.

Basics of Electricity

Because of the dangers inherent when working with electricity, only certified technicians should perform electrical maintenance in the printing plant. Nevertheless, since electricity is the primary means of powering our printing equipment, it is necessary that we understand the basic principles in order to purchase equipment and communicate better with service personnel.

When an electrical charge is produced by friction and remains fixed or stationary on bodies commonly electrified, or charged, it is called ***static electricity.*** For example, on a printing press, static electricity is caused by a combination of low humidity and high running speeds, which cause the generation of electrical charges. A common solution to this problem is handing tinsel from a printing press. Superior methods are available today to neutralize the forces that create static problems.

The most common electrical energy is of a dynamic nature—i.e., charges that are in motion. We measure the dynamics of electricity in voltage, current, and resistance:

Voltage. The pressure that forces current to flow through a circuit is called *voltage.* The amount of current depends directly upon the amount of voltage, which is measured in units call *volts* (V).

Current. The flow of electrons is called *current,* which performs work or powers a motor or any other electrical device. Current, which is measured in units called *amperes* (amps, A), is directly related to the amount of voltage; if voltage increases, so does the current, and if voltage decreases, so does the current.

Direct current (d.c.) is current in which the electrons flow directly from the negative terminal of the source, through the electrical circuit (e.g., wire, switches, lamps, motors), and back to the positive terminal of the source. As long as the voltage of the source and the resistances of the circuit elements remain constant, the current remains at a single level. The most common form of direct current is that produced with a battery, which produces a constant voltage.

Electricity supplied to us by power companies is alternating current. Most of the motors that drive small fans, pencil sharpeners, and small refrigeration units use alternating current to operate. *Alternating current* (a.c.) is a current in which the electrons alternate direction. The force that causes the electrons to move in alternate directions is called *alternating voltage.* With a battery, the voltage is essentially constant, except for the decrease in voltage due to its being discharged slowly over a long period of time. Alternating voltage, on the other hand, is changing constantly—following a cyclic pattern—over time, between a maximum positive level and a maximum negative level. The cycle is repeated continuously, and the number of such cycles per second (Hertz, Hz) is the frequency; in the United States, there are typically 60 cycles per second. The average value of voltage during any cycle is zero. When the voltage is at its maximum positive level, the rate of flow of electrons (current) is at its highest level in one direction. Conversely, when the voltage is at its maximum negative level, the rate of flow of electrons is at its highest level in the opposite direction. The rate of flow of electrons also follows a cyclic pattern.

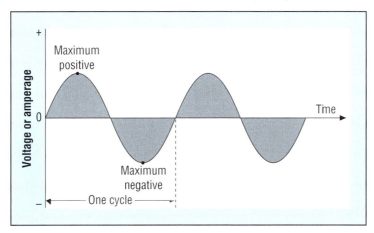

Figure 22-24. Alternating voltage or alternating current. If the voltage was 60 Hz, the cylce would repeat every 1/60 of a second and the direction of electron flow would change every 1/120 of a second.

Resistance. *Resistance,* which is measured in units called *ohms* (Ω), works against voltage by trying to stop current. The amount of opposition—resistance—depends on the type of material used as a conductor and the length and thickness of the conductor. The more resistance there is, the less current will flow in the circuit.

Ohm's Law. G. S. Ohm, a German physicist, developed a series of basic expressions showing the interrelationship among voltage, current, and resistance. These mathematical expressions are collectively called *Ohm's Law.* Voltage (E) expressed in volts is equal to current (I) expressed in amperes multiplied by resistance (R) expressed in ohms:

$$E = I \times R$$

Power. *Power* is the rate at which heat is released or the rate at which work is done. Power is usually measured in watts (or kilowatts) and horsepower.

In electrical work, the most common measurement of this heat release or work is the *watt* (W). Examples involving heat release are the 60-watt bulb and the 1,000-watt film dryer. Electrical power (P) in watts is equal to the voltage (E) expressed in volts multiplied by the current (I) expressed in amperes. Thus a current of 1 A and a voltage of 1 V produce 1 W of power. The following formula shows this relationship:

$$P = E \times I$$

Because the watt is a fairly small unit of power, we generally talk in terms of the *kilowatt* (kW), which is equivalent to 1,000 W. The electric bill shows that a printing plant's electric power usage is determined using the *kilowatt-hour* (kWh), which is equivalent to an average usage of 1,000 W, or 1 kW, for 1 hr. The following formula can be used to determine the cost for using any piece of equipment when the wattage is known:

$$kWh = \frac{W \times H}{1,000}$$

where W is the wattage of the printing equipment, H is the number of hours used, and 1,000 is a factor to convert watts into kilowatts.

When electric current is sent through an electric motor the electric energy is converted, by means of the motor, into mechanical energy. This energy or power when utilized by the motor is called *horsepower* (h.p.). By scientific definition, a 1-h.p. motor can lift a weight of 550 lb. (250 kg) at a rate of 1 ft./sec. (0.3 m/sec).

The relationship between horsepower and watts is important, as we generally are charged by the kilowatts for the electricity used by motor:

$$746 \text{ W} = 1 \text{ h.p.}$$

Basics of Electric Motors

An electric motor derives its energy from an outside source and converts this electricity to mechanical energy. Most motors and generators are similar, the basic difference being that a generator takes energy and converts it to electrical energy whereas motors take their energy from an electrical input and convert it to mechanical energy.

Electric motors have four basic parts, which are listed below:

- **Field magnets,** which are stationary magnets made to fit the round interior of the motor's case. There are normally two in each motor and they are of equal magnetism to keep the revolving action of the armature smooth.

- **Armature,** which is a magnet wrapped with wire and connected to the commutator. Rotating between the field magnets, this part is of opposite polarity from the field magnets.

- **Commutator,** which is composed of copper disks formed around the armature shaft and not connected in any way except for the contact of the brushes and armature. The commutator rotates in the same direction as the armature and changes the polarity of the electrical charge so as to permit it to rotate in one direction.

- **Brushes,** devices that, with the force of springs, rest on the commutator as the motor is in action and connect the electrical supply to the commutator and armature.

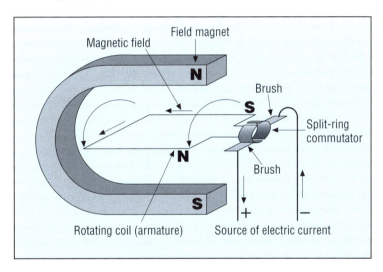

Figure 22-25. A simple motor.

Direct-current (d.c.) motor. When electricity is induced into a current carrying wire, it creates a magnetic field. This electricity is applied to the armature through the commutator and brushes. At the instant the positive and negative poles become exactly opposite, the commutator reverses the current in the armature, making the poles alike (negative to negative, positive to positive), and the armature is forced to revolve further. This continues as long as there is current.

Direct current motors and generators are used where it is necessary to control and change the speed on a piece of printing equipment. Applications of the d.c. motor can be found on web presses and where electroplating is done for multimetal litho plates and gravure cylinders.

Alternating-current (a.c.) motor. The a.c. electric motor operates on the principle that when the armature is rotated in a magnetic field, the current changes its direction every half turn. There are two alternations of current for each revolution. This kind of motor converts electrical energy into mechanical energy by inducing a current into the armature so that mechanical force is affected as in the d.c. motor. These motors are used on most printing equipment.

Polyphase motor. A polyphase motor (or generator) is a motor utilizing a combination of two or three armatures on one shaft, revolving in one magnetic field. This type of motor is very common in the printing industry, as it operates very smoothly and efficiently. Because of the multiple armatures, the shaft can move from armature to armature over a shorter distance, therefore creating a smoother action as it revolves.

Motor enclosures. Motors for printing applications are either open or totally enclosed. The three totally enclosed types are the pipe-ventilated, fan-cooled, and explosion-proof. Even open designs, though, do have some form of enclosure, with variations in their location and the shape of vent openings. Drip-proof, splash-proof, and weather-protected motors are among the more popular of the open types.

Troubleshooting motor problems. There are many problems that a maintenance person must be aware of in resolving motor problems. Three distinct problems that seem to recur when dealing with motors used in the printing industry are brushes sparking, motor noise, and bearings running hot. The possible solutions to these problems are intended for people who have a great deal of knowledge about electrical work. It can be dangerous and expensive for the novice to take a motor apart. Make sure the power is locked off before opening a motor up for inspection.

If the brushes are sparking, a number of items should be checked. If you are working with a generator, check for overload with an ammeter to make sure current is not above the rating of the generator. Check the electrical loads from the generator to make sure they do not exceed the rated output. Carefully check for grounding and any other possible electrical leaks using a voltmeter. If the motor is belted, check to make sure it is not excessively tight on the drive side, a problem sometimes indicated by slipping or squeaking. If the field of the motor is weak, the motor can run excessively fast and the sparking will be worse when first starting up. Sometimes this is due to an incorrect connection. Poor brush contact can also create sparking. Check the commutator to make sure it is clean, smooth, and chocolate brown in color. Do not use emery cloth on the commutator. Check to see that the brushes contact the commutator evenly. If they don't, sandpaper them until they do. Adjust springs to see if this eliminates sparking. A faulty armature coil can also create sparking. Short-circuit coils get very warm after running a short time. If an open-circuit coil exists, the sparking will be enormous and will occur at one spot on the commutator. This generally requires replacement of the coils.

All motors and generators make some noise. The following discussion, however, defines some specific problem sounds. Rattles are generally caused by loose nuts or other parts. Bumping sounds can be caused by the armature as it turns in its bearing. Sometimes resetting the collar to allow more endplay will alleviate this problem. Check the motor for alignment with the device it is driving. Squeaking can be caused

---

Safety Precautions for Motors

- Paper scraps on, under, or near motors can create fires.
- Motors should be placed on machines, on floors, or in locations where they will not interfere with normal movement.
- Nonenclosed motors should be in areas free from dust, moisture, or corrosive vapors.
- When motors are isolated, all power transmission lines should be guarded.
- Motors may be protected from floor moisture by mounting on elevated platforms.
- Care must be exercised in handling liquids around electrical motors. For example, wetting motors during cleanup operations can cause electrical failure and set up shock hazards. Pressrooms where solvents are used to clean floors also present this problem.
- When motors must be repaired or maintained and the operation of the electrical equipment can be started by remote control, the circuit should be opened at the switch box and locked in the off position.
- Always consider explosion-proof motors where there is any possibility of solvent or ink fumes. The motors and conduit are not difficult or expensive to install and are more than worth the small cost for the safety gained.

---

by poor brush tension. If the brushes are new, allow for a run-in period before using sandpaper to shape them to the curvature of the armature. Rubbing and pounding sounds can be caused by the armature rubbing on the pole faces. This results from worn bearings or armature windings working loose. Excessive vibration can be caused by the printing machine itself. Change the machine's speed to see if the vibration is alleviated. Sometimes armature imbalance or worn armature bearings can create this problem.

If the bearings are running hot and can't be held or if there is a smell of burning oil, the following may be causes:

- Dirt contamination in the bearings
- Motor and bearings are poorly lubricated
- Bearing is too tight
- Motor is out of alignment with equipment's drive section
- Shaft is crooked or bent
- No endplay in motor's shaft
- Commutator temperature is too high, heat is transferring through shaft and bearing

Pneumatic Systems Maintenance

Pneumatics is that branch of physics that deals with the properties of air and other gases. In the printing industry we are concerned with compressed air, which serves in a number of different capacities within the plant.

Some of the more common places that we see pneumatics at work in the printing industry are the vacuum and air blast systems on sheetfed presses, air shafts and chucks on webfed presses, air tables in the bindery, vacuum frames in platemaking, pneumatic maintenance tools, and general cleaning aids.

Basic Laws of Pneumatics

Boyle's Law. In the 1662 Robert Boyle, an Englishman, discovered the exact mathematical relationship between the pressure and the volume of gases such as air. Through experiments he determined that the pressure of a gas multiplied by its volume equaled 100 (constant), and that when the original pressure (P_1) was doubled the original volume (V_1) was halved. This relationship held true through a series of experiments where pressures and volumes were varied. However, it was discovered that the temperature must be held constant, because heating air will cause it to expand and cooling it will cause it to contract, thereby changing the pressure/volume relationship. This change causes problems with air compressors that use holding tanks without an automatic pressure-release valve.

Boyle's Law states that the volume of a given mass of gas (air) varies inversely with the pressure, providing the temperature of the gas does not change. Boyle's Law can also be expressed mathematically:

$$P_1 \times V_1 = P_2 \times V_2$$

where P_1 is the original pressure; P_2, the new pressure; V_1, the original volume; and V_2, the new volume.

Charles's Law. Jacques Charles, a French scientist (1746–1823), experimented with gases and developed ***Charles's Law,*** which states that all gases expand or contract with changes in temperature at the same rate, provided that the pressure remains unchanged. This law is also called Gay-Lussac's Law in honor of Joseph Gay-Lussac (1778–1850). Charles's Law can be expressed mathematically:

$$\frac{V_1}{V_2} = \frac{T_1}{T_2}$$

where T_1 is the original temperature; V_1, the original volume; T_2, the new temperature; and V_2, the new volume.

Combined Gas Law. Boyle's and Charles's Laws can be combined to calculate a new volume of gas when both temperature and pressure are changed. A change in pressure and/or temperature can also be calculated by using this formula:

$$\frac{V_1 \times P_1}{T_1} = \frac{V_2 \times P_2}{T_2}$$

where V_1 is the original volume; P_1, the original pressure; T_1, the original temperature; V_2, the new volume; P_2, the new pressure; and T_2, the new temperature.

Compressor Types

A ***compressor*** is a machine that increases air pressure by compressing it, i.e., reducing its volume. The discharge pressure from a compressor is higher than the initial intake pressure. (A vacuum pump is a compressor in reverse.) The printing industry uses two distinctly different types of compressors for its many needs: reciprocating and rotary.

Reciprocating compressors. A *reciprocating compressor* consists of a piston that compresses air from an inlet and expels it through an outlet at an increased pressure. The air that is compressed and expelled for short intervals, i.e., pulses, rather than as a continuous flow.

A disadvantage of this type of compressor is that it has a tendency to introduce oil vapor from the cylinder lubricant into the air being compressed. Oil vapor can stain the paper or other printing substrates and leave oil on flats, film, and plates.

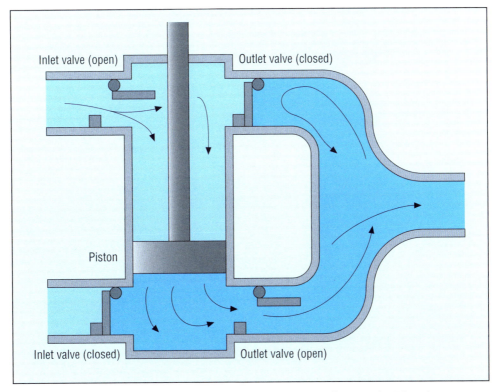

Figure 22-26. *Double-acting, single-stage reciprocating compressor. Compression takes place every stroke, both downard and upward.*

Rotary compressors. A *rotary compressor* increases air pressure by compressing the air into an ever-decreasing space (vane compressor) or by throwing air outward by centrifugal force through an impeller (centrifugal compressor). Rotary compressors are most commonly found where a consistent air blast or vacuum is required, as in vacuum frames. They are normally suitable only for low compression ratios and for small amounts of compressed air or vacuum delivery.

When the impeller of a centrifugal compressor is well-balanced there is a minimum of mechanical vibration; however, there is sometimes an objectionable high-frequency noise. Friction and leakage losses are great with rotary compressors. Their parts are subject to higher wear than those of reciprocating compressors, and they require precision machining. However, these compressors do not introduce oil vapor into the air being condensed, because there is no internal lubrication. The only lubrication necessary is to the bearings.

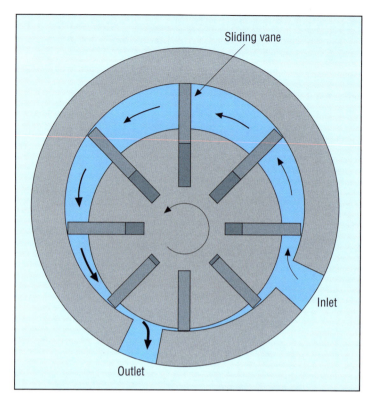

Figure 22-27. Vane compressor. Pressure is created by constricting the air into an ever-decreasing volume.

Rotary compressors are able to handle adequate volumes for most printing operations where pressures are relatively low. As a rule, single-stage vane compressors are used for pressures up to 50 lb./sq.in. (psi). Because the air is supplied in a steady stream from the compressor, a holding tank is not required.

Compressor Problem Troubleshooting

Problem	Possible Cause
Compressor motor does not want to run	• Blown fuse • Overload protector cutout • Motor problems
Compressor inoperative but motor runs	• Check valve closed • Broken compressor
Compressor runs continuously	• Leaks in system • Undersized compressor • Pressure cutout switch set wrong or malfunctioning
Excessive noise	• Lubrication failure • Compressor not bolted or balanced • Worn compressor • Leaks

Note: Before any maintenance work is done on a compressor, the electrical power to the machine must be locked off or disconnected.

Compressed Air Systems

Most printing plants have some form of **centralized compressed air system,** consisting of at least one large holding tank—called a **receiver**—and a reciprocating compressor. The compressed air is piped through the plant for purposes of cleaning and operating equipment. With a **decentralized compressed air system,** each machine has its own compressor.

The piping for a centralized compressed air system can be made of steel, reinforced rubber, or plastic. Whichever material is used, it must be able to withstand air pressure without breaking. Pressure losses occur in transporting the air throughout the plant due to friction and leaks. To compensate for these losses, it is necessary to increase the pressure in the holding tank. Therefore, the compressor must operate more to compensate for the loss. (It is not uncommon for leaks to account for 20% loss of pressure.) For these reasons, piping should be inspected every three months. Large leaks are apparent because of the hissing sound of escaping air. Soapsuds placed along a pipe can be used to detect small leaks or leaks in places where the hissing sound is inaudible because of the general noise of the printing plant; the soapsuds will bubble wherever there is an air leak. Piping should not be underground, as it is difficult to detect leaks.

In general, a centralized system works best where high volume and high pressures are needed. However, decentralized systems have advantages. When each machine has its own compressor, there is no need to be concerned about piping leaks throughout the plant causing pressure and volume drops. If an individual compressor breaks down it does not shut down the whole manufacturing operation.

Disadvantages to the decentralized compressor system include the need to inspect and do preventive maintenance on many small compressors. Also, when printing equipment with small decentralized compressors is being purchased, it is advisable to request from the manufacturer that adequate storage area be provided for the compressor within the printing machine. For example, a soundproof or sound-muffling compartment that is out of the way and easily accessible for maintenance functions would be optimal. Otherwise the small compressor will be underfoot where people can trip on the hoses or will be sitting in a corner making noise and collecting dust.

The lack of a lubricant in the compressed air system is a distinct problem. It can be rectified by injecting a lubricant into the compressed air supply beyond the compressor and as close to the end use as feasible. This will keep rubber gaskets and other materials from becoming hard and cracked, as well as lubricating metal parts and reducing dust buildup. It is advisable to check the oil level daily as it is expelled into the compressed air system. In printing it is imperative not to use excessive amounts of oil because of its potential for soiling the printed product.

Types of Valves

Valves for pneumatic systems are most commonly designed for pressures under 150 psi if there is no other indication of pressure rating. Valves can be categorized by three distinct functions:

- *Direction control valve,* which is used to open and close sections of the air system so that the air can be directed where needed.
- *Pressure control valve,* which is used to reduce maximum line pressure where reduced pressure is needed for printing equipment.
- *Flow control valve,* which is generally inserted directly into the air line to restrict the amount of air flowing beyond that point. A butterfly flow control valve is also used in dampening of high-velocity air in web dryers, and it should be noted that the control is only at the extreme end of the closing cycle. Needle valves are more accurate for pneumatic systems where throttle control to reduce volume is necessary.

Maintenance of Pneumatic Systems

The following maintenance checklist is intended to be general in nature and should not be considered complete. More complete maintenance procedures are contained in compressor maintenance manuals.

The importance of opening and draining moisture traps on a regular basis cannot be stressed enough. Perhaps because it must be done regularly, it seems to be forgotten or overlooked, but it is essential.

Before any maintenance work is done on a compressor, the electrical power to the machine must be locked off or disconnected.

Maintenance Checklist for a Pneumatic System

Daily:
- Check for noise and vibration.
- Drain all condensation traps.
- Check oil level.

Weekly:
- Clean air filter (if not self-cleaning).
- Check relief and safety valves for sticking.

Monthly:
- Check entire system for leaks.
- Check oil for contamination and change if necessary.
- If there is a drive belt, check tension.

Hydraulic Systems Maintenance

Hydraulics is the use of liquids to transfer power from one area to another. The word "hydraulic" was originally Greek and means "water pipe." However, our concern in the printing industry is with oil hydraulics as applied to force amplification. Therefore, a hydraulic power system is a means of transmitting power by the use of a relatively incompressible fluid. This system transfers energy from one location to another and converts that energy to useful work.

Hydraulic systems are used in several areas within the printing plant, including the clamp on paper cutters, plate and blanket impression units on large lithographic presses, and hydraulic lifts on presses and roll handling trucks.

A General Hydraulic System

Fluid. Any fluid used in a hydraulic system must transmit the power effectively and provide lubrication to the moving parts. To be effective, the fluid must be kept free from contaminants, such as paper dust and antisetoff spray. The majority of industrial hydraulic systems employ refined mineral oil blended with additives to ensure stability and avoid corrosion.

Reservoir. The reservoir or holding tank for hydraulic fluids also acts as a means of dissipating heat and filtering out dirt from the oil and serves as a mounting for pump and pressure gauges. A system with a large reservoir is preferable as it can transfer heat from the oil to its sides and thereby keep the hydraulic fluid fairly cool.

Oil lines must discharge the fluid below the oil surface. Otherwise foaming, frothing, and general aeration of the oil will occur, leading to pressure losses as the oil becomes more compressible. It is important to remember that oil is relatively incompressible, whereas air is a compressible gas. It is important, therefore, to keep air out of a hydraulic system because it can cause excessive wear, eventual pump failure, and spongy and erratic component action.

Filters and strainers. A filter in either the pressure line or the return line or both is necessary to remove fine contamination from hydraulic oil. The removal of coarse contaminants require strainers, most which are often used in inlet lines and reservoir fill ports. Strainers are usually classified by the degree of fineness of the weave or by meshes per linear inch, the average mesh being 50. Filters remove much smaller particles from the oil and are referred to in terms of the particle size they can trap. Fiber filters, which are disposable, consist of wool or cellulose impregnated with resin. Some filters have magnetic assemblies to collect metallic particles that do not settle to the bottom but remain suspended in the hydraulic fluid.

In the maintenance process it is important to develop a regular sequence for cleaning both filters and strainers. The hydraulic oil gets contaminated not only with abrasive particles from the printing plant but also from hydraulic pump wear and general flaking of the internal workings of the system. These particles are caught in the filters and can plug them completely, causing a loss in pressure because the hydraulic oil can't get through.

Piping. Piping for hydraulic systems comes in many different materials, which may be solid, semirigid, or flexible. For solid piping, extraheavy-service annealed seamless steel tubing is specified. Copper piping is not recommended because of fatigue failures—i.e., it wears out from use.

Flexible hoses are used where a certain amount of movement is necessary between pressure input and the activating or work portion of the piece of equipment.

Examples of such applications would be hydraulic lifts attached to web presses, fork-lifts, and grab trucks. In the replacement of flexible hoses, be sure that you replace the old hose with one having correct pressure rating.

Pumps. A *hydraulic pump* moves fluid by converting mechanical energy into hydraulic force, which is measured in pounds per square inch (psi). A pump that delivers a fixed output at a constant speed is called a *fixed-displacement pump,* and a pump that delivers a variable output at a constant speed is called a *variable-displacement pump.* Most pumps in the printing industry are of the fixed-displacement type. There are three basic types of pumps available:

- *Gear pumps.* With a gear pump, two gears mesh together displacing fluid; at least one of the gears must rotate. Gear pumps are used for low- to medium-high-pressure (0–1,200 psi) applications, but rarely for high-pressure (1,200–3,000-psi) applications.

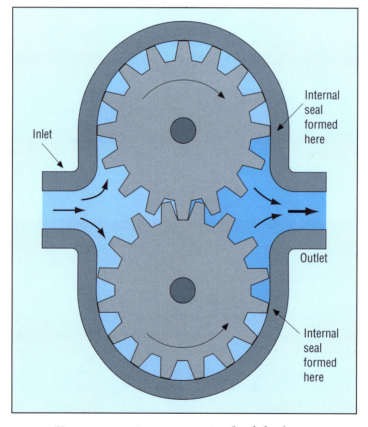

Figure 22-28. External gear pump.

- *Vane pumps.* A vane pump is a fixed-displacement pump in which radial vanes produce the pumping action. The vanes slide in and out of a rotating hub and maintain contact with an outer ring. Gear pumps are used for low- to medium-high-pressure applications, but rarely for high-pressure applications.
- *Piston pumps.* With a piston pump, a piston moves back and forth in a cylinder bore, displacing fluid. A piston pump can be used in hand-pumping,

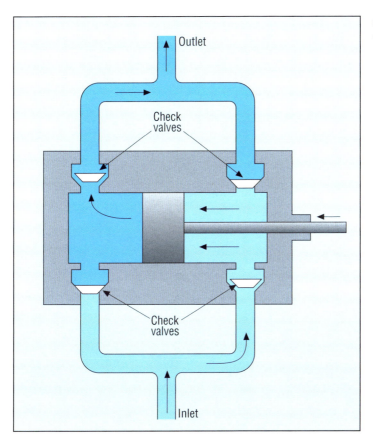

Figure 22-29. Reciprocating piston pump.

low-pressure (0–1,000 psi) applications, but is most often used for high-pressure applications and sometimes for extra-high-pressure (above 3,000 psi) applications.

Valves. In a hydraulic system the flow of oil is guided and directed by *valves.* Their function is to start and stop the flow, direct the flow, and control the volume and/or pressure of oil. One or more such valves are found in hydraulic systems even though they may be hidden from view. Valves can be controlled manually, mechanically or electrically, and entirely automatic hydraulic circuits can be designed. Following are the basic types of hydraulic valves:

- *Pressure relief valve.* A relief valve is a protective device that operates when the pressure in the system reaches a dangerous or predetermined level.

- *Pressure reducing valve.* A pressure reducing valve keeps pressure in a secondary circuit lower than in the main circuit. When the pressure is normal, the valve remains open. When the pressure in the second circuit increases, the valve closes partially and restricts flow through the valve.

- *Volume control valve.* A volume control valve regulates the amount of hydraulic fluid flowing through a circuit. The simplest volume control valve resembles a faucet: when closed, no fluid flows; as it is opened, the rate of flow increases. Spring-loaded volume control valves are sensitive to pressure changes and can react to maintain the outlet pressure at a constant level.

- *Sequence valve.* A sequence valve sends hydraulic fluid to a motor or cylinder in a series before it sends the fluid to another part of the circuit.
- *Directional control valve.* There are two kinds of direction control valves: check valves and spool valves. A *check valve* allows fluid flow in only one direction. The check valve is spring-loaded and only allows fluid flow when the pressure in one direction exceeds a level sufficient to counteract the strength of the spring. A *spool valve* is a device that opens and closes inlets and outlets to alter the path that hydraulic fluid takes through a circuit. An example would be a spool valve that controls a cylinder: if the spool is pushed in one direction, the cylinder extends; if pushed in the other direction, the cylinder retracts.

Cylinders and motors. Hydraulic cylinders and motors convert flow into mechanical motion.

A *hydraulic cylinder* consists of a piston (and/or piston rod) housed in a cylinder with limited inlets and outlets for the hydraulic fluid. The piston moves (extends or retracts) when hydraulic fluid flows through the cylinder. There are two basic kinds of hydraulic cylinder: single-acting and double-acting. With a *single-acting cylinder,* hydraulic pressure is used to extend the piston and gravity or a spring is used to retract the piston. With a *double-acting cylinder,* hydraulic pressure is used to both extend and retract the piston.

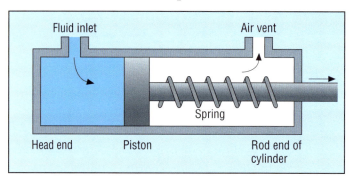

Figure 22-30. *Single-acting cylinder, in which hydraulic fluid enters at the head end of the cylinder and forces the piston to extend. The piston retracts because of gravity or a spring (as shown) between the piston and the rod end of the cylinder.*

A *hydraulic motor* is similar in many respects to a hydraulic pump but works in reverse; i.e., the fluid pressure is converted into mechanical energy to turn a drive shaft. All hydraulic motors are positive-displacement motors because any fluid entering the motor must turn the drive shaft. The two common types of hydraulic motors are gear motors and piston motors.

Maintenance of Hydraulic Systems

A planned preventive maintenance program minimizes operational failures. Since, hydraulic fluid is generally a petroleum-based oil, refill drums should be stored in a dry, cool place. The spout of the drum should always be cleaned before oil is withdrawn, and portable filters should be used when pouring fluid from the original drum to the hydraulic reservoir. While the oil is in the hydraulic system, it must be kept clean by filtration. Only filters of the correct mesh size for your operation should be used, and they should be kept clean or replaced on a regular, predetermined schedule. The reservoir and piping should be emptied, flushed, and thoroughly cleaned periodically.

Failure to follow such a maintenance program will bring unwanted problems: the system can be expected to fail to deliver the proper volume of fluid at the required pressure; and noise will become apparent, indicating improper internal lubrication, a possibility of excessive internal wear, and even the possible breakage of internal parts.

Sometimes the original reservoir is too small, a situation creating excessive heat in the system because there is not enough time for fluid to cool down sufficiently between cycles. If this occurs, the reservoir should be replaced with a larger one. Other causes of excessive heat are a poorly functioning air cooler (if the system has one); leakage from pump, valves, and cylinders; incorrect clearances between pump parts; relief valve that is set for too high a pressure; clogged pipes or replacement pipes that are too small; and oil of incorrect viscosity.

These problems can be solved by cleaning, checking, and setting pump, valves, and oil pipes and by replacing oil with a more suitable one.

All installations should be bled periodically to eliminate any air in the system and thus to ensure more effective operation without a loss in constant pressure.

Maintenance Checklists

As explained earlier in the chapter, preventive maintenance is simply an organized program of maintenance that will keep facilities and equipment in the best possible condition to suit the needs of production, and a preventive maintenance program has four areas of concern: lubrication, inspection, cleaning, and parts adjustment and repair. This section describes some of the major preventive maintenance functions performed on the various components of a web offset press.

Effective preventive maintenance programs must have various components for success.

1. All the press equipment's operation manuals must be stored in a maintenance library for the basis of maintenance checklist development.

2. Preventive maintenance checklists must be developed in a professional team atmosphere. The team members must consist of press management, maintenance management, press operators, and maintenance staff technicians. Everyone must understand each step in the checklist and why they must be performed.

3. Develop a clear and concise malfunction report form and routing procedures for the autonomous maintenance inspection activities carried out by the equipment operators.

4. Maintenance activity checklists must be based on different specified time requirements.

Paster and Infeed

- Lubricate the drive on a multiarmed paster, the spindles and shafts of all rollers and cylinders, and the drive gears on the acceleration belts. If the paster has festoon rollers, each of these should be lubricated.

- Lubricate the mounting of the dancer roller. If the linkage to the roll-stand brake is mechanical, lubricate appropriate points along the linkage.
- On flying pasters, frequently inspect and clean the mechanisms that sense the roll during the paster cycle (photoelectric eyes, butt switches, or electro-mechanical sensing fingers).
- Periodically inspect and clean the rollers or brushes that are used to apply pressure to the splice.
- Periodically inspect the web-severing knife for sharpness and the acceleration belts for wear. Replace as needed.
- Periodically inspect and lubricate all rollers in the infeed.
- Check the roller settings of infeed metering unit. Light settings may allow slippage and reduce tension control.

Printing Units

- Lubricate the mountings of all rollers according to the manufacturer's specifications.
- Regularly check and adjust all inking and dampening rollers, especially the form rollers. Improper settings can cause ink and water distribution problems and undue plate wear, and generally impair print quality.
- Regularly perform reconditioning washups of each inking system, using multi-step solvents to remove dried gum and ink, as well as to recondition roller surfaces (soft and hard). Out-of-the-way places should be thoroughly cleaned.
- Thoroughly clean each dampening system. Scrub built-up chemicals from the fountain pan. Remove grease, gum, and other materials from the roller surfaces. If a recirculating system is in use, flush the entire system with water. Replace or clean the filters as required.
- Frequently clean and inspect the cylinder bearers. Foreign particles will prevent smooth rolling contact.
- Periodically check the bearer pressures.
- Clean cylinder bodies each time plates and blankets are changed.
- Inspect and clean all gear trains; remove traces of old grease and oil. Relubricate the entire train. Inspect gear trains that are sealed in oil baths to ensure that no grit has infiltrated the system.

Dryer

- Regularly inspect and clean all nozzles in the dryer to prevent clogging. Clogged nozzles cannot evenly dry the web; subsequently, marking occurs on the folder. Remove any ink buildup from hot air orifices. These deposits are made by a fluttering web.
- Gas dryers should be inspected for leaks in the gas lines.
- Periodically check the exhaust system. Insufficient exhaust allows solvent vapors to saturate the air inside the dryer. Greater heat is required to evapo-

rate solvent vapors, thus reducing dryer efficiency. Maintain negative air pressure inside the dryer to prevent solvent vapors from exhausting into the pressroom. Antipollution systems that recycle clean hot air back into the dryer must be maintained in top operating condition.

Chill Rolls

- Lubricate the main drive into the chill rolls, the variable-speed drive, and the chill roller mountings.
- Thoroughly clean the chill roll surfaces with a strong solvent to remove ink buildup.
- Periodically flush out the chill rollers. Systems that constantly filter and recirculate water inhibit mineral buildup. Consult the manufacturer's instructions for the proper flushing procedure.
- Check the refrigeration system to make sure that it is operating efficiently.

Folder

- Regularly and completely lubricate the entire folder. This includes the gear train from the press drive; the mountings of wheels, rollers, and cylinders; the tucker blades and folding jaws; the actuating mechanism on the chopper blade; and the drive rollers on the delivery belts.
- Regularly inspect the trolley wheels, slitter wheels, cross perforators, cutoff knives, impaling pins, tucker blades and jaws, chopper blade, and delivery belts. Replace as needed.
- Regularly inspect the mechanisms used to hold the signature and to ensure folding accuracy (e.g., cams, brushes, and wheels). Replace as required.

Glossary

addendum. The portion of the gear tooth between the pitch circle and the outside diameter.

additive. (1) Any compound which, when combined with another, reduces or improves flow (workability), or otherwise changes the composition of a lubricant to a predetermined state. (2) A substance added to another in relatively small amounts to impart or improve desirable properties or suppress undesirable properties. In printing, these substances may be added to ink, paper, and dampening solutions.

afterburner. In incinerator technology (such as that found on web presses), a burner located so that the combustion gases are forced to pass through its flame in order to remove smoke and odors. It may be attached to or separated from the incinerator proper.

air pollution control device. Mechanism or equipment that cleans emissions generated by an incinerator by removing pollutants that would otherwise be released to the atmosphere.

air shaft. A special roller in the roll stand of a web press that uses air-actuated grippers to hold the core of the roll of paper.

alcohol, isopropyl. An organic substance added to the dampening solution of a lithographic printing press to reduce the surface tension of water. *Alternative term:* isopropanol.

alcohol substitute. A chemical used in a lithographic dampening solution in place of isopropyl alcohol (isopropanol).

amperage. The unit of measurement of current.

angle bar. A metal bar at a 45° angle horizontal to the direction of the printing press. It is used to turn the web when feeding from the side, or to bypass turning it in ribbon printing. The angle bar is usually filled with cooled air and perforated to reduce the friction resulting from web travel.

anticorrosion additive. An additive to circulating oils and some types of greases that retards corrosion and rusting of metal surfaces exposed to moist air or water.

antifoaming agent. A substance that prevents the buildup of foam in a dampening solution.

antiskinning agent. A material added to ink to prevent a rubbery layer from forming on its surface when it is exposed to air.

auxiliary equipment. Web guides, ink circulating systems, antistatic devices, and other such things that are not standard on presses but are often incorporated for better control of the substrate.

back cylinder. See *impression cylinder.*

back pressure. The force between the blanket cylinder and the impression cylinder that facilitates the transfer of the image from the blanket to the printing substrate. *Alternative term:* impression pressure.

backing away. A condition in which an ink does not flow under its own weight or remain in contact with the fountain roller. It "backs away" and is not transferred to the ductor roller. Eventually, the prints become uneven, streaky, and weak. A conical ink agitator, which applies a finite amount of force to the ink, keeps it flowing or prevents it from backing away in the fountain while automatic ink leveling keeps the fountain full.

backlash. The free movement at the pitch circle of one gear in the direction of the circumference; the play between teeth.

backlash gear. A thin second gear bolted to the spur gear to reduce play between gears.

back-trap mottle. Blotches and streaks in the solids and tones of an overprinted ink film on a press sheet due to the transfer of a printed ink film from the paper to the blanket of a subsequent printing unit. This trap problem occurs almost exclusively on sheetfed presses with four or more printing units.

backup registration. Correct relative position of the printing on one side of the sheet or web and the printing on the other side.

bareback roller. A form or ductor roller in a conventional dampening system that operates without cloth or paper covers.

basic size. Sheet size in inches for a particular type of paper.

basis weight. Weight, in pounds, of a ream of paper cut to its basic size, in inches.

Baumé scale. Unit for measuring the density or specific gravity of the liquids used in the printing processes.

bearer. A hardened metal ring attached to the cylinder body or journal of the plate and blanket cylinders.

bearer pressure. The force with which the bearers of opposed cylinders contact each other on a sheetfed offset press.

bearer-to-bearer. The cylinder arrangement in which the bearers of the plate and blanket cylinders contact each other.

bearing. A machined element in which another part (e.g., axle or shaft) turns or slides and that reduces friction between moving and stationary parts of machinery.

belt press. A printing press that uses two continuous tracks for printing books in an in-line operation from a paper roll to a delivered book, ready for its binding at the end of the press.

bending fixture. See *plate bending device*.

bending jig. See *plate bending device*.

bible paper. A very thin, lightweight, bright, strong, opaque paper made from rag and mineral fiber pulp.

bladeless ink fountain. A disposable sheet of polyester foil that is held in contact with the fountain roller by a series of small cylinders lying parallel to it.

blanket. A fabric coated with synthetic or natural rubber that transfers the image from the printing plate to the substrate.

blanket, compressible. A blanket with a specially manufactured layer designed to "give" or compress under pressure from the plate and impression cylinder.

blanket, conventional. A hard, noncompressible blanket that bulges out on either or both sides of a nip under pressure.

blanket compressibility. The extent to which blanket thickness reduces under pressure, such as during the printing impression.

blanket compression set. The permanent reduction in thickness of a blanket or any of its component parts.

blanket creep. The slight forward movement or slip of the part of the blanket surface that is in contact with the plate or paper.

blanket cylinder. The cylinder that carries the printing blanket and has two primary functions: (1) to carry the offset rubber blanket into contact with the inked image on the plate cylinder and (2) to transfer, or offset, the ink film image to the paper (or other substrate) carried by the impression cylinder.

blanket-to-blanket. A setup on a perfecting press whereby two blankets, each acting as an impression cylinder for the other, simultaneously print on both sides of the paper passing between them.

blister. In printing, an oval-shaped, sharply defined, bubblelike formation that bulges out on both sides of the web.

body. The relative term describing the consistency of an ink, referring mainly to the stiffness or softness of an ink, but implying other things including length and thixotropy.

bonding. The elimination of a difference in electrical potential between objects.

bridge roller. A roller in a combination continuous-flow dampening system that contacts the dampening form roller and the first ink form roller and transports dampening solution from the dampening system into the inking system.

buffer. A substance capable of neutralizing acids and bases in solutions and thereby maintaining the acidity or alkalinity level of the solution.

bushing. A case-hardened core that prevents metal from rubbing metal, or a machined element in which another part (e.g., axle or shaft) turns or slides and that reduces friction between moving and stationary parts of machinery.

bustle wheel. A mechanical device used to compensate for paper stretching in web offset printing.

butt. The unusable portion of a roll of paper on a web press.

butt splice. The end-to-end joining of two similar materials, such as webs of paper.

cage. A device that keeps the rolling elements in a bearing properly spaced and prevents adjoining rolling elements from touching.

caliper. The thickness of a sheet of paper or other material measured under specific conditions. Caliper is usually expressed in mils or points, both ways of expressing thousandths of an inch.

cam. A machine part that either rotates or slides back and forth to produce a prescribed motion in a contacting element, the follower.

catalyst. A substance that alters (initiates or accelerates) the velocity of a reaction between two or more substances without changing itself in chemical composition.

catch-up. The condition that occurs when insufficient dampening causes the nonimage areas of the plate to become ink-receptive and print as scum or when excessive ink reaches the plate. *Alternative term:* dry-up.

chalking. Poor adhesion of ink to the printing surface. This condition results when the substrate absorbs the ink vehicle too rapidly. The ink dries slowly and rubs off as a dusty powder.

chemicals, hazardous. An EPA designation for any hazardous material requiring a Material Safety Data Sheet (MSDS) under OSHA's Hazard Communication Standard. Such substances are capable of producing fires and explosions or adverse health effects like cancer and dermatitis. Hazardous chemicals are distinct from hazardous waste.

chill rolls. On a web offset press, the section located after the drying oven where heatset inks are cooled below their setting temperature.

chuck. The mechanism on a paper roll stand that centers and grips the roll core.

Clean Air Act. U.S. federal guidelines for air quality and emission controls affecting all manufacturing concerns.

Clean Water Act. U.S. federal guidelines for water quality and controls affecting all manufacturing concerns.

closed loop. A process in which all control functions have been automated, including sensing output errors and correcting the input to compensate for the error.

coating. (1) An unbroken, clear film applied to a substrate in layers to protect and seal it, or to make it glossy. (2) Applying waxes, adhesives, varnishes, or other protective or sealable chemicals to a substrate during the converting process. (3) The mineral substances (clay, blanc fixe, satin white, etc.) applied to the surface of a paper or board during manufacture.

collating mark. A distinctive, numbered symbol printed on the folded edge of signatures to denote the correct gathering sequence.

color, HiFi. A special high-fidelity color reproduction process that uses cyan, magenta, yellow, and black plus additional special colors to expand the color gamut of printing. With HiFi color based on the Küppers model, seven basic colors are used: cyan, yellow, magenta, orange, green, violet, and black. Because seven colors are used, color separations are made using stochastic screening technology to prevent moiré, which would occur if conventional halftone screening technology was used.

color bar. A device printed in a trim area of a press sheet to monitor printing variables such as trapping, ink density, dot gain, and print contrast. It usually consists of overprints of two- and three-color solids and tints; solid and tint blocks of cyan, magenta, yellow, and black; and additional aids such as resolution targets and dot gain scales. *Alternative terms:* color control strip; color control bar.

color match. Condition resulting when no significant difference in hue, saturation, and lightness can be detected between two color samples viewed under standard illumination.

color sequence. The order in which colors are printed on a substrate as indicated by the order that the inks are supplied to the printing units of the press. Color sequence determines how well inks trap on a substrate. *Alternative term:* color rotation.

color strength. An ink's color power as determined by its pigment concentration.

color temperature. The degree (expressed in Kelvins) to which a blackbody must be heated to produce a certain color radiation. For example, 5,000 Kelvin is the graphic arts viewing standard.

color variation. Changes that occur in the density of a color during printing as a result of deviations in the amount of ink accepted by paper or the amount of ink fed to the paper.

colorimeter. An instrument that measures and compares the hue, purity, and brightness of colors in a manner that simulates how people perceive color.

common impression cylinder. A drum-like cylinder that contacts several blanket cylinders, permitting multicolor printing on one side of the sheet.

compression set. A permanent reduction in blanket thickness or the thickness of any of its component parts.

console. The computer system workstation where operators perform specific tasks by executing commands through a keyboard. Modern presses have consoles that control inking, dampening, and plate register moves. With some systems, the results of the operator's commands can be reviewed on a nearby monitor.

corrosion. The deterioration—e.g., rusting—of metal due to oxidation of unlubricated metal parts or by the contact of developers and fountain solutions with printing equipment that has not been protected by a lubricant.

corrosion inhibitor. An additive to the dampening solution to prevent it from reacting with the plate.

creep. (1) Movement of the blanket surface or plate packing caused by static conditions or by the squeezing action that occurs during image transfer. (2) The displacement of each page location in the layout of a book signature as a result of folding the press sheet.

cross-perforation. A series of holes or slits pierced at a right angle to the direction of web travel to prevent the signature from bursting during folding. Cross-perforating also prevents gusseting.

cross-web. The position at a right angle to the grain or machine direction of a web of flexible material.

crystallization. The drying of an ink to form a hard, impervious surface that interferes with dry trapping.

cutoff length. (1) The distance between corresponding points of repeated images on a web. (2) The circumference of the plate cylinder.

cylinder, hydraulic. A cylinder that converts fluid flow into mechanical motion through the linear movement of a piston.

cylinder guide marks. Lines on an offset press plate that match corresponding lines on the press's plate cylinder, ensuring that each plate will be positioned accurately on press.

cylinder undercut. The difference between cylinder body radius and bearer radius.

dampener covers. Molleton, paper, or fiber sleeves placed over dampening rollers in a conventional dampening system that aid in carrying the dampening solution.

dampening solution. A mixture of water; gum arabic; an acid, neutral, or alkaline etch; and isopropyl alcohol or an alcohol substitute used to wet the nonimage areas of the lithographic printing plate before it is inked. *Alternative term:* fountain solution.

dampening system. A group of rollers that moistens the nonimage areas of a printing plate with a water-based dampening solution that contains additives such as acid, gum arabic, and isopropyl alcohol or other wetting agents.

dampening system, brush. A system on a lithographic press that uses a rotating brush to transfer small amounts of water from the fountain to the dampener roller train.

dampening system, continuous-feed. A ductorless dampening system in which there is a continuous flow of dampening solution from fountain roller to form roller. *Alternative term:* continuous-flow dampening system.

dampening system, conventional. A dampening system consisting of a fountain, fountain pan roller, ductor roller, oscillator roller, and one or more covered or uncovered form rollers. The ductor roller transfers dampening solution from the fountain roller to the oscillator roller. *Alternative term:* intermittent-flow dampening system.

dampening system, inker-feed. An integrated, continuous-feed dampening system that delivers dampening solution to an ink form roller.

dampening system, plate-feed. A continuous-feed dampening system that applies dampening solution directly to the plate using dampening form rollers, rather than indirectly using the first inking form roller.

dampening system, spray bar. A variation of the inker-feed continuous-flow dampening system that applies a very fine mist of dampening solution directly to the inking system rollers.

dancer roller. A weighted or spring-tensioned controlled roller positioned between a paper roll and a press unit on a web printing press. It senses and removes web slack by controlling the paper reel brake.

deadweight micrometer. A device that uses the deadweight of an anvil to obtain repeatable measurements on plate, blankets, and packing. *Alternative terms:* bench micrometer, blanket thickness gauge, Cady gauge.

deionization. A chemical water purification process that uses two ion exchange resins to remove minerals from water.

delamination. The continuous splitting, or separation, of the paper's surface caused by the tack of the ink and the rubber blanket.

densitometer, reflection. An instrument for measuring the optical density of a photographic print or press sheet.

desensitization. In platemaking, the making of a nonimage area less receptive to ink by the application of a gum solution.

dimensional stability. How well an object maintains it size. The extent to which a sheet maintains its dimensions with changes in its moisture content or applied stressing.

direct lithography. A lithographic process in which the plate and printing surface are brought into contact.

dished roll. A web of paper with progressive concave or convex edge misalignment, which is noticeable immediately after the roll is unwrapped.

distillation. A water purification method where water is boiled in a still. The steam that rises from the boiling water is cooled through condenser coils where it is converted back to mineral-free liquid water.

dot gain. The optical increase in the size of a halftone dot during prepress operations or the mechanical increase in halftone dot size that occurs as the image is transferred from plate to blanket to paper in lithography.

double-sixteen. A folder that takes a thirty-two-page form and folds it as two separate or inserted sixteen-page forms.

double-thirty-two. A folder that takes a sixty-four-page form and folds it as two inserted or separate thirty-two-page forms.

drier. An ink additive, such as a salt of cobalt or manganese, that acts as a catalyst to convert a wet ink film to a dry ink film.

drive, mechanical. A mechanical system that transmits power and movement to equipment; usually consists of some combination of chains and sprockets, belts and pulleys, cams, and gears.

drum. (1) An oscillating metal ink distribution roller. (2) A synonym for "cylinder" in many press applications.

dry dusting. A preliminary pass of the sheet under pressure through the press to remove excessive spray powder, surface material, or other debris.

dry offset. Printing from relief plates by transferring the ink image from the plate to a rubber surface and then from the rubber surface to the paper. Printing with this process on an offset press eliminates the need to use water. *Alternative terms:* indirect letterpress, letterset, relief offset.

dry-back. An optical loss of density and color strength that may occur while an ink is setting. To achieve the proper dry density, the ink is printed with a wet density slightly higher than the projected dry density.

dryer. A unit on a heatset web press that evaporates the solvent ingredient in the heatset ink.

drying agent. An ink additive, such as a salt of cobalt or manganese, that acts as a catalyst to convert a wet ink film to a dry ink film.

drying section. Section of a papermaking machine where water is removed by passing the web over hot drying cylinders.

drying stimulator. A substance—e.g., cobalt chloride—that complements the drier in the ink.

dry-up. The problem that occurs when ink appears in the nonimage area due to insufficient dampening of the plate. *Alternative term:* catch-up.

ductor. (1) A small-diameter cylinder that alternately contacts the ink fountain roller and the first roller of the ink train. (2) In a conventional dampening system, a small-diameter cylinder that alternately contacts the dampening fountain roller and the oscillator. *Alternative term:* ductor roller.

ductor shock. The vibration sent through the inking system when the ductor first contacts the oscillating roller.

ductorless inking systems. Also called an *anilox inking system.* This ink-metering system transfers ink in a continuous flow from the fountain using an engraved fountain roller. A doctor blade squeegees the surface of the rotating fountain roller, leaving ink in the cells that is picked up by the next roller in the system.

duplicator. Any sheetfed press smaller than 11×17 in. (279×432 mm) without bearers (hardened metal disks attached to the ends of the cylinder or to the cylinder's journal).

durometer, type-A. Instrument used in printing to measure the hardness of roller compounds.

dwell. (1) The length of time that the ductor roller contacts the fountain roller. (2) The condition that occurs when the follower remains stationary for some period of rotation of the cam.

electrostatic-assisted drying. The use of electrostatic energy within the hot air dryer to break up the solvent-laden laminant air on the surface of the web, aiding dryer efficiency and reducing air turbulence requirements.

emulsification. (1) Condition that occurs when an ink picks up too much dampening solution and prints a weak, snowflaky pattern. (2) The mixing of two mutually insoluble liquids—e.g., water and a lubricant—in which one is dispersed as droplets throughout the other.

endplay. Undesirable lateral movement of a roller due to poor fit between roller shaft and roller bracket.

fan delivery, web. A rotary unit with blades that form pockets which transfer individual folded signatures or newspapers from the folder to the conveyors that carry them to the delivery on a web press.

fan-out. An expansion of the sheet near the tail edge.

feeler gauge. A thin strip of steel ground to precise thickness and marked accordingly. It is used to adjust clearances between various press mechanisms.

felt side. The top side of the paper formed on the paper machine wire. It is the preferred side for printing. See also: *wire side.*

festoon. A method of storing a relatively large amount of paper on a series of rollers so that the press can continue to operate while a roll is spliced and accelerated to press speed.

fiber cut. A short, straight, fairly smooth slice in the web caused by a fiber bundle catching as paper passes through the calender.

fiber puffing. Surface roughening of a coated paper containing groundwood fibers. Condition occurs during heatset drying.

fire point. The temperature at which a vapor will ignite when a flame or spark is applied and burn steadily without an additional spark.

flagging. (1) Indicating a web splice so that the spliced product can be removed from the press folder and discarded. (2) Marking printed matter to indicate a change or correction. (3) Inserting small strips of paper into a skid of press sheets as needed to indicate segments of defective printed sheets.

flash point. The temperature at which vapors will ignite momentarily when a flame or spark is applied.

flying paster. An automatic device that attaches a new roll of paper to an expiring roll without a press stop, while the paper is running at press speed and without the use of a festoon.

folder, jaw. Three cylinders in the in-line finishing area of a web press that make one or two parallel folds at right angles to the direction of web travel. *Alternative terms:* parallel folder; tucker folder.

folder, ribbon. A folder on a web press used for publication work. It slits the web into multiple strips of the width required by the desired product size. Each ribbon is turned over an angle bar and guided into position so that all ribbons align with each other ahead of the jaw-folding section. The ribbons of paper are collated and brought down to the cutoff knives and folding jaws in either one or two streams. The press then simultaneously delivers either one or two sets of signatures of the same size.

follower. A device that follows the contoured cam surface or that moves when contacted by a cam.

form roller. A device, riding in contact with the printing plate, that transfers dampening solution or ink from an oscillator roller to the printing plate. Presses typically have one or two *dampening* form rollers and three to five *inking* form rollers.

form roller, oscillating. A special roller substituted for the first and, sometimes, fourth (last) form rollers of a press to reduce ghosting on a job.

former. A smooth, triangular-shaped, metal plate over which a printed web passes prior to entering an in-line folder. The former folds the moving web in half lengthwise. *Alternative term:* former board.

former fold. First fold given paper coming off a web press. The former fold is made in the direction of web travel, thus parallel to the grain.

fountain. A reservoir for the dampening solution or ink that is fed to the plate.

fountain blade. A spring steel plate, steel segment, or plastic angled against the fountain roller and forming the bottom of the ink fountain. Moving the blade closer or farther from the fountain roller controls the thickness of the ink film across the roller.

fountain cheeks. Vertical metal pieces contacting the edges of the fountain roller and blade to form an ink-tight trough.

fountain height monitor. A sensing device, usually mechanical or ultrasonic, that checks the height of ink moving over the agitator.

fountain keys. A series of thumb screws or motor-driven screws or cams behind the blade that provide for variable ink flow across the fountain.

fountain roller. A metal roller that rotates inter-mittently or continuously in the ink or dampening fountain and carries the ink or dampening solution on its metal surface.

fountain solution. See *dampening solution*.

fountain solution concentrate. A mixture of chemicals (compounded acids and gums) that, when combined with water and alcohol or another wetting agent, form dampening solution.

fountain splitter. A device that divides the ink fountain so that two or more inks can be used in the same ink fountain. Each ink will print a differ-ent section of the press sheet; e.g., red on the left side and blue on the right side.

fungicide. A substance that prevents the formation of mildew and the growth of fungus and bacteria in the dampening system.

furnish. Mixture of fibrous and nonfibrous materials like colorants, fillers, and sizing in a water suspen-sion from which paper or paperboard is made.

gear. A mechanism that transmits mechanical rotary power from one shaft to another or from a shaft to a slide (rack) by means of successively engaging projections of teeth.

gear streaks. Alternating light and dark marks that appear as bands in halftones and solids parallel to the gripper edge of the sheet. The distance between the marks is the same as the interval between the gear teeth on a cylinder.

ghosting. The appearance of faint replicas of an image in undesirable places, caused by mechanical or chemical processes, other than setoff or show-through. *Mechanical ghosting* is caused by ink starvation. *Chemical ghosting* is the appearance of gloss or dull ghosts of images that are printed on the reverse side of the sheet and is caused by the chemical-activity influence that inks have on each other during their critical drying phases.

glaze. A combination of oxidized roller surface, embedded ink pigment, dried ink vehicle, and gum from dampening solution on an inking roller.

gloss. High reflectance of light from a smooth surface.

grain direction. In papermaking, the alignment of fibers in the direction of web travel. In printing, paper is *grain-long* if the grain direction parallels the long dimension of the paper and *grain-short* if it parallels the short dimension.

grammage. Weight in grams of a single sheet of paper having an area of one square meter (1 m^2).

grater rollers. Textured press cylinders that support a web before drying, reducing smearing and marking.

grayness. The amount of unwanted gray that appears to reflect from a process color ink film (yellow, magenta, cyan). Measured as a percent-age of reflectance.

grease. Oil that has been mixed with soap to thicken it so it will stand up under greater heat and pressure.

grounding. The elimination of a difference in electrical potential between an object and the ground.

groundwood. Mechanical pulp used in papermak-ing produced by forcing bark-free logs against a revolving, abrasive grinding stone in the presence of water.

guide roller. A cylinder on the roll stand between the roll of paper and the dancer roller. It is used to compensate for slight paper variations. *Alternative term:* cocking roller.

hair cut. A smooth, curved slice in a paper web that usually occurs because a piece of felt from the manufacturing process is embedded in the web and passes through the calender.

hairline register. A standard for accuracy in which the maximum deviation between printing colors is 0.003 in. (0.08 mm).

Hazard Communication Standard. An OSHA regulation that requires chemical manufacturers, suppliers, and importers to assess the hazards of the chemicals that they make, supply, or import,

page 426 of Web Offset Press Operating

and to inform employers, customers, and workers of these hazards through Material Safety Data Sheets (MSDS).

helical gear. A gear that has teeth cut at an angle.

hickey. An imperfection in printing due to a particle on the blanket or, sometimes, the plate. A *doughnut hickey* consists of a small, solid printed area surrounded by a white halo, or unprinted area. A *void hickey* is a white, unprinted spot surrounded by printing.

hickey-picking roller. A roller that has synthetic fibers embedded in its surface to help it remove hickeys from the surface of an offset printing plate or to fill in the white ring on the plate surface. This roller replaces one of the ink form rollers.

horsepower. The energy or power utilized by a motor; equivalent to the force necessary to move 550 lb. a distance of 1 ft. in 1 sec.

hot-weather scumming. The tendency of ink to print in nonimage areas when the dampening feed rate is too low.

hue error. The degree to which a process color ink film (yellow, magenta, cyan) varies from a perfect hue. Measured as a percentage of impurity.

hydrophilic. Water-receptive, as in the nonimage areas of the printing plate.

hydrophobic. Water-repellent, as in the image areas of the printing plate.

hygroscopic. The ability of paper or other substrates to absorb or release moisture and, in so doing, expand or contract.

image area. On a printing plate, the area that has been specially treated to receive ink.

image fit. The agreement in distance between the register marks on each color from the gripper to the tail edge of the press sheet.

impaling pins. Sharp pieces of metal that maintain web control within a folder. The web is punctured by the pins, just behind the web cutoff point, to control and pull the web around the folder cylinder, releasing it when the fold is started.

impression. (1) The printing pressure necessary to transfer an inked image from the blanket to the substrate. (2) A single print.

impression cylinder. A large-diameter cylinder that transports the press sheet and forces the paper or other substrate against the inked blanket.

impression cylinder pressure. The force of the impression (or back) cylinder against the blanket cylinder.

infeed. The set of rollers controlling web tension ahead of the first unit on a web press.

ink, heatset. An ink used in high-speed web offset printing that dries primarily by evaporation while the job is still on the press. The inked web passes through a high-velocity, hot-air dryer, which drives the solvents out of the ink, and then over chill rolls that cool and set the ink before the substrate is transferred to the folder.

ink absorbency. The extent that an ink penetrates the paper.

ink agitator. A revolving cone-shaped device that moves from one end of the fountain to the other keeping the ink soft and flowing.

ink drying. Process by which an ink is transformed from an original semifluid or plastic state to a solid.

ink feed. The amount of ink delivered to the ink form rollers.

ink fountain. The trough on a printing press that holds the ink supply to be transferred to the inking system. The operator controls ink volume from adjustment screws or keys on the fountain or from a remote console.

ink holdout. The extent to which paper resists or retards the penetration of the freshly printed ink film.

ink setting. (1) The increase in viscosity or body (resistance to flow) that occurs immediately after the ink is printed. (2) An adjustment that the press operator makes to the inking system to control ink volume.

ink vehicle. A complex liquid mixture in which pigment particles are dispersed.

ink-dot scum. On aluminum plates, oxidation characterized by scattered pits that print sharp, dense dots, or ink material trapped in the grain.

inking control console. A computerized device that enables the press operator to control inking and a variety of other functions without leaving the inspection table.

inking system. A series of rollers that apply a metered film of ink to a printing plate.

ink/water balance. In lithography, the appropriate amounts of ink and water required to ink the image areas of the plate and keep the nonimage areas clean.

in-line converting. Converting done directly from the last printing station or drying unit into the converting machinery in one continuous operation.

in-line finishing. Manufacturing operations such as numbering, addressing, sorting, folding, die-cutting, and converting that are performed as part of a continuous operation right after the printing section on a press.

insert, free-standing (FSI). A four-page, eight-page, etc., self-contained signature typically added to a newspaper.

interface. The electronic device that enables one kind of equipment to communicate with or control another. The common boundary between systems or parts of systems.

isopropanol. See *alcohol, isopropyl.*

journal. A shaft, axle, roll, or spindle.

Kelvin temperature scale. A scale developed by the physicist Sir William Thomson (better known as Lord Kelvin) to measure absolute temperature; 0 on the scale being equivalent to absolute zero; used to convert Celsius readings to the absolute scale. *Alternative term:* absolute temperature scale.

kiss impression. The minimum pressure at which proper ink transfer from the blanket to the substrate is possible.

knife rollers. Small-diameter hard rollers that help to keep the ink system clean by picking up ink skin particles, lint, etc. *Alternative term:* lint roller.

lay. Position of the printed image on the sheet.

lay sheet. The first of several sheets run through the press to verify lineup, register, type, and non-image areas.

lint. Loosely bonded paper surface fibers and dust that accumulates on an offset plate or blanket and interferes with print quality. *Alternative terms:* linting, fluffing.

liquid drier. A drier in which metal salts are suspended in liquids such as a petroleum solvent.

lithography, offset. A planographic printing process that requires an image carrier in the form of a plate on which photochemically produced image and nonimage areas are receptive to ink and water, respectively.

loupe. An adjustable-focus magnifier incorporating a precise measuring scale, with or without a self-contained light source. It is used to inspect fine detail.

makeready. All of the operations necessary to get the press ready to print the current job.

masstone. The color of ink in bulk, such as in a can, or a thick film of ink. It is the color of light reflected by the pigment and often differs from the printed color of the ink.

Material Safety Data Sheet. A product specification form used to record information about the hazardous chemicals and other health and physical hazards employees face in an industrial workplace, along with guidelines covering exposure limits and other precautions. Employers are required to compile and maintain files of this information under the OSHA Hazard Communication Standard set forth by the U.S. federal government.

metering nip. The line of contact between the two rollers of an inker-feed dampening system.

milking. A coating buildup on the nonimage areas of the offset blanket that usually occurs when the coating softens because it does not adequately resist water.

misregister. Incorrectly positioned printed images, either in reference to each other or to the sheet's edges.

misregister, random. Misregister that varies from sheet to sheet.

misting. Flying ink that forms fine droplets or filaments that become diffused throughout the pressroom.

moiré. An undesirable, unintended interference pattern caused by the out-of-register overlap of two or more regular patterns such as dots or lines. In process-color printing, screen angles are selected to minimize this pattern. If the angles are not correct, an objectionable effect may be produced.

molleton. A thick cotton fabric, similar to flannel, with a long nap and used to cover form rollers in conventional lithographic dampening systems.

mottle. Irregular and unwanted variation in color or gloss caused by uneven absorbency of the paper.

multicolor printing. The printing of two or more colors, often one over another.

nip. The line of contact between two cylindrical objects, such as two rollers on an offset press.

nonimage area. The portion of a lithographic printing plate that is treated to accept water and repel ink when the plate is on press. Only the ink-receptive areas will print an image.

offset blanket. See *blanket.*

ohm. The unit of measurement of resistance.

OK sheet. A press sheet that closely matches the prepress proof and has been approved by the customer and/or production personnel. It is used as a guide to judge the quality of the rest of the production run.

oleophilic. Oil-receptive, as in the image areas of the printing plate.

oleophobic. Oil-repellent, as in the dampened nonimage areas of the printing plate.

opacity. (1) The ability of a printed ink film to hide what is underneath. (2) The extend to which light transmission is obstructed.

oscillator. A driven inking or dampening roller that not only rotates but also moves from side to side, distributing and smoothing out the ink film and erasing image patterns from the form roller. *Alternative terms:* oscillating drum, oscillating roller, or vibrator.

overpacking. Packing the plate or blanket to a level that is excessively above the level of the cylinder bearer.

overrun. The quantity of printed copies exceeding the number ordered to be printed. Trade custom allows a certain tolerance for overruns and underruns.

packing. (1) The procedure for setting the pressure between the plate and blanket cylinders. (2) The paper or other material that is placed between the plate or blanket and its cylinder to raise the surface to printing height or to adjust cylinder diameter to obtain color register in multicolor printing.

packing gauge. A device for measuring the height of the plate or blanket in relation to its cylinder bearers.

paper conditioning. Bringing the paper's temperature to equilibrium with the temperature or atmosphere of the pressroom without removing its wrapping or exposing it to atmospheric and humidity changes.

paper sizes, international. The common paper sizes used in Europe and Japan. They are: A3 (11.7×16.5 in.); A4 (8.3×11.7 in.); A5 (5.8×8.3 in.); B4 (10.1×14.3 in.); B5 (7.2×10.1 in.); and B6 (5.1×7.2 in.). See also: *basis weight.*

paste drier. A highly viscous drier prepared by grinding the inorganic salts of manganese or other metals in linseed oil varnishes.

paster. (1) A device used to apply a fine line of paste on either or both sides of the web to produce finished booklets directly from the folder without saddle stitching. The paste is applied from a stationary nozzle as the web passes underneath it. (2) An automatic web splicer on a press. (3) The rejected web with a splice in it.

perfecting. The printing of at least one color on both sides of a sheet in a single pass through a press.

perfector. A press that can print at least one color on both sides of a sheet in a single pass. A blanket-to-blanket web offset press is an example of a perfector.

perforating. Punching a row of small holes or incisions into or through a sheet of paper to permit part of it to be detached; to guide in folding; to allow air to escape from signatures; or to prevent wrinkling when folding heavy papers. A perforation may be indicated by a series of printed lines, or it may be blind; in other words, without a printed indication on the cutline. *Alternative term:* perf.

pH. A measure of a solution's acidity or alkalinity, specifically the negative logarithm of the concentration (in moles/liter) of the hydrogen ions in a solution. Measured on a scale of 0 to 14, with 7 as the neutral point.

pick resistance. Ability of a paper to resist a force applied perpendicularly to its surface before picking or rupturing occurs.

picking. The delamination, splitting, or tearing of the paper surface that occurs when the tack of the ink exceeds the surface strength of the paper.

pigment. Finely divided solid particles derived from natural or synthetic sources and used to impart colors to inks. They have varying degrees of resistance to water, alcohol, and other chemicals and are generally insoluble in the ink vehicle.

piling. A buildup of paper, ink, or coating on the offset blanket, plate, or rollers in such quantity that it interferes with print quality.

pinless folder. A folder that uses a set of belts (rather than impaling pins), one on each side of the cut signature, to grip and guide the cut signature through the folding mechanism.

pipe rollers. Small-diameter hard rollers that help to keep the ink system clean by picking up ink skin particles, lint, etc.

pitch diameter. The working diameter of the gear attached to the cylinder journal.

planography. A printing process that uses a flat image carrier, such as the lithographic printing plate, which has no relief images and has image and nonimage areas on the same level (or plane).

plate. A flexible image carrier with ink-receptive image areas and, when moistened with a water-based solution, ink-repellent nonimage areas. A thin metal, plastic, or paper sheet serves as the image carrier in many printing processes.

plate, bimetal. A negative-working multimetal printing plate that usually consists of copper electroplated on a base metal such as aluminum or stainless steel.

plate, negative-working. A printing plate that is exposed through a film negative. Plate areas exposed to light become the image areas.

plate, positive-working. A printing plate that is exposed through a film positive. Plate areas exposed to light become the nonimage areas.

plate, presensitized. A sheet of metal or paper supplied to the user with the light-sensitive material already coated on the surface and ready for exposure to a negative or positive.

plate, subtractive. A printing plate in which the light-sensitive coating also contains an image-reinforcing material.

plate, surface. A printing plate in which a light-sensitive coating applied to the plate surface is made ink-receptive in the image areas during exposure and processing, while in the nonimage areas it is removed or converted to a water-receptive layer.

plate, trimetal. A positive-working multimetal plate consisting of a top layer of chromium (the nonimage metal) and a bottom layer of copper (the image metal) electroplated to a base metal.

plate, waterless. A presensitized negative- or positive-working planographic image carrier that uses ink-repellent silicone rubber, instead of a water-based dampening solution, to keep ink from adhering to the nonimage areas of the plate.

plate bending device. A device that creases a printing plate in such a way that it fits precisely into the clamps of a press's plate cylinder. *Alternative terms:* bending fixture, bending jig.

plate blinding. The loss of ink receptivity in the image area due to an excessively acidic fountain solution.

plate clamp. A device that grips the lead and trailing edges of the plate and pulls it tight against the cylinder body.

plate cylinder. A cylinder that carries the printing plate. It has four primary functions: (1) to hold the lithographic printing plate tightly and in register, (2) to carry the plate into contact with the dampening rollers that wet the nonimage area, (3) to bring the plate into contact with the inking rollers that ink the image area, and (4) to transfer the inked image to the blanket carried by the blanket cylinder.

plate scanner. A device that measures all of the various densities in a plate's image area at selected increments across the printing plate before it is mounted on the press. The press operator then sets the ink fountain keys to match the ink densities indicated by the plate scanner's measurements before beginning to print the job.

plate scumming. The pickup of ink in nonimage areas of the plate.

polymerization. A chemical reaction—usually carried out with a catalyst, heat, or water, and often under high pressure—in which a large number of relatively simple molecules combine to form a chain-like macro-molecule. Some printing inks dry by polymerization (a chemical reaction between the binder and solvent leaves a tough and hard ink deposit on the substrate).

preloaded pressure. The amount of force required to hold the plate and blanket cylinder in firm contact when the cylinders are overpacked to create the recommended squeeze pressure.

press, blanket-to-blanket. A perfecting press in which the blankets from two printing units are in contact, with the paper passing between the two blankets. Since each blanket acts as the impression cylinder for the other, no impression cylinder is needed.

press, offset lithographic. A mechanical device that dampens and inks a planographic printing plate and transfers the inked image to the blanket and then to the printing substrate.

press, perfecting. A press that can print at least one color on both sides of a sheet in a single pass.

press, sheetfed offset. A printing press that feeds and prints on individual sheets of paper (or other substrate) using the offset lithographic printing method. Some sheetfed presses employ a rollfed system in which rolls of paper are cut into sheets before they enter the feeder; however most sheetfed presses forward individual sheets directly to the feeder.

press, single-color. A press consisting of a single printing unit, with its integral inking and dampening systems, a feeder, a sheet transfer system, and a delivery. It can also be used for multicolor printing by changing the ink and plate and running the paper through the press again.

press, small offset. Any press smaller than 11×17 in. (279×432 mm) without bearers (hardened metal disks attached to the ends of the cylinder or to the cylinder's journal).

press, web offset. A offset lithographic press that prints on a continuous web, or ribbon, of paper fed from a roll and threaded through the press.

press section. Section of papermaking machine where water is removed from the web by suction and applied pressure.

press sheet. A single sheet of paper selected for the job to be printed on the press.

pressrun. (1) The total of acceptable copies from a single printing. (2) Operating the press during an actual job.

presswork. All operations performed on or by a printing press that lead to the transfer of inked images from the image carrier to the paper or other substrate. Presswork includes makeready.

preventive maintenance. An organized program of maintenance that will keep facilities and equipment in the best possible condition to suit the needs of production.

print contrast. The ratio of the difference in the density of a 75% (three-quarter) tone and a solid print to the density of the saturated solids on the press sheet. This densitometric measurement indicates how well the three-quarter tone to shadow areas of an image are reproducing on press.

printing pressure. The force, in pounds per square inch, required to transfer the printed image to the substrate. In lithography, this includes the pressure between the plate and blanket, the blanket and the impression cylinder, and the impression cylinder and the substrate.

printing unit. The section of the offset lithographic press that houses the components for reproducing an image on the substrate. With an in-line web offset press, a printing unit includes the inking and dampening systems and the plate and blanket cylinders.

proof. A prototype of the printed job made photomechanically from plates (a press proof), photochemically from film, or digitally from electronic data (prepress proofs). Prepress proofs serve as samples for the customer and guides for the press operators. Press proofs are approved by the customer and/or plant supervisor before the actual pressrun.

proof, direct digital color. A type of prepress proof in which digital information is used to directly image the color proofing material. Various technologies, such as ink jet and dye sublimation, are used to image the color proofing material. No film intermediates are required.

proof, overlay. A type of photochemical prepress proof used in multicolor or process-color printing where pigmented or dyed sheets of plastic (for each process color and black) are exposed to a halftone negative or positive from a set of color separation films, processed, registered to each other, and taped or pin-registered to a base.

proof, prepress. A simulation of the printed piece that is made digitally from electronic data or photochemically using light-sensitive papers (principally to proof single-color printing), colored films, or photopolymers. *Alternative term:* off-press proof.

proof, single-sheet. A type of photochemical prepress proof used for multicolor or process-color proofing where the printing colors are built up on a base through lamination, exposure to a halftone negative or positive from a set of color separation films, and toning or other processing.

proof press. A printing machine used to produce photomechanical proofs. It has most of the features of a true production machine but is not meant for long pressruns.

proofing. Producing simulated versions of the final reproduction from films and dyes or digitized data (prepress proofing) or production trial images directly from the plate (press proofing).

pull. A group of inspection sheets removed from the delivery of the press.

quality control. Systematically planning, measuring, testing, and evaluating the combination of staff resources, materials, and machines during (and directly after) manufacture with the objective of producing a product that will satisfy established standards and profitability of an enterprise.

ream. With a few exceptions, 500 sheets of paper.

reducer. An ink additive that softens the ink and reduces its tack.

reelroom. At a web printer, newspaper printers in particular, the separate area where roll stands are sometimes housed.

refiner mechanical pulp. Papermaking pulp produced by passing wood chips through a disk refiner instead of pressing the wood against an abrasive grinding stone.

refractive index. Measure of the ability of a material, such as a pigment particle, to bend or refract light rays. The result is expressed as the ratio of the speed of light in one medium to the speed of light in another medium (usually air or a vacuum).

register. The accurate positioning of images— either in relation to images on other press sheets or in relation to an image already printed on that press sheet.

register, circumferential. The alignment of successive ink films on top of each other on the printed sheets, usually accomplished on a rotary printing press by moving the plate cylinder toward the gripper or tail.

register marks. Small reference patterns, guides, or crosses that aid in color registration and correct alignment of overprinted colors on press sheets.

release. The readiness of the blanket to give up the paper after it leaves the nip.

remote control console. A computerized device that enables the press operator to control a variety of functions without leaving the inspection table. Among the functions controlled are inking, dampening, and lateral and circumferential image register.

repeat flaws. In an ink system, any imperfection in the roller surface that creates a flaw in the ink film. This imperfection is passed to the next roller and can continue from roller to roller to the plate.

Roller oscillation and varying roller diameters help to eliminate this problem.

resilience. The ability of a blanket to regain its thickness after pressure on its surface has been removed.

reverse osmosis. A water purification method where water is filtered through a membrane to remove impurities.

reverse slip nip. The point of contact where two rollers are rotating in opposite directions in a dampening system.

rewind. Rerolling a web onto a new core after printing.

roll. Paper or cardboard produced in a continuous strip and wound uniformly around a central shaft or hollow core.

roll set curl. Paper curl that occurs because the web has been stored in roll form long enough to cause its curved condition to become permanent. *Alternative term:* wrap curl.

roll stand. The mechanism that supports the roll of paper as it unwinds and feeds into the press.

roll stand, auxiliary. An extra roll stand mounted on top of another roll stand. This reduces downtime by permitting one stand to be reloaded while the other is still unwinding. The auxiliary roll stand cannot be used to feed two webs at the same time unless it is converted to a dual roll stand.

roll stand, dual. A support for two rolls of paper, one stacked above the other, to feed two webs at the same time, or to reduce reloading time if a single web is used.

roller, intermediate. A friction- or gravity-driven roller between the ductor and form roller that transfers and conditions the ink. It is called a *distributor* if it contacts two rollers and a *rider* if it contacts a single oscillating drum.

roller cover. Absorbent cloth or paper that covers the rollers and helps to provide more continuous dampening by increasing the solution-carrying and solution-storing capacity of the rollers.

roller stripping. Condition that occurs in lithography when ink oscillators fail to accept ink because they have been desensitized by dampening solution.

roller-setting gauge. A device that shows the amount of pressure exerted when the press operator pulls a metal feeler strip between the two rollers being set.

roller-stripe gauge. A device that is marked with stripes of specified widths and used to visually determine the width of an ink stripe on a roller or plate.

roll-fed. A printing press or converting machine that receives paper as continuous webs from rolls, instead of as sheets. See also: *press, web.*

roll-to-roll printing. Printing webs of substrates and then rewinding them directly onto another roll core after printing.

scanning densitometer. A computerized quality control table that measures and analyzes press-sheet color bars using a densitometer.

scumming. The problem that occurs when a permanent ink image—usually dots—appears in the nonimage area.

sensitization. In platemaking, the making of an image area more ink-receptive.

sequestering agent. A substance that prevents the calcium and magnesium compounds in the dampening solution from precipitating.

setoff. Condition that results when wet ink on the surface of the press sheets transfers or sticks to the backs of other sheets in the delivery pile. Sometimes referred to as "offset," a term that is reserved for the offset method of printing.

shaftless press. A press lacking a drive shaft, driven by individual, independent AC servo motors attached to each driven element in the press, including printing couples, reel stand, chill rolls, and folder. Each electric motor is controlled via a fiber optic link to a central controller.

sheeter. A specific web press delivery unit that cuts the printed web into individual sheets.

shortening compound. An ink additive that reduces ink flying, or misting.

showthrough. A term used to describe the visibility of printed material from the opposite side of the sheet. This characteristic is proportional to transparency of the substrate and the oiliness of the ink.

signature. One or more printed sheets folded to form a section of a book or pamphlet.

slabbing. The practice of removing several layers of paper from the outside of a new roll prior to inspection. *Alternative term:* stripping.

slip compound. An ink additive that improves scuff resistance of the printed ink film.

slip sheet. A sheet of paper placed between freshly printed sheets to prevent setoff or blocking.

smash-resistance. The ability of a blanket to recover from being momentarily subjected to excessively high pressure.

smoother. A device that helps to keep the sheet flat on the feedboard.

snowflaking. The tiny, white, unprinted specks that appear in type and solids if the ink is excessively emulsified.

specific gravity. Ratio of the weight of one material to the weight of an equal volume of water.

Specifications for Newsprint Advertising Production (SNAP). Originally called Specifications for Non-Heatset Advertising Printing, a set of standards for color separations and proofing developed for those printing with uncoated paper and newsprint stock in the United States.

Specifications for Web Offset Publications (SWOP). A set of standards for color separation films and color proofing developed for those involved in publications printing. The SWOP standards help magazine printers achieve accuracy when color separations from many different sources are printed on one sheet.

spectrophotometer. An instrument used to measure the relative intensity of radiation throughout the spectrum as reflected or transmitted by a sample.

spectrophotometry. The science of measuring color by analyzing the reflection or transmission of samples at specified points across the electromagnetic spectrum. The spectrophotometric curve is the most precise means for specifying colors since metameric pairs can be distinguished.

splice. The area where two paper rolls are joined to form a continuous roll.

split fountain. A divided ink fountain, or the use of dividers, to provide separate sections capable of holding two or more colors of ink, to permit the printing of two or more colors, side by side, in one pass through the press.

sprocket. A toothed wheel engineered to engage a chain.

spur gear. A gear that has teeth cut straight across.

squeeze. Printing pressure between the plate and blanket cylinders. It is expressed as the combined height of the plate and blanket over their respective bearers on a *bearer-contact press* and as the combined height of the plate and blanket over their respective bearers minus the distance between the bearers on a *non–bearer-contact press.*

stacker. A device attached to the delivery conveyor of a web press that collects, compresses, and bundles printed signatures.

stacker, compensating/counter. A machine that alternates the layering of a stack of printed products by turning them 180° to offset the uneven thickness between face and spine.

standard viewing conditions. A set of American National Standards Institute (ANSI) specifications that dictate the conditions under which originals (transparencies and reflection prints), proofs, and reproductions are viewed. For the graphic arts, the standard specifies a color temperature of 5000 K, a light level of approximately 200 footcandles, a color-rendering index of 90, and, for viewing transparencies, a neutral gray surround. Large format transparencies must be viewed with 2–4 in. of white surround and should never be viewed with a dark surround. It is also necessary to view the original or reproduction at an angle to reduce glare.

start-of-print line. A horizontal line that indicates the limit of the printing area. It is often engraved in the gutters about an inch behind the plate cylinder's leading edge.

static eliminator. A printing press attachment that attempts to reduce the amount of static developing on a press because of low relative humidity and the movement of paper over metal surfaces. It can also be helpful in eliminating ink setoff or paper feeding problems. *Alternative term:* antistatic device.

stereotype. Early method of imaging cylinders for letterpress web presses. Involved pressing a flat metal relief plate against a papier-maché mold (called a "flong"). The mold was wrapped inside a cylindrical carrier, and molten metal was poured into the mold to form the relief cylinder.

substrate. Any base material with a surface that can be printed or coated.

supercalendering. Finishing operation in papermaking where the web of paper passes between a series of hard metal rollers and soft, resilient rollers that impart varying degrees of smoothness and gloss to the paper.

surface strength. Ability of a paper to resist a force applied perpendicularly to its surface before picking or rupturing occurs.

tack. Resistance of a liquid to splitting. It is measured by determining the force required to split an ink film between two surfaces.

tail-end hook. A curl in the paper that develops at the back edge of the sheet away from the printed side.

temperature conditioning. Process of allowing paper to reach pressroom temperature before unwrapping the paper.

thermomechanical pulp. Papermaking pulp produced by preheating wood chips with steam prior to passing them through a disk refiner.

thixotropy. Characteristic of a material that causes it to change consistency on being worked.

through drier. A slow-acting drier that solidifies the ink film throughout and does not form a hard surface.

tight-edged paper. A paper whose exposed edges have given up moisture to the atmosphere and shrunk.

tinting. The bleeding of ink pigment particles into the dampening solution. *Alternative term:* toning.

top drier. Drier that gives a very hard surface to the ink.

trap. A densitometric measure of how well one ink film prints over another ink film. Expressed as a percentage of efficiency.

trapping. (1) Printing a wet ink over a previously printed dry or wet ink film. (2) How well one color overlaps another without leaving a white space between the two or generating a third color where they overprint.

trapping, dry. (1) The ability of a dry, printed ink film to accept a wet ink film over it. (2) Printing overprints, or one color on top of another, when the first color is already dry. Printing multicolor work on a single-color press is an example of dry trapping.

trapping, wet. (1) The ability of a wet, printed ink film to accept another wet ink film printed over it. (2) Printing overprints, or one color on top of another, when the first color is not dry. Printing multicolor work on a web press is an example of wet trapping.

true rolling. A term often used to describe the condition when there is no slip in the printing nip.

tucker blade. A reciprocating knife used to force signatures into jaws to produce a jaw fold, or between rollers to produce a chopper fold.

undercut. The difference between the radius of the cylinder body and the radius of the cylinder bearers.

undertone. The color of a thin film of ink. It is the color of light reflected by the paper and transmitted through the ink film.

unitack. A series of printing inks that have the same tack rating.

viscoelastic. A material, like an offset ink, that behaves as both a fluid and an elastic solid.

warp. The direction of maximum strength on a blanket.

washup. The process of cleaning the inking systems and blankets of a press with specially formulated cleaning solutions to remove all ink as required at the end of the operating day or whenever an ink color change is necessary.

water pan. A device that holds the dampening solution to be fed to the plate. *Alternative term:* water fountain.

water stop. One of a series of devices that are set against the surface of the dampening fountain roller; commonly used to reduce the amount of solution reaching heavily inked areas of the printing plate.

waterless lithography. A planographic printing process that does not require the use of a water-based dampening solution to prevent ink from adhering to nonimage areas of the printing plate. It requires special inks, presensitized waterless plates, and temperature-controlled inking systems.

watt. The most common measurement of heat release or work in electrical work; equal to the voltage (E) expressed in volts multiplied by the current (I) expressed in amperes.

wavy-edged paper. A paper whose exposed edges have absorbed moisture and become wavy.

web. A roll of any substrate that passes continuously through a printing press or converting or finishing equipment.

web lead. The continuous strip of paper passing from supply roll, over various rollers, through press units, to the folder.

web lead rollers. Any of the rollers used to support the paper web as it is fed through a web press.

web offset. A lithographic printing process in which a press prints on a continuous roll of paper instead of individual sheets.

webfed. A printing press that prints on a continuous roll of paper instead of individual sheets.

weft. The direction of minimum strength on a blanket.

wet printing. See *trapping, wet.*

wettability. The ease with which a pigment can be completely wet by the ink vehicle.

wetting agent. (1) In inkmaking, an additive that promotes the dispersion of pigments in the vehicle. (2) A substance, such as isopropanol or an alcohol substitute, found in a dampening solution, that decreases the surface tension of water and water-based solutions.

wire side. Side of the paper that is in contact with the paper machine's wire during papermaking.

zero-speed splicer. An automatic device that attaches a new roll of paper to an expiring roll without a press stop. The device is used in conjunction with a festoon to permit the expiring roll to come to a complete stop just before the splice is made and then to accelerate the new roll up to press speed.

Index

About the Author

Daniel G. Wilson, Doctor of Industrial Technology, is an associate professor and coordinator of graphic communications at Illinois State University, where he teaches a variety of graphic communications courses. He has taught in higher education for 17 years and has also worked in the printing industry as a lithographic press operator and technical trainer. While working for Baker Perkins, Inc., he developed training for web offset press operators.

Dr. Wilson, the coauthor of the textbook *Lithographic Technology in Transition*, has written a number of publications for PIA/GATF*Press*, including the *Lithography Primer*, *Web Offset Press Operating*, *PrintScape: A Crash Course in Graphic Communications*, and *The Bindery Training Curriculum*. He lives in Hudson, Illinois with his wife and their three children.

About PIA/GATF

The Printing Industries of America/Graphic Arts Technical Foundation (PIA/GATF), along with its affiliates, delivers products and services that enhance the growth, efficiency, and profitability of its members and the industry through advocacy, education, research, and technical information.

The 1999 consolidation of PIA and GATF brought together two powerful partners: the world's largest graphic arts trade association representing an industry with more than 1 million employees and $156 billion in sales and a nonprofit, technical, scientific, and educational organization dedicated to the advancement of the graphic communications industries worldwide.

Founded in 1924, the Foundation's staff of researchers, educators, and technical specialists help members in more than 80 countries maintain their competitive edge by increasing productivity, print quality, process control, and environmental compliance and by implementing new techniques and technologies. Through conferences, Internet symposia, workshops, consulting, technical support, laboratory services, and publications, PIA/GATF strives to advance a global graphic communications community.

In continuous operation since 1887, PIA promotes programs, services, and an environment that helps its members operate profitably. Many of PIA's members are commercial printers, allied graphic arts firms such as electronic imaging companies, equipment manufacturers, and suppliers. To serve the unique needs of specific segments of the print and graphic communications industries, PIA developed special industry groups, sections, and councils. Each provides members with current information on their specific segment, helping them to meet the business challenges of a constantly changing environment. Special industry groups include the Web Offset Association (WOA), Label Printing Industries of America (LPIA), and Binding Industries of America International (BIA). The sections include Printing Industry Financial Executives (PIFE), Sales & Marketing Executives (S&ME), *EPS—the Digital Workflow Group* (EPS), Digital Printing Council (DPC), and the E-Business Council (EBC).

PIA/GATF*Press* publishes books on nearly every aspect of the field; training curricula; audiovisuals (CD-ROMs and videocassettes); and research and technology reports. It also publishes *GATFWorld*, a bimonthly magazine providing articles on industry technologies, trends, and practices, and *Management Portfolio*, a bimonthly magazine that provides information on business management practices for printers; economic trends, benchmarks, and forecasts; legislative and regulatory affairs; human and industrial relations issues; sales, marketing, and customer service techniques; and management resources.

For more information about PIA/GATF, special industry groups, sections, products, and services, visit www.gain.net.

PIA/GATF*Press:* Selected Titles

Basics of Print Production. Hardesty, Mary.

Bindery Training Curriculum.

Chemistry for the Graphic Arts. Eldred, Nelson R.

Color and Its Reproduction. Field, Gary G.

Encyclopedia of Graphic Communications. Romano, Frank J. and Richard M. Romano.

Field Guide to Color Reproduction. Field, Gary G.

Fundamentals of Lithographic Printing, Mechanics of Printing. MacPhee, John.

Guide to Troubleshooting for the Sheetfed Offset Press. Destree, Thomas M. (ed.)

Guide to Troubleshooting for the Web Offset Press. Oresick, Peter M. (ed.)

Imaging Skills Training Curriculum.

Handbook of Graphic Arts Equations. Breede, Manfred.

*Lithography Primer.** Wilson, Daniel G.

Materials Handling for the Printer. Geis, A. John & Paul L. Addy.

Paper Buying Primer. Wilson, Lawrence.

The PDF Print Production Guide. Shaffer, Julie, and Joseph Marin

PrintScape: A Crash Course in Graphic Communications. Wilson, Daniel G., Deanna M. Gentile, and PIA/GATF Staff.

*Sheetfed Offset Press Operating.** DeJidas, Lloyd P. and Thomas M. Destree.

Sheetfed Offset Press Training Curriculum.

Total Production Maintenance: A Guide for the Printing Industry. Rizzo, Kenneth.

Web Offset Press Operating. Wilson, Daniel G. and PIA/GATF Staff.

Web Offset Press Training Curriculum.

Web Offset Problem-Solving Training Program. PIA/GATF Staff.

*What the Printer Should Know about Ink.** Eldred, Nelson R.

*What the Printer Should Know about Paper.** Wilson, Lawrence A.

---

*Also available in Spanish

Colophon

Web Offset Press Operating was edited, designed, and printed at PIA/GATF, head-quartered in Sewickley, Pennsylvania. The manuscript was written using Microsoft Word, and the files were emailed to PIA/GATF. The edited files were imported into QuarkXPress 4.1 running on an Apple Power Macintosh G3. The primary typefaces for the interior are New Caledonia and Helvetica Condensed. Line drawings were created in Adobe Illustrator and Macromedia FreeHand, and photographs were scaled and cropped in Adobe Photoshop. Adobe Acrobat PDFs and page proofs produced on a Xerox Regal color copier with Splash RIP were used for author approval.

Upon completion of the editorial/page layout process, the illustrations were transmitted to PIA/GATF's Robert Howard Center for Imaging Excellence, where all images were adjusted for the printing parameters of PIA/GATF's in-house printing department and proofed.

The preflighted pages were printed to Agfa's Apogee production system. Agfa's Sherpa 43 was used to produce the digital imposed proofs for customer approval. Creo Preps was used to impose the pages, and then the book was output to a Creo Trendsetter 800 Quantum platesetter. The interior of the book was printed on PIA/GATF's 26×40-in., four-color Heidelberg Speedmaster Model 102-4P sheetfed perfecting press, and the cover was printed two-up on PIA/GATF's 20×28-in., six-color Komori Lithrone 28 sheetfed press with coater. Finally, the book was sent to a trade bindery for case binding.